T0189016

Behavioral Synthesis for Hardware Security

Srinivas Katkoori • Sheikh Ariful Islam
Editors

Behavioral Synthesis for Hardware Security

Editors
Srinivas Katkoori
Department of Computer Science
and Engineering
University of South Florida
Tampa, FL, USA

Sheikh Ariful Islam
Department of Computer Science
University of Texas Rio Grande Valley
Edinburg, TX, USA

ISBN 978-3-030-78843-8 ISBN 978-3-030-78841-4 (eBook)
https://doi.org/10.1007/978-3-030-78841-4

This Springer imprint is published by the registered company Springer Nature Switzerland AG
The registered company address is: Gewerbestrasse 11, 6330 Cham, Switzerland

To my beloved wife, Hemalatha.
To my beloved family.

Preface

Today, behavioral synthesis tools are widely used in industry to routinely design billion-transistor complex integrated circuits. While there are several knobs available in these tools for area, power, and performance optimization, none exist for automatically incorporating security in a design. The primary reason for this state of affairs is that the problem of hardware security is not yet well understood. The problem inherently is a difficult problem, for example, there is no metric to measure the level of security, unlike those available for area, power, and performance.

There is a large and active community of hardware security researchers addressing the secure design problem at lower levels of design abstraction, namely at physical and gate levels. Relatively speaking, the number of research groups working at higher levels of abstractions is low. The main reason is that the knowledge base built at the lower levels is quintessential to address the problem reliably at the higher levels (as was the case with the area, power, and performance optimization).

This edited book provides a snapshot of the ongoing hardware security research at the register-transfer level and above. The intended audience are the academic researchers actively conducting research in the area as well as those who intend to work in this area. Further, the designers in the semiconductor industry can benefit by understanding how one could address the problem at higher levels. The book can also be used in undergraduate or graduate advanced hardware security elective. We take this opportunity to thank the contributing authors who have enthusiastically contributed to this book. We also thank Springer for providing us with an opportunity to publish this research book. We thank Ruchitha Chinthala, Omkar Dokur, Rajeev Joshi, Kavya Lakshmi Kalyanam, and Md Adnan Zaman from our research group for providing invaluable feedback.

Tampa, FL, USA
Edinburg, TX, USA
February 2021

Srinivas Katkoori
Sheikh Ariful Islam

Contents

Contributors

Richa Agrawal Synopsys Inc., Mountain View, CA, USA

Armaiti Ardeshiricham UC San Diego, La Jolla, CA, USA

Hannah Badier Lab-STICC, ENSTA Bretagne, Brest, France

Swarup Bhunia Department of Electrical and Computer Engineering, University of Florida, Gainesville, FL, USA

Mike Borowczak University of Wyoming, Laramie, WY, USA

Rajat Subhra Chakraborty Department of Computer Science and Engineering, Indian Institute of Technology, Kharagpur, West Bengal, India

Deming Chen Department of Electrical and Computer Engineering, University of Illinois at Urbana-Champaign, Champaign, IL, USA

Philippe Coussy Lab-STICC, Universite de Bretagne Sud, Lorient, France

Siddharth Garg New York University, New York, NY, USA

Tony Givargis Center for Embedded Systems and Cyber-Physical Systems (CECS), University of California, Irvine, CA, USA

Guy Gogniat Lab-STICC, Universite de Bretagne Sud, Lorient, France

Wei Hu Northwestern Polytechnical University, Xi'an, China

Sheikh Ariful Islam Department of Computer Science, University of Texas Rio Grande Valley, Edinburg, TX, USA

Robert Karam University of South Florida, Tampa, FL, USA

Ramesh Karri New York University, New York, NY, USA

Ryan Kastner UC San Diego, La Jolla, CA, USA

Srinivas Katkoori Department of Computer Science and Engineering, University of South Florida, Tampa, FL, USA

S. T. Choden Konigsmark Department of Electrical and Computer Engineering, University of Illinois at Urbana-Champaign, Champaign, IL, USA

Jean-Christophe Le Lann Lab-STICC, ENSTA Bretagne, Brest, France

Matthew Lewandowski Department of Computer Science and Engineering, University of South Florida, Tampa, FL, USA

Chen Liu Intel Corp., Hillsboro, OR, USA

Steffen Peter Center for Embedded Systems and Cyber-Physical Systems (CECS), University of California, Irvine, CA, USA

Christian Pilato Politecnico di Milano, Milan, Italy

Mahendra Rathor Indian Institute of Technology Indore, Indore, India

Francesco Regazzoni ALaRI Institute, Università della Svizzera italiana, Lugano, Switzerland

Wei Ren Department of Electrical and Computer Engineering, University of Illinois at Urbana-Champaign, Champaign, IL, USA

Pranesh Santikellur Department of Computer Science and Engineering, Indian Institute of Technology, Kharagpur, West Bengal, India

Benjamin Carrion Schafer Department of Electrical and Computer Engineering, The University of Texas at Dallas, Richardson, TX, USA

Donatella Sciuto Politecnico di Milano, Milan, Italy

Anirban Sengupta Indian Institute of Technology Indore, Indore, India

Nandeesh Veeranna ST Micro, Singapore, Singapore

Ranga Vemuri University of Cincinnati, EECS Department, Cincinnati, OH, USA

Martin D. F. Wong Department of Electrical and Computer Engineering, University of Illinois at Urbana-Champaign, Champaign, IL, USA

Lingjuan Wu Huazhong Agricultural University, Wuhan, China

Chengmo Yang Department of Electrical and Computer Engineering, University of Delaware, Newark, DE, USA

Farhath Zareen University of South Florida, Tampa, FL, USA

Acronyms

3PBIP	Third-Party Behavioral Intellectual Property
ACO	Ant-Colony Optimization
ADPCM	Adaptive Differential Pulse-Code Modulation
AES	Advanced Encryption Standard
AIS	Artificial Immune Systems
ALM	Adaptive Logic Module
ARF	Auto-Regression Filter
AST	Abstract Syntax Tree
ATPG	Automatic Test Pattern Generation
BFOA	Bacterial Foraging Optimization Algorithm
BIP	Behavioral Intellectual Property
BMC	Bounded Model Checking
CAD	Computer-Aided Design
CDFG	Control Data Flow Graph
CE	Consumer Electronics
CIG	Colored Interval Graph
CS	Control Step
CSA	Clonal Selection Algorithm
DAG	Directed Acyclic Graph
DBT	Dual Bit Type
DCT	Discrete Cosine Transform
DDL	Dynamic Differential Logic
DDoS	Distributed Denial-of-Service
DFS	Design for Security
DfT	Design-for-Trust
DPA	Differential Power Analysis
DSE	Design Space Exploration
DSP	Digital Signal Processing
DWT	Discrete Wavelet Transform
ECC	Elliptic Curve Cryptography
EDA	Electronic Design Automation

FDS	Force-Directed Scheduling
FIR	Finite Impulse Response
FPU	Floating-Point Unit
FSM	Finite State Machine
FSMD	Finite-State Machine with Data Path
FU	Functional Unit
GA	Genetic Algorithm
HD	Hamming Distance
HDL	Hardware Description Language
HLS	High-Level Synthesis
HLT	High-Level Transformations
HT	Hardware Trojan
HW	Hamming Weight
IC	Integrated Circuit
ICC	International Chamber of Commerce
IDCT	Inverse Discrete Cosine Transform
IFT	Information Flow Tracking
IG	Interval Graph
ILP	Integer Linear Programming
IoT	Internet-of-Things
IP	Intellectual Property
IR	Intermediate Representation
KaOTHIC	Key-based Obfuscating Tool for HLS in the Cloud
LLVM	Low Level Virtual Machine
MI	Mutual Information
ML	Machine Learning
MPEG	Moving Picture Experts Group
MPSoC	Multiprocessor System-on-Chip
MSE	Mean Square Error
MSMA	Macro Synchronous Micro Asynchronous
MTD	Measurements To Disclosure
NoC	Network-on-Chip
OCM/ OEM	Original Component/Equipment Manufacturer
PCA	Principle Component Analysis
PCB	Printed Circuit Board
PCC	Proof-Carrying Code
PDK	Process Design Kit
PI	Primary Input
PSO	Particle Swarm Optimization
PUF	Physical Unclonable Function
QoR	Quality of Results
RCDDL	Reduced Complementary Dynamic and Differential Logic
RE	Reverse Engineering
ROBDD	Reduced Ordered Binary Decision Diagram
RTL	Register Transfer Level

SA	Simulated Annealing
SaaS	Software-as-a-Service
SAIF	Switching Activity Interchange Format
SDDL	Simple Dynamic Differential Logic
SDML	Secure Differential Multiplexer Logic
SFLL	Stripped-Functionality Logic Locking
SMT	Satisfiability Modulo Theories
SoC	System-on-Chip
SPA	Simple Power Analysis
STG	State Transition Graph
STPs	State Transition Probabilities
SVA	System Verilog Assertion
SVM	Support Vector Machine
WDDL	Wave Dynamic Differential Logic

Chapter 1
Introduction and Background

Srinivas Katkoori and Sheikh Ariful Islam

In today's information age, it has become a commonplace to hear about security breaches almost on a daily basis in various contexts such as financial companies, social media, defense companies, etc. In a truly globally connected world through the modern Internet, bad actors have ready access to systems to steal sensitive information and cause serious damage.

While attacks on a computer system via software have been around for more than five decades in the form of computer virus, malware, Trojans, etc., attacks via system hardware are relatively new, perhaps, in the last 10–15 years. Hardware that was taken as the *root of trust* is no longer the case. One of the main reasons for this paradigm shift is the global off-shoring due to the fabless semiconductor manufacturing model [1–5], which exposes the design flow to potential attackers.

Hardware attacker has easy and low-cost access to complex resources such as cloud computing, parallel computing hardware (e.g., GPUs), etc., to analyze the victim hardware and craft sophisticated attacks. The types of attacks on hardware are varied such as hardware Trojans and backdoors, side-channel attacks (power, delay, and electro-magnetic), SAT-based attacks, tampering, and counterfeiting. In response to these attacks, the hardware design research community has been actively proposing countermeasures such as hardware obfuscation, logic locking, and Physically Unclonable Functions (PUFs).

To tackle the billion-transistor complexity of modern Integrated Circuits (ICs), Electronic Design Automation (EDA) tools are routinely used in all phases of the design cycle. Today, automated synthesis of fabrication-ready layouts from high-

S. Katkoori (✉)
Department of Computer Science and Engineering, University of South Florida, Tampa, FL, USA
e-mail: katkoori@usf.edu

S. A. Islam
Department of Computer Science, University of Texas Rio Grande Valley, Edinburg, TX, USA
e-mail: sheikhariful.islam@utrgv.edu

S. Katkoori, S. A. Islam (eds.), *Behavioral Synthesis for Hardware Security*,
https://doi.org/10.1007/978-3-030-78841-4_1

level behavioral descriptions is quite common in practice. While logic and layout synthesis tools are mature in the industry, behavioral synthesis tools have gained traction only in the last decade or so.

Using an automated tool chain, the designer can automatically optimize chip area, clock speed, and power consumption. However, security as an optimization goal is not available to the designer yet. Many research groups across the world are actively conducting research in robust high-level synthesis tools for security optimization. This research monograph is an attempt to bring together the latest research results in this area.

The rest of this chapter is organized as follows. Section 1.1 provides background on hardware security. Section 1.2 presents a brief overview of a typical ASIC synthesis flow in practice today. Section 1.3 introduces behavioral synthesis flow and design model. Section 1.4 motivates the need and advantages of addressing security optimization during behavioral synthesis. Finally, Sect. 1.5 provides a bird's eye view of the monograph.

1.1 Hardware Security

Security attacks on hardware have become prevalent in more than a decade [6–8]. In this section, we will briefly discuss the major driving forces that reshaped the hardware design landscape. For a detailed discussion on hardware security issues, the interested reader is referred to in-depth survey paper [4, 9–16] and research monograph [17]. We identify the following eight (8) driving forces that have made hardware vulnerable (Fig. 1.1) to security attacks:

- *Intellectual Property (IP) Usage* To handle the billion-transistor system design complexity, the usage of pre-designed third-party IP blocks has become the norm. Complex building blocks such as processors, memories, caches, memory controllers, Digital Signal Processing (DSP) cores, etc., are now readily available to rapidly build complex System-on-Chip (SoC) in every electronics customer segment [18–21]. Soft and hard IP blocks provide considerable design flexibility to perform quick or detailed trade-off analyses for a given time-to-market constraint. The profit model is based on the per-instance usage of the IP block. The global semiconductor IP market is estimated [22] to be USD 5.6 billion in 2020 and is projected to reach USD 7.3 billion in 2025. This financial incentive motivates attackers to steal IP through reverse engineering, illegal over-production, etc. Typically, IP blocks are encrypted (black-box model) due to proprietary reasons. This provides avenues for attackers to insert hidden malicious hardware into the system with the IP block as a host.
- *Hardware Complexity* While the semiconductor industry can rightfully boast an astonishing growth of IC capacity at an exponential rate in terms of the number of transistors, the flip side of the coin is the unmanageable complexity of the design

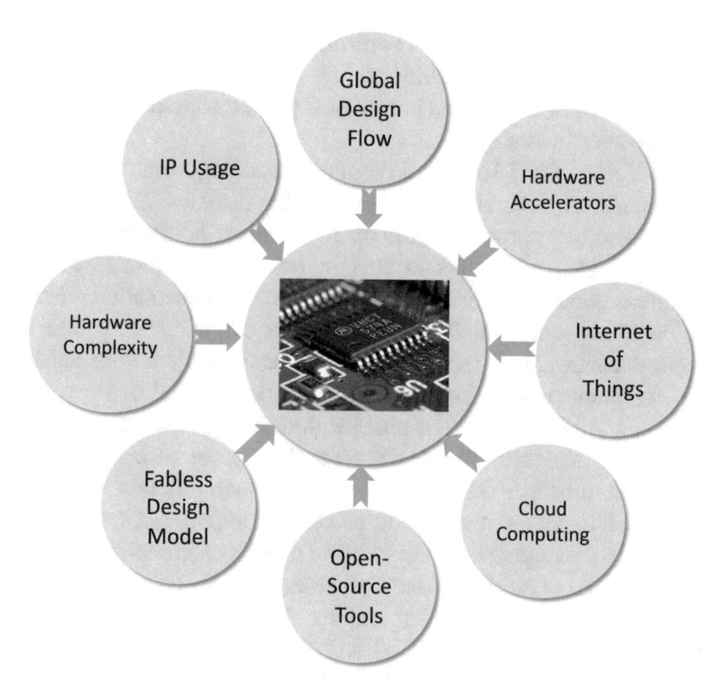

Fig. 1.1 Forces on hardware security (Courtesy: IC image from wikimedia.org)

verification problem. Hardware attackers can thus embed malicious artifacts into billion-transistor IC designs that can go undetected during the verification stage.

- *Fabless Companies* Similar to the software industry, in order to maximize the profit margin, semiconductor companies moved to an offshore design and manufacturing model. For low-complexity ICs, design is done in-house and the IC manufacturing is off-shored. In such a *fabless* paradigm, the design is naturally exposed to attackers in the fabrication phase. The fabless approach has given rise to a new problem that needs to be addressed: *Is the fabricated IC a faithful implementation of the design provided?*
- *Global Design Flow* Two decades ago, the ASIC design flow consisting of design, fabrication, and test tasks was executed completely within a company. To handle the design complexity as well as maximize the profit margin, the design flow is now distributed across multiple companies located across multiple continents. While fabless companies only outsource the fabrication task to a third party, in this approach, most of the tasks such as RTL design, physical implementation, fabrication, testing, packaging, etc., are outsourced. The more the number of tasks outsourced, the more the exposure of the design to the attackers.
- *Hardware Accelerators* With the capability to integrate a billion transistors on the same chip, coupled with EDA tools to automatically synthesize hardware from C/C++ descriptions, *porting* software to hardware has become easier.

A hardware accelerator is a hardware implementation of a function that was traditionally implemented as software. As it is now common to have several hardware accelerators integrated into an SoC implementation of a system, attacks on hardware have increased.

- *Internet-of-Things (IoT)* IoT is an ongoing technology transition where *things* can be humans, objects, and creatures, besides machines. IoT technology has a great promise in improving our daily lives. Smart home, smart health, smart transportation, etc., are IoT concepts in home automation, healthcare, and transportation domains wherein "edge" devices can constantly collect data and send to the cloud where intelligent applications can analyze the data and make intelligent decisions. IoT implies embedding every object with hardware/software to incorporate computing and communication capabilities.
- *Cloud Computing Resources* The attacker does have to own a powerful network of computers to analyze victim hardware and stage sophisticated attacks. Cloud computing based on per-usage model provides such a network at a fraction of the price. Such inexpensive access to powerful computing resources will enable smart attackers to stage reverse engineering attacks on complex ICs.
- *Powerful Computing Platforms and Open-Source Tools* Personal computers with multi-core computing capabilities and specialized parallel platforms such as Graphical Processing Units (GPUs) are other powerful computing platforms available to the attackers. A technology-savvy attacker can build a local cluster consisting of instances of these computers. Advanced open-source software is also readily available. For example, an attacker can run pattern detection algorithms to reconstruct layers of an IC using OpenCV. The recent breakthrough of machine learning algorithms and easy access to their hardware accelerators through the cloud make the attacker armed to the teeth.

1.2 ASIC Synthesis Flow

A simplified ASIC synthesis flow from specification to layout implementation is shown in Fig. 1.2. A high-level executable specification in VHDL, Verilog, C, or C++ is synthesized into an RTL design consisting of behavioral controller and structural data path. The RTL design is synthesized into logic-level netlist using logic synthesis tools. Finally, an optimized layout is generated with a place-and-route tool.

Input to the Design Flow The behavioral input should capture the input–output behavior of the design. An algorithm written in high-level languages such as behavioral VHDL, Behavioral Verilog, System Verilog, C, C++, etc. is used as an input. Typically, a subset of the language constructs is identified as a synthesizable subset. Some language constructs, for example, exact timing constructs, cannot be synthesized. The user must be familiar with the synthesizable subset to ensure successful implementation of the behavioral input.

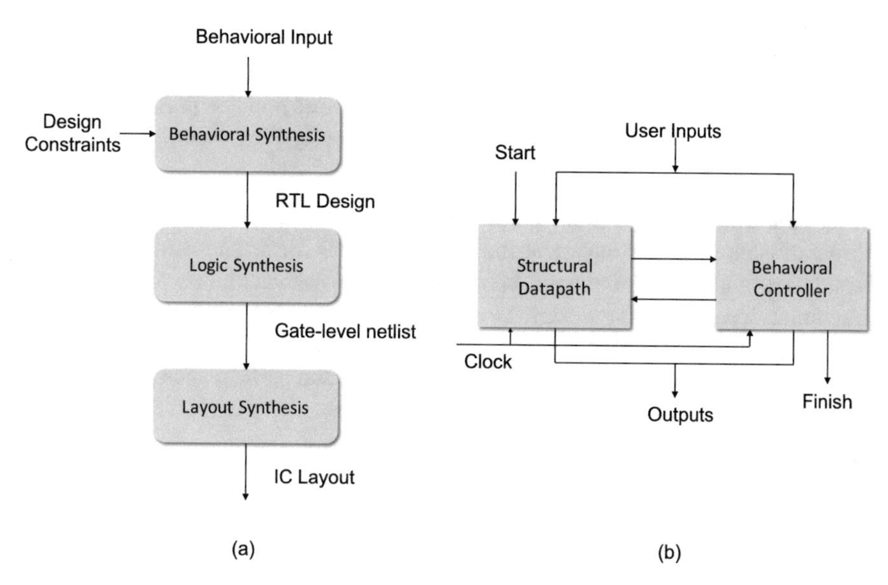

Fig. 1.2 (**a**) A simplified top-down ASIC design flow and (**b**) Register-Transfer Model

Design Description and Simulation Languages There are many design languages used during the ASIC design flow. Hardware Description Languages (HDLs) are specifically designed to describe hardware at various levels of abstraction. VHDL and Verilog are popular HDLs used in the industry. For behavioral, RTL, and gate-level descriptions, these languages can be used. The descriptions can be simulated for correctness with the commercial simulators for these languages. For the layout-level description, the standard description format is GDSII. Transistor-level descriptions extracted from the layout can be simulated using SPICE.

Design Goals and Constraints The design has goals and/or constraints on area, power, and/or performance. All synthesis tools have the capability to meet the design goals within the specified constraints. The traditional design goals/constraints are area and performance (usually clock speed). In the last 10–15 years, the synthesis tools have been extended to handle the power constraint at all abstraction levels. Security as a goal/constraint is not yet handled in the standard design flow.

Component Libraries One of the key inputs in the design flow is the component library. The library consists of basic Boolean logic gates, adders, multipliers, multiplexers, registers, register files, etc., described at various levels of design abstraction. It is possible to have soft or hard IPs as part of the library, which can be instantiated in the design. Often third-party libraries are readily available that are compatible with the synthesis tools so that the designer can use them out of the box. Pre-characterization of the components in terms of area, power, and performance is required so that the optimization engines in the synthesis tools can use them to accurately estimate the design-level attributes.

Target Technology For custom and semi-custom ASICs, the ASIC design flow targets a given technology node such as 32 nm or 18 nm. The design flow is equally applicable to other target platforms. The most popular platform is Field Programmable Gate Arrays (FPGAs) due to rapid prototyping capability as well as reconfigurability.

Design and Timing Verification As the design is synthesized down to the layout, the output of each synthesis tool is verified for functional correctness and equivalence. Simulation and formal methods are two ways to carry out this task. Besides, functional correctness the designers are interested in are timing correctness and power consumption. Design corner checks for nominal speed, and slow corners are carried out to ensure that the design meets the power and performance goals. In case the design fails to meet the goals, engineering change order is carried out by making local changes until the goals are met.

1.3 What Is Behavioral Synthesis?

Behavioral synthesis or high-level synthesis is the process of automatically generating a register-transfer level design from an input behavioral description. The RTL design consists of a structural data path and a behavioral controller. The data path is built with instances of functional units (adders, subtractors, ALUs, etc.), while the controller is a finite-state machine. The data path consists of the necessary data processing paths that can be invoked by the controller to implement the input behavior. Figure 1.2b shows the traditional Glushkovian model of the RTL architecture with a common clock. The data path generates appropriate conditional flags using which the controller will generate consequent control signals; typically, multiplexer selects values and register write_enable signals.

A simplified behavioral synthesis flow is shown in Fig. 1.3. The HDL input is converted into a control data-flow graph (CDFG) and stored in an internal format amenable for processing by the synthesis algorithms. The three fundamental steps are operation scheduling, resource allocation, and resource binding. Given resource and/or latency constraints, the operation scheduling algorithm analyzes the CDFG for data dependencies and time stamps the operations such that the dependencies are satisfied and constraints are met. The resource allocation algorithm allocates necessary functional, storage, and interconnect resources. The resource binding algorithm maps the high-level operations and edges to allocated resources and storage units, respectively. In principle, the three fundamental steps can be performed in any order or simultaneously. The most popular design flow is to perform scheduling, allocation, and binding sequentially in that order, as shown in Fig. 1.3.

A large body of the literature [23, 24] exists replete with algorithms for operation scheduling [25], resource allocation and binding [26], register optimization [27, 28], resource estimation [29, 30], area minimization [30, 31], delay minimization, early power estimation [32], leakage power minimization [33–36], dynamic power min-

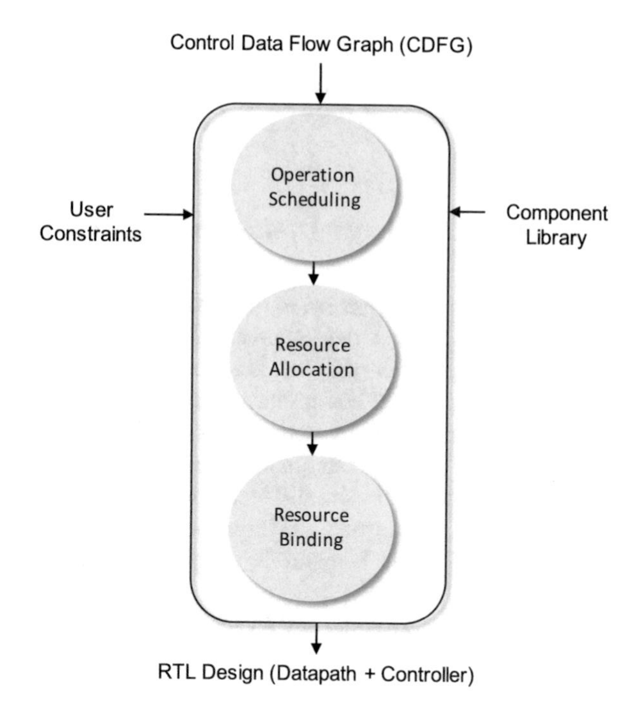

Fig. 1.3 A simplified sequential HLS synthesis flow with three key steps

imization [37–41], cross-talk minimization [42–48], thermal optimization [49, 50], interconnect optimization [51–53], design space exploration [37, 53], etc. Some of the popular HLS heuristics are based on force [25, 40, 41], simulated annealing [36, 45, 47, 48, 50, 54], genetic algorithm [50, 55, 56], etc. As with many VLSI optimization problems, the high-level synthesis steps are amenable to casting as graph optimization problems. The general formulations of scheduling, allocation, and binding problems are computationally hard problems, and thus most of the proposed optimization approaches are heuristic in nature. Exact formulations are applicable to small problem sizes and hence are only of pedagogical interest.

One of the main advantages of high-level synthesis is the early design space exploration. For the given input algorithm and user constraints, many feasible RTL designs with area, delay, and speed trade-offs are possible, thus providing a rich set of design alternatives. Often this feasible space is very large. The design space can be automatically explored especially focusing on Pareto-optimal [57–59] fronts. Early design space exploration techniques are proposed based on general search algorithms such as simulated annealing [36, 45, 47, 48, 50, 54] and genetic algorithm [19, 50, 56].

Behavioral synthesis tools can be targeted to ASIC or FPGA technologies. While a platform-agnostic RTL design can be produced, knowing the specific target technology, the tool can perform platform-specific optimizations. Over the last three decades or so, several academic and industrial HLS tools have been proposed. Some academic HLS tools are AUDI [35], Bambu [60], GAUT [61], and LegUp [62].

Some widely used commercial HLS tools are VivadoHLS [63], CatapultC [64], and Stratus HLS [65]. For recent detailed survey papers on HLS tools, the interested reader is referred to [59, 66].

1.4 Need for Secure Design During Behavioral Synthesis

EDA tools have always come to the rescue in handling the billion-transistor design complexity. A typical development trajectory for the EDA toolchain has been bottom-up, i.e., first the problem is addressed at the layout and gate levels, followed by RT level and above. For instance, this trend can be clearly seen in case of IC area minimization, performance optimization, and power optimization and management, and FPGA tool development. In the case of IC security, we can expect the same kind of growth trajectory. In fact, significant secure design work has already been done at that layout and gate levels. The in-depth understanding and the knowledge at these abstraction levels can be gainfully employed in building secure-aware behavioral synthesis tools.

The following are some key reasons for HLS tools to incorporate secure design techniques:

- *Design Freedom and Less Design Data* The amount of design freedom available at higher levels of abstraction provides great opportunities to incorporate security into a design. For example, a combination of algorithmic transformations can be applied to the input algorithms (say 1000 lines of C code) quickly due to less design information. Due to manageable design complexity at the higher levels, we can get an understanding of the global impact of a transformation on the design security besides the area, power, and performance overheads.
- *Early Design Space Exploration* Early automatic design space exploration can also be leveraged to identify highly secure design alternatives. However, this requires early assessment of the level of security. Thus, while HLS can provide early avenues to optimize security, it also poses the challenge of accurately and meaningfully "estimating" the security level. Reliable security metrics amenable for early evaluation are itself an open topic of research.
- *HLS is Common in Today's Design Toolchain* Today's ICs are increasingly built with behavioral synthesis tools starting with an algorithm rather than with an RT-level design. As security assessment is not yet considered, it is possible that the tools may incorporate back doors in the designs that could be exploited by attackers. As the current HLS tools are not security-aware, the designers would have to resort to manually inserting the security defenses. Needless to say, this is tedious and error-prone. Further, as the design may be optimized for area, power, or latency or a combination thereof, there may be very little room for incorporating security techniques.
- *Designer's Lack of Knowledge on Security Attacks and Defenses* The current status is such that the security solutions are largely in research phase and the

practicing engineer is not sufficiently knowledgeable in the security defenses so that they can incorporate them in their designs. As it takes a considerable amount of time to educate all the designers and given that security is a serious problem, automated secure HLS tools can fill this gap. By providing security-aware HLS tools, the designer is instantly equipped with a variety of security defenses.
- *Keeping up with the Attackers* Just as in the software world wherein computer antivirus software is constantly updated to keep up with new viruses, HLS tools have the potential to help to keep up with the new hardware vulnerabilities on the FPGA platform. On detection of a vulnerability, HLS tool can resynthesize the design to "patch up" the vulnerability and the FPGA can be reconfigured in the field.

1.5 Organization of This Book

This book consists of fifteen chapters, organized into four parts, discussing the role of behavioral synthesis in hardware security. In Part I, we start with hardware IP protection through various obfuscation techniques. Then, in Part II, we shift focus to hardware IP protection through watermarking and state encoding. In Part III, we present techniques to defend against hardware Trojans. Finally, in Part IV, we cover techniques for side-channel defense.

1.5.1 Part I: Hardware IP Protection Through Obfuscation

- Chapter 2 describes key-based obfuscation mechanism through the (automated) introduction of multiplexer during high-level synthesis. Particularly, the authors demonstrate how non-critical paths can be used to prevent an attacker from reusing IP for malicious purposes. Also, the resiliency of the obfuscation framework has been demonstrated in terms of variable key lengths.
- Chapter 3 presents the impact of obfuscation on design parameters (area, performance, power, etc.). The approach modifies several programming constructs of the source code using genetic algorithm. The approach is evaluated on the SystemC benchmark suite.
- Chapter 4 discusses an approach on how to modify RTL components (data path and controller) with the key while maintaining the original functionality. It includes an additional FSM to provide two modes of operations by judiciously modifying conditional assignments. The chapter ends with evaluating security metrics.
- Chapter 5 presents the role of cloud-based HLS service in protecting hardware IP. The chapter discusses transient logic locking mechanism and introduces obfuscation (de-obfuscation) as an additional step to traditional HLS. During

obfuscation, the attention is focused on bogus code insertion and key-based control flow flattening.
- Chapter 6 proposes algorithm-level obfuscation by extending the capability of HLS algorithms while obfuscating sensitive information of the input program. It also discusses how the correct functionality of obfuscated IC can be restored with the original key stored in tamper-proof memory.

1.5.2 Part II: Hardware IP Protection Through Watermarking and State Encoding

- Chapter 7 introduces watermarking as a basic protection mechanism of Digital Signal Processing IP. It looks at different watermarking methods (e.g., single phase, multi-phase, binary encoding, etc.) along with optimal algorithm to introduce such in behavioral synthesis.
- Chapter 8 presents a resilient FSM design approach against side-channel attacks. The authors focus on the reachability metric (Hamming Distance) to derive graded security with minimal increase in power consumption. Further, they describe secure FSM encoding and analyze the trade-offs between power and security.
- Chapter 9 discusses hybridized genetic algorithm to embed (and reproduce) digital watermark on sequential designs. The scheme also allows to pinpoint the root causes of intolerable overhead due to deliberate watermarking and proposes to modify state codes in favor of low-performance overhead.

1.5.3 Part III: Hardware Trojans: Modeling, Localization, and Defense

- Chapter 0 presents results from an empirical study on bit width to localize potential Hardware Trojan. The study provides useful insights on how the bit width of the arithmetic module can be broken into pieces based on the dual-bit type model. Experiments show higher accuracy on analytical model with lower mean square error compared to empirical results.
- Chapter 11 proposes "diversity" to detect or mute the malicious actions during runtime caused by collusion among third-party untrusted IP vendors. The overall scheme applies Integer Linear Programming and heuristics to determine their effectiveness on MPSoC.
- Chapter 12 describes the role of Artificial Immune Systems to distinguish between safe and unsafe behavior of third-party IP. The framework captures the IP behavior from RTL description to develop and train machine learning model, which is later used to classify vulnerable IP from benign IP.

1.5.4 *Part IV: Side-Channel Defense via Behavioral Synthesis*

- Chapter 13 proposes selective annotations to high-level program descriptions in order to generate efficient hardware IP with minimal (or no) information leakage. The authors primarily focus on resilience against Differential Power Attack by detecting confidential operations (variables) with cycle-accurate simulations.
- Chapter 14 discusses the secure FSM encoding and synthesis to thwart power-based side-channel attacks by minimizing mutual information (entropy) between source and model. The authors examine the constraint programming in the encoding phase.
- Chapter 15 examines the timing attack and provides defenses against it by balancing security-critical paths and timing annotations to basic blocks of the program. It validates timing invariance for proven security of hardware IP.
- Chapter 16 presents a framework to integrate information flow tracking method across different blocks of IP during HLS. It covers the model formalization and verification technique to maintain different security policies, including isolation, timing side channels, etc.

References

1. Kumar, R.: Fabless Semiconductor Implementation. McGraw-Hill Education, New York (2008)
2. Hackworth, M.L.: Fabless or IDM? What the future holds for both. IEEE Des. Comput. **20**(6), 76–85 (2003). https://doi.org/10.1109/MDT.2003.1246166
3. Edwards, C.: Fabless future for venture capital. IEE Rev. **51**(6), 22 (2005). https://doi.org/10.1049/ir:20050601
4. Kumar, R.: Simply fabless! IEEE Solid-State Circuits Mag. **3**(4), 8–14 (2011). https://doi.org/10.1109/MSSC.2011.942448
5. Shelton, J., Pepper, R.: Look, Ma–No Fabs! IEEE Solid-State Circuits Mag. **3**(4), 25–32 (2011). https://doi.org/10.1109/MSSC.2011.942451
6. Adee, S.: The hunt for the kill switch. IEEE Spectr. **45**(5), 34–39 (2008). https://doi.org/10.1109/MSPEC.2008.4505310
7. Collins, D.R.: Trust in integrated circuits. https://apps.dtic.mil/dtic/tr/fulltext/u2/a482032.pdf (2008)
8. DARPA: Integrity and reliability of integrated circuits (IRIS). https://www.darpa.mil/program/integrity-and-reliability-of-integrated-circuits (2012)
9. Bhunia, S., Abramovici, M., Agarwal, D., Bradley, P., Hsiao, M.S., Plusquellic, J., Tehranipoor, M.: Protection against hardware trojan attacks: towards a comprehensive solution. IEEE Des. Test **30**, 6–17 (2013)
10. Tehranipoor, M., Koushanfar, F.: A survey of hardware trojan taxonomy and detection. IEEE Des. Test Comput. **27**(1), 10–25 (2010). https://doi.org/10.1109/MDT.2010.7
11. Koushanfar, F.: Hardware metering: a survey. In: Introduction to Hardware Security and Trust, pp. 103–122. Springer, New York (2012)
12. Quadir, S.E., Chen, J., Forte, D., Asadizanjani, N., Shahbazmohamadi, S., Wang, L., Chandy, J., Tehranipoor, M.: A survey on chip to system reverse engineering. J. Emerg. Technol. Comput. Syst. **13**(1) (2016). https://doi.org/10.1145/2755563

13. Rostami, M., Koushanfar, F., Karri, R.: A primer on hardware security: models, methods, and metrics. Proc. IEEE **102**(8), 1283–1295 (2014). https://doi.org/10.1109/JPROC.2014.2335155
14. Xiao, K., Forte, D., Jin, Y., Karri, R., Bhunia, S., Tehranipoor, M.: Hardware trojans: lessons learned after one decade of research. ACM Trans. Des. Autom. Electron. Syst. **22**(1) (2016). https://doi.org/10.1145/2906147
15. Valea, E., Da Silva, M., Di Natale, G., Flottes, M., Rouzeyre, B.: A survey on security threats and countermeasures in IEEE test standards. IEEE Des. Test **36**(3), 95–116 (2019). https://doi.org/10.1109/MDAT.2019.2899064
16. Chang, C., Zheng, Y., Zhang, L.: A retrospective and a look forward: fifteen years of physical unclonable function advancement. IEEE Circuits Syst. Mag. **17**(3), 32–62 (2017). https://doi.org/10.1109/MCAS.2017.2713305
17. Tehranipoor, M., Wang, C.: Introduction to Hardware Security and Trust. Springer, Berlin (2011)
18. Cheng, T.: IEEE Std 1500 enables core-based SoC test development. IEEE Des. Test Comput. **26**(1), 4–4 (2009). https://doi.org/10.1109/MDT.2009.11
19. Benini, L., De Micheli, G.: Networks on chips: a new SoC paradigm. Computer **35**(1), 70–78 (2002). https://doi.org/10.1109/2.976921
20. Bergamaschi, R.A., Bhattacharya, S., Wagner, R., Fellenz, C., Muhlada, M., White, F., Daveau, J., Lee, W.R.: Automating the design of SOCs using cores. IEEE Des. Test Comput. **18**(5), 32–45 (2001). https://doi.org/10.1109/54.953270
21. Saxby, R., Harrod, P.: Test in the emerging intellectual property business. IEEE Des. Test Comput. **16**(1), 16–18 (1999). https://doi.org/10.1109/54.748800
22. Saxby, R., Harrod, P.: Semiconductor Intellectual Property (IP) Market with COVID-19 Impact Analysis by Design IP, IP Core, IP Source, End User, Vertical (Consumer Electronics, Telecom & Data Centers, Automotive, Commercial, Industrial), and Geography - Global Forecast to 2025. https://www.researchandmarkets.com/reports/5185336/semiconductor-intellectual-property-ip-market. October 2020
23. Gajski, D., Dutt, N., Wu, A.C-H., Lin, S.Y-L.: High Level Synthesis: Introduction to Chip and System Design. Kluwer Academic, Boston (1992)
24. De Micheli, G.: Synthesis and optimization of digital circuits. McGraw Hill, New York (1994)
25. Paulin, P.G., Knight, J.P.: Force-directed scheduling for the behavioral synthesis of ASICs. IEEE Trans. Comput. Aided Des. Circuits Syst. **8**(6), 661–679 (1989)
26. Gopalakrishnan, C., Katkoori, S.: Resource allocation and binding approach for low leakage power. In: Proceedings of 16th International Conference on VLSI Design, pp. 297–302 (2003). https://doi.org/10.1109/ICVD.2003.1183153
27. Katkoori, S., Roy, J., Vemuri, R.: A hierarchical register optimization algorithm for behavioral synthesis. In: Proceedings of 9th International Conference on VLSI Design, pp. 126–132 (1996). https://doi.org/10.1109/ICVD.1996.489471
28. Vemuri, R., Katkoori, S., Kaul, M., Roy, J.: An efficient register optimization algorithm for high-level synthesis from hierarchical behavioral specifications. ACM Trans. Des. Autom. Electron. Syst. **7**(1), 189–216 (2002). https://doi.org/10.1145/504914.504923
29. Katkoori, S., Vemuri, R.: Accurate resource estimation algorithms for behavioral synthesis. In: Proceedings Ninth Great Lakes Symposium on VLSI, pp. 338–339 (1999). https://doi.org/10.1109/GLSV.1999.757449
30. Natesan, V., Gupta, A., Katkoori, S., Bhatia, D., Vemuri, R.: A constructive method for data path area estimation during high-level VLSI synthesis. In: Proceedings of ASP-DAC '97: Asia and South Pacific Design Automation Conference, pp. 509–515 (1997). https://doi.org/10.1109/ASPDAC.1997.600319
31. Fernando, P., Katkoori, S.: An elitist non-dominated sorting based genetic algorithm for simultaneous area and wirelength minimization in VLSI floorplanning. In: 21st International Conference on VLSI Design (VLSID 2008), pp. 337–342 (2008). https://doi.org/10.1109/VLSI.2008.97
32. Katkoori, S., Vemuri, R.: Simulation based architectural power estimation for PLA-Based Controllers. In: Proceedings of 1996 International Symposium on Low Power Electronics and Design, pp. 121–124 (1996). https://doi.org/10.1109/LPE.1996.547492

33. Pendyala, S., Katkoori, S.: Interval arithmetic based input vector control for RTL subthreshold leakage minimization. In: 2012 IEEE/IFIP 20th International Conference on VLSI and System-on-Chip (VLSI-SoC), pp. 141–146 (2012). https://doi.org/10.1109/VLSI-SoC.2012.7332091
34. Gopalakrishnan, C., Katkoori, S.: Tabu search based behavioral synthesis of low leakage datapaths. In: IEEE Computer Society Annual Symposium on VLSI, pp. 260–261 (2004). https://doi.org/10.1109/ISVLSI.2004.1339548
35. Gopalakrishnan, C., Katkoori, S.: KnapBind: an area-efficient binding algorithm for low-leakage datapaths. In: Proceedings 21st International Conference on Computer Design, pp. 430–435 (2003). https://doi.org/10.1109/ICCD.2003.1240935
36. Pendyala, S., Katkoori, S.: Self similarity and interval arithmetic based leakage optimization in RTL datapaths. In: 2014 22nd International Conference on Very Large Scale Integration (VLSI-SoC), pp. 1–6 (2014). https://doi.org/10.1109/VLSI-SoC.2014.7004171
37. Kumar, N., Katkoori, S., Rader, L., Vemuri, R.: Profile-driven behavioral synthesis for low-power VLSI systems. IEEE Des. Test Comput. **12**(3), 70 (1995). https://doi.org/10.1109/MDT.1995.466383
38. Katkoori, S., Kumar, N., Vemuri, R.: High level profiling based low power synthesis technique. In: Proceedings of ICCD '95 International Conference on Computer Design. VLSI in Computers and Processors, pp. 446–453 (1995). https://doi.org/10.1109/ICCD.1995.528906
39. Gopalakrishnan, C., Katkoori, S.: Power optimization of combinational circuits by input transformations. In: Proceedings First IEEE International Workshop on Electronic Design, Test and Applications '2002, pp. 154–158 (2002). https://doi.org/10.1109/DELTA.2002.994605
40. Gupta, S., Katkoori, S.: Force-directed scheduling for dynamic power optimization. In: Proceedings IEEE Computer Society Annual Symposium on VLSI. New Paradigms for VLSI Systems Design. ISVLSI 2002, pp. 75–80 (2002). https://doi.org/10.1109/ISVLSI.2002.1016878
41. Katkoori, S., Vemuri, R.: Scheduling for low power under resource and latency constraints. In: 2000 IEEE International Symposium on Circuits and Systems (ISCAS), vol. 2, pp. 53–56 (2000). https://doi.org/10.1109/ISCAS.2000.856256
42. Gupta, S., Katkoori, S.: Intrabus crosstalk estimation using word-level statistics. In: Proceedings of 17th International Conference on VLSI Design, pp. 449–454 (2004). https://doi.org/10.1109/ICVD.2004.1260963
43. Gupta, S., Katkoori, S., Sankaran, S.: Floorplan-based crosstalk estimation for macrocell-based designs. In: 18th International Conference on VLSI Design Held Jointly with 4th International Conference on Embedded Systems Design, pp. 463–468 (2005). https://doi.org/10.1109/ICVD.2005.100
44. Gupta, S., Katkoori, S.: Intrabus crosstalk estimation using word-level statistics. IEEE Trans. Comput. Aided Des. Integr. Circuits Syst. **24**(3), 469–478 (2005). https://doi.org/10.1109/TCAD.2004.842799
45. Sankaran, H., Katkoori, S.: Bus binding, re-ordering, and encoding for crosstalk-producing switching activity minimization during high level synthesis. In: 4th IEEE International Symposium on Electronic Design, Test and Applications (delta 2008), pp. 454–457 (2008). https://doi.org/10.1109/DELTA.2008.114
46. Sankaran, H., Katkoori, S.: On-chip dynamic worst-case crosstalk pattern detection and elimination for bus-based macro-cell designs. In: 2009 10th International Symposium on Quality Electronic Design, pp. 33–39 (2009). https://doi.org/10.1109/ISQED.2009.4810266
47. Sankaran, H., Katkoori, S.: Simultaneous scheduling, allocation, binding, re-ordering, and encoding for crosstalk pattern minimization during high level synthesis. In: 2008 IEEE Computer Society Annual Symposium on VLSI, pp. 423–428 (2008). https://doi.org/10.1109/ISVLSI.2008.95
48. Sankaran, H., Katkoori, S.: Simultaneous scheduling, allocation, binding, re-ordering, and encoding for crosstalk pattern minimization during high-level synthesis. IEEE Trans. Very Large Scale Integr. Syst. **19**(2), 217–226 (2011). https://doi.org/10.1109/TVLSI.2009.2031864
49. Krishnan, V., Katkoori, S.: Simultaneous peak temperature and average power minimization during behavioral synthesis. In: 2009 22nd International Conference on VLSI Design, pp. 419–424 (2009). https://doi.org/10.1109/VLSI.Design.2009.78

50. Krishnan, V., Katkoori, S.: TABS: temperature-aware layout-driven behavioral synthesis. IEEE Trans. Very Large Scale Integr. Syst. **18**(12), 1649–1659 (2010). https://doi.org/10.1109/TVLSI.2009.2026047
51. Katkoori, S., Alupoaei, S.: RT-level interconnect optimization in DSM regime. In: Proceedings IEEE Computer Society Workshop on VLSI 2000. System Design for a System-on-Chip Era, pp. 143–148 (2000). https://doi.org/10.1109/IWV.2000.844543
52. Alupoaei, S., Katkoori, S.: Net-based force-directed macrocell placement for wirelength optimization. IEEE Trans. Very Large Scale Integr. Syst. **10**(6), 824–835 (2002). https://doi.org/10.1109/TVLSI.2002.808453
53. Krishnan, V., Katkoori, S.: Clock period minimization with iterative binding based on stochastic wirelength estimation during high-level synthesis. In: 21st International Conference on VLSI Design (VLSID 2008), pp. 641–646 (2008). https://doi.org/10.1109/VLSI.2008.85
54. Gopalan, R., Gopalakrishnan, C., Katkoori, S.: Leakage power driven behavioral synthesis of pipelined datapaths. In: IEEE Computer Society Annual Symposium on VLSI: New Frontiers in VLSI Design (ISVLSI'05), pp. 167–172 (2005). https://doi.org/10.1109/ISVLSI.2005.46
55. Krishnan, V., Katkoori, S.: A genetic algorithm for the design space exploration of datapaths during high-level synthesis. IEEE Trans. Evol. Comput. **10**(3), 213–229 (2006). https://doi.org/10.1109/TEVC.2005.860764
56. Lewandowski, M., Katkoori, S.: A Darwinian genetic algorithm for state encoding based finite state machine watermarking. In: 20th International Symposium on Quality Electronic Design (ISQED), pp. 210–215 (2019). https://doi.org/10.1109/ISQED.2019.8697760
57. Ferretti, L., Kwon, J., Ansaloni, G., Guglielmo, G.D., Carloni, L.P., Pozzi, L.: Leveraging prior knowledge for effective design-space exploration in high-level synthesis. IEEE Trans. Comput. Aided Des. Integr. Circuits Syst. **39**(11), 3736–3747 (2020). https://doi.org/10.1109/TCAD.2020.3012750
58. Chen, J., Carrion Schafer, B.: Exploiting the benefits of high-level synthesis for thermal-aware VLSI design. In: 2019 IEEE 37th International Conference on Computer Design (ICCD), pp. 401–404 (2019). https://doi.org/10.1109/ICCD46524.2019.00062
59. Schafer, B.C., Wang, Z.: High-level synthesis design space exploration: past, present, and future. IEEE Trans. Comput. Aided Des. Integr. Circuits Syst. **39**(10), 2628–2639 (2020). https://doi.org/10.1109/TCAD.2019.2943570
60. Pilato, C., Ferrandi, F.: Bambu: a modular framework for the high level synthesis of memory-intensive applications. In: 2013 23rd International Conference on Field Programmable Logic and Applications, pp. 1–4 (2013). https://doi.org/10.1109/FPL.2013.6645550
61. de Bretagne-Sud, U.: GAUT - high level synthesis tool. http://hls-labsticc.univ-ubs.fr/. Accessed 23 Feb 2021
62. Canis, A., Choi, J., Aldham, M., Zhang, V., Kammoona, A., Anderson, J.H., Brown, S., Czajkowski, T.: LegUp: high-level synthesis for FPGA-based processor/accelerator systems. In: Proceedings of the 19th ACM/SIGDA International Symposium on Field Programmable Gate Arrays, FPGA '11, pp. 33–36. Association for Computing Machinery, New York, NY (2011). https://doi.org/10.1145/1950413.1950423
63. Xilinx: Vivado Design Suite - HLx Editions. https://www.xilinx.com/products/design-tools/vivado.html. Accessed 23 Feb 2021
64. Seimens: Catapult - High-Level Synthesis. https://eda.sw.siemens.com/en-US/ic/ic-design/high-level-synthesis-and-verification-platform. Accessed 23 Feb 2021
65. Cadence: Stratus High-Level Synthesis. https://www.cadence.com/en_US/home/tools/digital-design-and-signoff/synthesis/stratus-high-level-synthesis.html. Accessed 23 Feb 2021
66. Nane, R., Sima, V., Pilato, C., Choi, J., Fort, B., Canis, A., Chen, Y.T., Hsiao, H., Brown, S., Ferrandi, F., Anderson, J., Bertels, K.: A survey and evaluation of FPGA high-level synthesis tools. IEEE Trans. Comput. Aided Des. Integr. Circuits Syst. **35**(10), 1591–1604 (2016). https://doi.org/10.1109/TCAD.2015.2513673

Part I
Hardware IP Protection Through Obfuscation

Chapter 2
Behavioral Synthesis of Key-Obfuscated RTL IP

Sheikh Ariful Islam and Srinivas Katkoori

2.1 Introduction

As Moore envisioned [1], the recent growth of advanced semiconductor manufacturing nodes has changed the computing history. Technology scaling has played a vital role in propelling the "silicon age" revolution and enabled many computing devices to be pervasive in our daily life. These electronic devices have become critical parts of larger systems, in diverse application domains such as transportation, communication, finance, health, and defense. As more transistors have been integrated within single silicon die, the complexity of Integrated Circuit(IC) design and manufacturing has also increased. This has led to a globalized horizontal business model. Technology constraints and human factors are also dominant factors leading to the reuse of pre-designed and pre-verified components. A not surprising consequence of this is the lack of centralized control of the multi-party trust. There are evidences of attacks on hardware as IC goes through a deeper supply chain. This includes, but is not limited to, overproduction/overbuilding, counterfeiting, side-channel analysis, etc., without any visible knowledge of the original Intellectual Property (IP) owner. Reverse Engineering (RE) is commonly employed to execute the variants of IP theft. Hardware RE can be destructive or nondestructive. Destructive RE is expensive and requires precise equipment

This work was part of the first author's doctoral dissertation research at USF.

S. A. Islam (✉)
University of Texas Rio Grande Valley, Edinburg, TX, USA
e-mail: sheikhariful.islam@utrgv.edu

S. Katkoori
University of South Florida, Tampa, FL, USA
e-mail: katkoori@usf.edu

S. Katkoori, S. A. Islam (eds.), *Behavioral Synthesis for Hardware Security*,
https://doi.org/10.1007/978-3-030-78841-4_2

(Scanning Electron Microscope) to analyze die features [2]. For nondestructive RE, an attacker may utilize a non-invasive method such as side-channel analysis.

Counterfeit electronics account for 5–7% of the worldwide economy [2]. In 2005, the US Department of Defense and military equipment manufacturers spent (unknowingly) $300M on purchasing counterfeit parts. In another instance, a certifying agent provided false certifications of flight-critical aircraft parts [2]. Similarly, to damage company image and sabotaging reliable systems, the attacker scraps the ICs from the Printed Circuit Board (PCB) and resells them, which degrades the system's lifetime. Remarking the ICs and selling them as higher grade products were most common among all reported incidents [3]. Related to this, the number of fake components that do not conform to the specification of the Original Component/Equipment Manufacturer (OCM/OEM) has also increased. The number of these counterfeit components being sold worldwide has quadrupled since 2009 [2]. Similarly, the International Chamber of Commerce (ICC) estimated that electronics counterfeiting and cloning would amount to $4.2 trillion by the end of 2022 [4].

In light of the above problems, we argue that IP protection against all counterfeiting types should be performed in the early design phase (i.e., Behavioral Synthesis) to overcome IP theft. Key-based obfuscation provides the first line of defense against tampering, counterfeiting (recycled/remarking/rebranding), overproduction, and other reverse engineering approaches (cloning) that attempt to compromise security. This anti-tamper mechanism should be easy to implement with the highest security guarantee. Further, the impact of obfuscation should have minimal design overhead and be carried out during the design cycle. Similarly, low-cost and early-on obfuscation techniques are a requirement to maintain design for security. By employing obfuscation, we transform the original IP into an equivalent design with a higher barrier to uncovering functional semantics without the correct key. Key bits and their locations are transparent to the user, and the design is functional when the correct key is applied. Hence, the potential to extract the key from obfuscated hardware should be extremely low. The additional hardware obfuscation requirements as a successful anti-tamper technique include indistinguishable key bearing logic from the rest of the design and algorithmic resistance by creating confusion.

Implementing key-based obfuscation by embedding a large number of state elements incurs significant design overhead. Hence, security primitives should be implemented with minimal design overhead. As a result, carefully determining obfuscation points is of paramount interest. This motivates us to pursue an approach to obfuscate RTL design from a high-level design description (C, C++, SystemC) at an early phase of the design cycle. In this chapter, we present an automated approach to incorporate a key-based obfuscation technique in an RTL design during High-Level Synthesis. Our proposed technique determines possible obfuscation points on non-critical paths of input Control Data-Flow Graph (CDFG). We require minimal modifications to an existing design for obfuscation logic insertion during the datapath generation. Furthermore, unlike traditional obfuscation and logic locking mechanisms, our technique does not require a key storage mechanism in obfuscated design.

The rest of this chapter is organized as follows. Section 2.2 provides detailed background on hardware obfuscation and presents related work on key-based IP obfuscation. Section 2.3 describes the threat model assumptions. Section 2.4 presents the obfuscation framework, time complexity of obfuscation algorithm, and resilience of key-obfuscated RTL. Section 2.5 describes experimental evaluations of three different key lengths under three design corners for different design parameters. Finally, Sect. 2.6 draws the conclusion.

2.2 Background and Related Work

Multiplexer-based logic locking at gate level is a well-known technique in hardware obfuscation. However, there is a need for an automatic generation of an obfuscated design with minimal overhead during High-Level Synthesis. This section provides terminology and related background work.

2.2.1 *Hardware Obfuscation*

Hardware obfuscation is a protection technique for IP/IC so that an attacker cannot decipher the implementation details. It increases the difficulty of reading, interpreting, and eventually modifying the structure of IP/IC without modifying the original functionality. Generally speaking, hardware obfuscation inserts simple logic blocks (e.g., multiplexer, XOR) into the combinational design that works as key inputs. With the embedded key, the designer can conceal the input–output relationships without affecting the original functionality. Similar obfuscation can be performed for sequential design by introducing additional decision nodes (states) before entering *normal mode*. In both cases, decompiling the obfuscated structure requires the correct key to enable the functionality. Such transformations are effective in hiding the data and control flow of the program.

2.2.2 *High-Level Obfuscation*

The generated RTL from HLS can be denoted by $f(dp, ctrl)$, where dp denotes the datapath containing ALUs, registers, and steering logic, and $ctrl$ denotes the controller with appropriate control flow logic. With obfuscation, any RTL design f can be transformed into $\langle f_1, f_2, \ldots, f_n \rangle$ where f_1 to f_n are obfuscated RTL designs depending on the type, number, and location of obfuscation logic, as shown

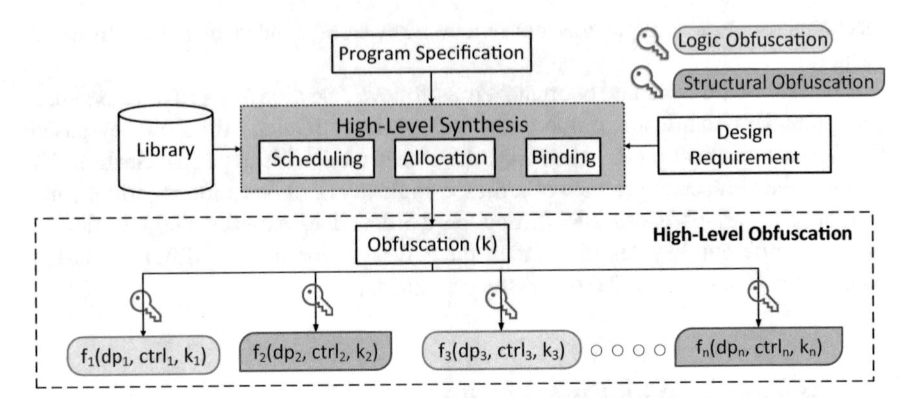

Fig. 2.1 High-level obfuscation technique for key-obfuscated RTL design

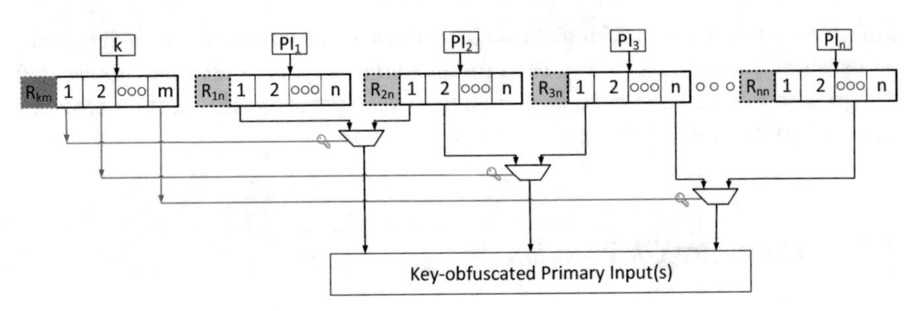

Fig. 2.2 Sample datapath obfuscation mechanism

in Fig. 2.1. Mathematically, $f \xrightarrow{\mathcal{T}(k)} \langle f_1, f_2, \ldots, f_n \rangle$, i.e., functional equivalence between f and any of the structurally different obfuscating transformation from f_1 to f_n exists for transformation function, $\mathcal{T}$, and correct key, k.

With HLS, we can perform obfuscation at functional (algorithmic) and structural levels. For the functional level, the input algorithm can be modified to obfuscate the functionality. For the structural level, the structural design being generated can be modified to hide the implementation details by inserting additional blocks (e.g., multiplexer, XOR). We can leverage the standard algorithmic transformations such as loop unrolling, tree-height reduction, folding/unfolding, interleaving, etc. for obfuscation purposes. One of the advantages of High-Level Synthesis is the ability to generate different structural implementations for the same function. This flexibility helps in synthesizing functionally equivalent but structurally obfuscated designs. Furthermore, we can have a better handle on the design overheads for the obfuscation.

Figure 2.2 illustrates the key obfuscation technique applied to primary inputs of an RTL datapath. Here, we have n Primary Inputs (PIs), each having n-bit width and obfuscation key (k) of length, m, that can be applied as a primary input. However, it is not a requirement that the key length should necessarily be the same as the

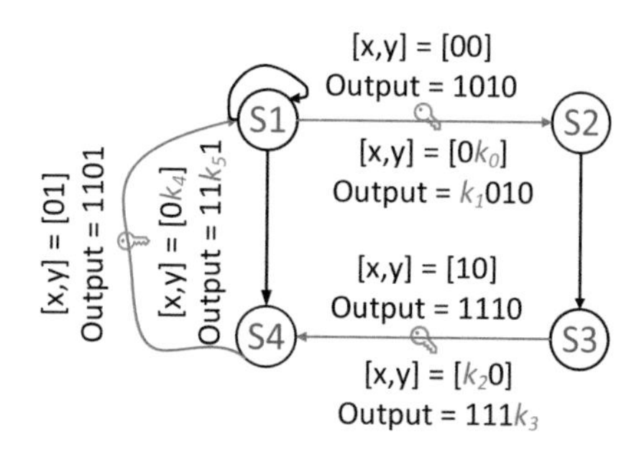

Fig. 2.3 Sample controller obfuscation mechanism

number of PIs or the bit width of any particular PI. According to constraints applied on PI whether to include it as an input to obfuscation element, the key size tends to be smaller than the PI bit width. In this example, k_1 is used to multiplex two primary input bits, $PI_1[1]$ and $PI_2[1]$. Similarly, $k_2 \ldots k_m$ are used as select lines. Note that many obfuscation approaches store k in a Non-Volatile Memory. In this approach, as the key is applied through a primary input, no key storage is required. We can also obfuscate internal signals and primary output(s) using the same technique.

Similar to datapath, we can also perform obfuscation in the controller by multiplexing status signals from datapath as shown in the illustrative example (Fig. 2.3). The controller is a Mealy machine with four states and two inputs. The initial state of the controller is $S1$. The value of input signals (x and y) can be replaced by obfuscation key bits, k_0, k_2, and k_4 for transitions $S1 \rightarrow S2$, $S3 \rightarrow S4$, and $S4 \rightarrow S1$, respectively. Furthermore, k_1, k_3, and k_5 substitute some of the output bits.

2.2.3 *Related Work*

The works in [5–7] introduce several ways to perform combinational logic obfuscation, while sequential logic obfuscation techniques are discussed in [8–11]. The designer can conceal the input–output relationships without affecting the original functionality, with either an embedded key or additional state elements. However, these logic obfuscation mechanisms do not minimize the design overheads. Furthermore, these obfuscation algorithms considered are applicable to either combinational designs (ISCAS85 [12]) or sequential designs (ISCAS89 [13]) only and the size of these designs is smaller.

There have been several works that perform structural obfuscation to protect DSP IPs from reverse engineering. External key-enabled reconfigurator is embedded within the design to enable High-Level Transformations (pipelining, interleaving, folding, etc.) [14]. Although the key and configuration data should be known *apriori* for correct functionality, the optimal choice of folding is not explored to expose unique HLS optimization opportunities. The works in [15, 16] performed High-Level Transformations (HLT) based on Particle Swarm Optimization (PSO) to obtain low-cost obfuscated design. The same transformations, along with multi-check-pointing, are explored for fault-secured obfuscated designs [17]. Although the authors in [15–17] evaluate area–power trade-offs with PSO Design Space Exploration (DSE) at the cost of runtime, they do not account for the relative cost of obfuscating each operation. Selective algorithm-level obfuscation of C language during HLS is proposed in [18]. Although the key-bit size comparable to masking all details of an algorithm grows, the overhead is as large as 30%. An accelerator-rich micro-architecture is proposed in [19] to track the information flow by embedding the taint tags to the variables. The tags are stored in additional taint registers without considering register sharing, hence incurring higher overhead. Similarly, the security analysis of these taint registers is missing against removal attack. The work in [15] is complementary to our work as HLT can be applied to the input prior to submitting it to our obfuscation technique.

At the gate and layout levels, there is a wealth of techniques in the literature for IP protection such as hardware metering [8, 20, 21], digital watermarking [22, 23], and fingerprint analysis [24, 25]. The underlying goal of key-based obfuscation methods [9, 10, 26–28] is to prevent RE of the state transition functionality extraction from Register-Transfer Level (RTL) and gate-level design. These approaches perform transformations to the original FSM by embedding additional states to introduce *modes of operation*; however, they incur significant design overhead. An attacker has to execute additional state transition function(s) to unlock the key or uncover structural variations. Such key-based approaches have two common characteristics: (a) system initialization with correct key value and (b) a large number of sequential elements to prevent brute-force attack. At the same time, state-of-the-art obfuscation methods protect only RTL IP netlist that does not contain both the datapath and the controller generated during HLS.

2.3 Threat Model

We assume an attack model as shown in Fig. 2.4, wherein the early design phase until RTL design is in the trusted zone and untrusted later on. To reverse engineer the key, we assume an attacker has access to a locked netlist and an oracle black-box IC. We also assume that the attacker does not have access to internal nets except primary input(s) and output(s). Under these assumptions, the adversary can apply any input

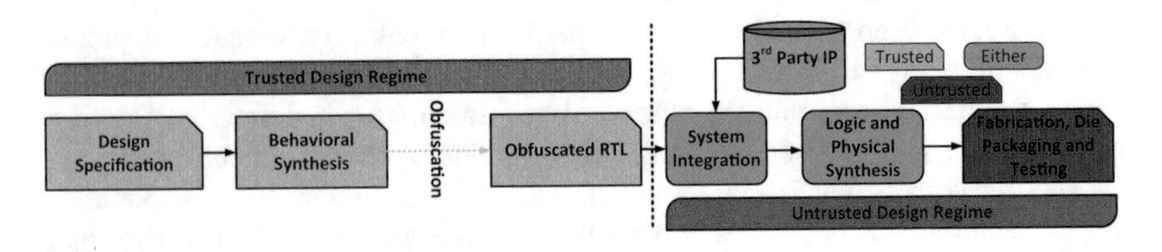

Fig. 2.4 Threat model assumption for RTL obfuscation

sequence, observe the output, and analyze the input–output behavior. Further, we assume that the design cannot be subjected to sophisticated micro-probing attacks. Though the design is locked at RTL, anyone (system integrator, layout designer, foundry, and distributor) possessing the IP in the electronics supply chain beyond the RTL stage can synthesize and simulate it. However, in our technique, we do not use any tamper-proof on-chip memory to store the key, which has to be activated and passed on to make obfuscated RTL IP functional. Instead, we apply the key through primary inputs.

The attacker's goal during pre-silicon is to understand the interaction between datapath elements (registers, functional units, and interconnection units) and control unit. From a foundry perspective, a rogue manufacturer can depackage, delayer, image, and annotate the layout images to infer the functionality of a locked module/block. This understanding would help the attacker to reconstruct the design and obfuscation key, which in turn can lead to overproduction of IP. Under latency constraint minimum resource problem, the proposed obfuscation of a subset of operations would shuffle the circuit output(s) signal in case of the invalid key, thus increasing the reverse engineering complexity and potential unsafe use by hiding the functionality of the design. Even under an oracle-guided attack on the structurally obfuscated RTL design as proposed in this chapter, the design would exhibit erroneous behavior due to various optimization heuristics applied during HLS.

2.4 Proposed RTL Obfuscation Approach

High-Level Synthesis performs scheduling, resource allocation, and binding on a given algorithmic/design description to generate RTL design. To limit the scope of reverse engineering and trading off performance with security, we propose an HLS framework (Fig. 2.5) where the additional synthesis objective is to embed obfuscation logic before generating RTL netlist. The framework accepts high-level design description, user constraints, and component library. Before performing scheduling, for a given key length, the overhead estimator provides revised area and clock frequency estimates. Based on these estimates, the constraints are revised and provided as an input to perform synthesis on CDFG.

2.4.1 Obfuscation Algorithm

We have modified the Force-Directed Scheduling (FDS) algorithm [29] to include obfuscation key and accordingly estimate design overhead. We have observed that delay of embedded multiplexer does not increase the latency of non-critical paths according to delay information available in Process Design Kit (PDK) library [30].

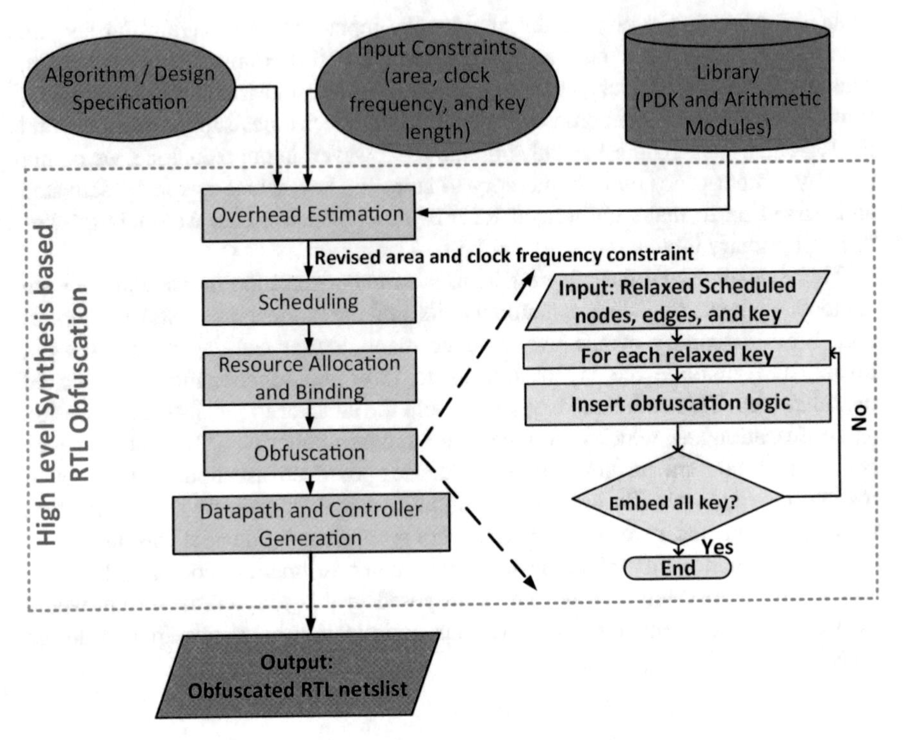

Fig. 2.5 Obfuscated RTL IP design flow during Behavioral Synthesis

Based on the estimated overhead and information about the force of each operation, we select an operation to schedule and embed multiplexor along its input and/or output line. We perform several runs of modified FDS and choose the one that provides scheduling information with best cost. The intermediate and modified CDFG is provided to obfuscation phase, which identifies possible obfuscation points and embeds obfuscation logic in addition to regular datapath elements and controller.

We provide the pseudocode of proposed RTL obfuscation in Algorithm 2.1. The input to the algorithm is CDFG of input design (scheduled, allocated, and bound) and key length (Lines 1–2). There are two stages in this algorithm before we can determine probable obfuscation points and embed obfuscation logic. In the first stage (scheduling phase of HLS), we analyze the mobility of each operation (i.e., the input and output) in the non-critical path and store the operations in $\mathcal{V}$ (Lines 4–8), and this has been shown in Fig. 2.6a. We calculate the mobility based on ASAP and ALAP scheduling metrics, and this choice does not degrade the performance of design. Then, we determine the cycle time of each operation that helps to further categorize operations and choose whether the selection of particular operation will lead to violation of timing slack. We omit those operations from further consideration (Lines 4–8). Remaining operations are sorted in ascending

Algorithm 2.1 RTL obfuscation approach

Initialization:

1: $G(V, E)$ = Scheduled, Allocated, and Bound graph;
2: $K = \langle K_1, K_2, \ldots, K_n \rangle$;
3: $O = \phi$;

Procedure:

4: **for** $v \in V$ **do**
5:     **if** $mobility(v) > 1$ and both inputs of v have zero slack **then**
6:         $\mathcal{V} \leftarrow \mathcal{V} \cup \{v\}$;
7:     **end if**
8: **end for**
9: $O_{sorted} \leftarrow$ Sort $\mathcal{V}$ in ascending order of v according to their slack;
10: **for** $k_i \in K$ **do**
11:     $v_j \leftarrow$ Select $v_j \in O_{sorted}$;
12:     Let v_j is mapped to FU_j;
13:     Let $\delta \leftarrow$ Register-to-Register delay of FU_j;
14:     Select input, x of $FU_j \ni$ Arrival time $[x] < \delta$;
15:     Embed obfuscation mux, M at input x;
16:     Set select(M) $= k_i$;
17:     $O_{sorted} \leftarrow O_{sorted} - \{v_j\}$;
18: **end for**

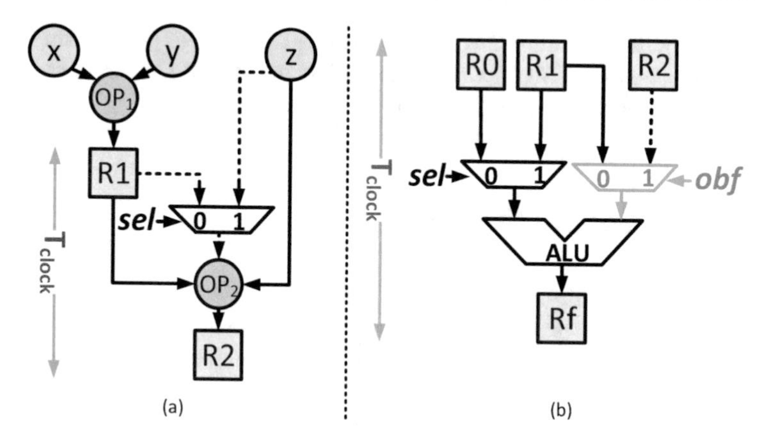

Fig. 2.6 Two possible stages to embed obfuscation logic in non-critical paths. (**a**) Scenario 1 (during HLS). **b** Scenario 2 (after HLS)

order based on their slack value (Line 9). For each operation in the sorted list to be obfuscated (Line 11), we first determine the register-to-register delay (δ) of the functional unit where the selected operation is going to be mapped (Line 13). Then, we select a particular input line whose arrival time is bound by δ (Line 14) and provide it as the input to multiplexer (Line 15). Choosing such input and embedding multiplexer do not increase register-to-register delay as shown in Fig. 2.6b. We iterate this technique until all keys are embedded (Line 10), and then we generate obfuscated RTL datapath. Both datapath and controller have to reverse-engineered for identifying key locations and value. Additionally, the synthesis information is kept confidential and known only to the designer.

2.4.2 Time Complexity of HLS-Based RTL Obfuscation

We compute the time complexity in two stages:

- Stage 1: First, we compute the mobility of each operation where CDFG is represented as Directed Acyclic Graph (DAG). For DAG, the worst-case scheduling complexity is $O(|V| + |E|)$, where V is the set of operations and E is the set of directed edges in the input graph, $G(V, E)$. Then, we remove operations in $O(P)$ time, where P is the number of possible obfuscating operations in $\mathcal{V}$. Sorting the remaining N operations in $\mathcal{V}$ can be completed in $O(|NlogN|)$.
- Stage 2: Estimating the register-to-register delay takes constant time, $O(1)$. The identification of an input to provide as an input to multiplexer is limited to a number of inputs per operation in CDFG. In our case, it is 2. Hence, such identification can be completed within $O(|M|)$, where $M = 2$.

In summary, the algorithm runs in $O(|V|+|E|+|NlogN|+2|N|)$. As the number of candidate operations (N) for obfuscation is less than the total operations (V), the overall time complexity is $O(|V| + |E|)$.

2.4.3 An Illustrative Example

Consider a simple multiplicative function, namely, $f = (a * b) * (c * d) * e$. We have to perform resource-constrained scheduling for f given two multipliers. We further assume all primary input values are available at the beginning of execution (control step, cs0). A possible schedule, resource allocation, and binding information for f are shown in Fig. 2.7a. The corresponding synthesized datapath is shown in Fig. 2.7b. We can observe that the critical path is along {a,t0,t1,f}; therefore, we exclude this path while choosing obfuscation points. As operation 4 can be scheduled in either control step 1 or 2, we can leverage this flexibility to perform obfuscation on the inputs of this operation.

Given the scheduling flexibility of operation 4 on the non-critical path, we show two possible schedules in Fig. 2.8. Figure 2.8a shows a schedule with three possible obfuscation paths (P0, P1, and P2), while Fig. 2.8b shows another with two obfuscation paths (P0 and P1). Along these paths, we can embed obfuscation logic (multiplexer). Figure 2.9a shows an obfuscated datapath where k_2, k_1, and k_0 refer to obfuscation key bits used as select signals of the embedded multiplexers for nodes $\{e, c, d\}$. For the sake of simplicity, we have used 2-input multiplexers in this example. However, multiplexers with large input sizes can be used as needed. Figure 2.9b shows another obfuscated datapath with two multiplexers. Table 2.1

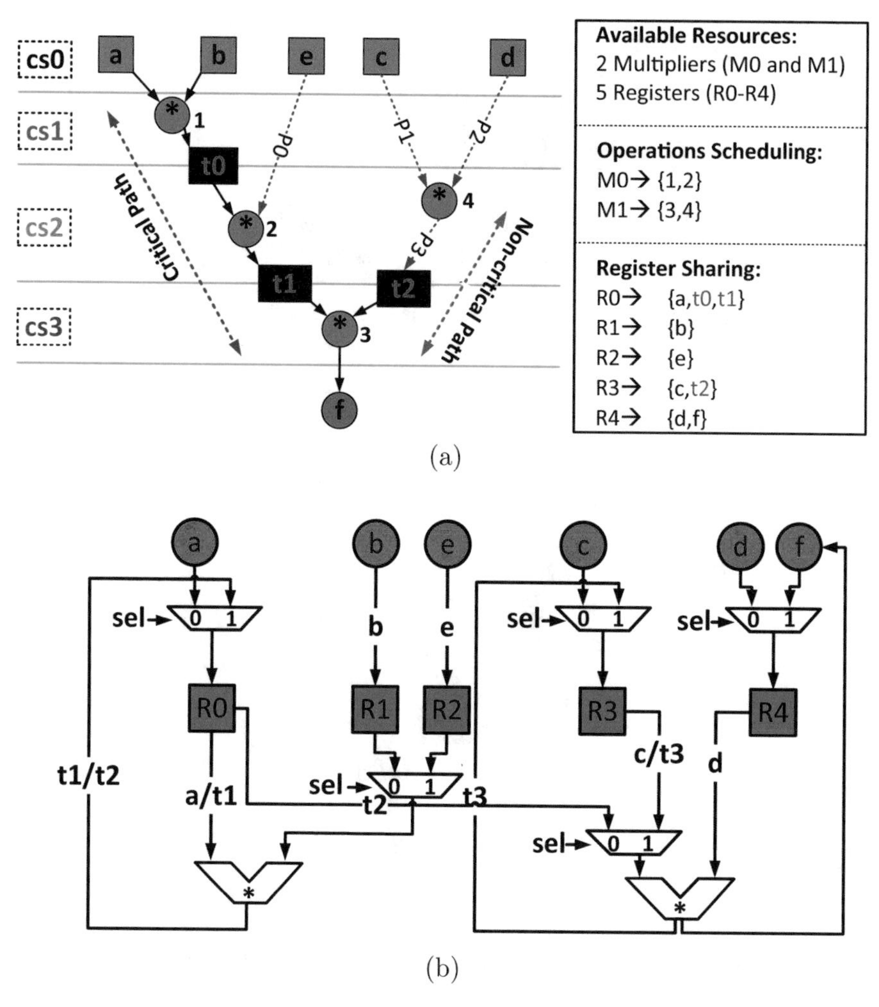

Fig. 2.7 Synthesized datapath for arithmetic function: $f = (a * b) * (c * d) * e$; (**a**) scheduled data-flow graph, operation mapping, and register mapping information; (**b**) synthesis datapath with four possible obfuscation paths (P0, P1, P2, and P3)

provides obfuscated functions and correct key values. This example demonstrates the obfuscation possibilities arising from the scheduling freedom.

2.4.4 Difficulty of Attack

Resilience refers to the withstanding capability of the obfuscated RTL IP against a brute-force attack. In general, for a given obfuscation key length (k) and the

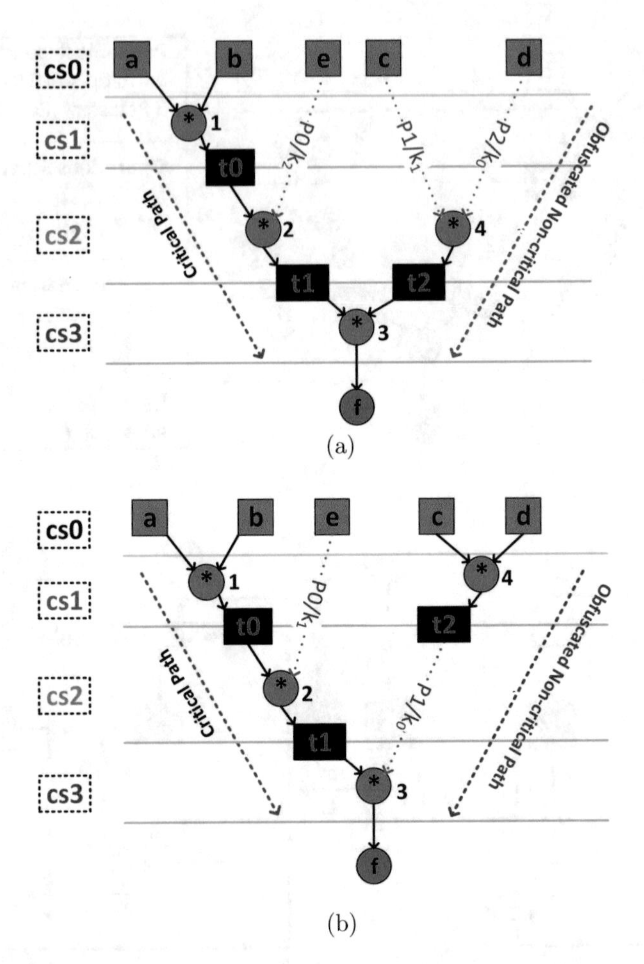

Fig. 2.8 Scheduling freedom for high-level obfuscation technique for function: $f = (a * b) * (c * d) * e$

number of registers (R) each with n-bit width, an attacker will have to look up $\sum_{k=1}^{R} \binom{R*n}{k}$ possibilities to search for correct key position and value. We can augment the search by rearranging k keys into $k!$ different ways for R registers. Hence, the design resilience can be defined as follows:

$$P(R, n, k) = \frac{1}{\sum_{k=1}^{R} \binom{R*n}{k} * k!}. \tag{2.1}$$

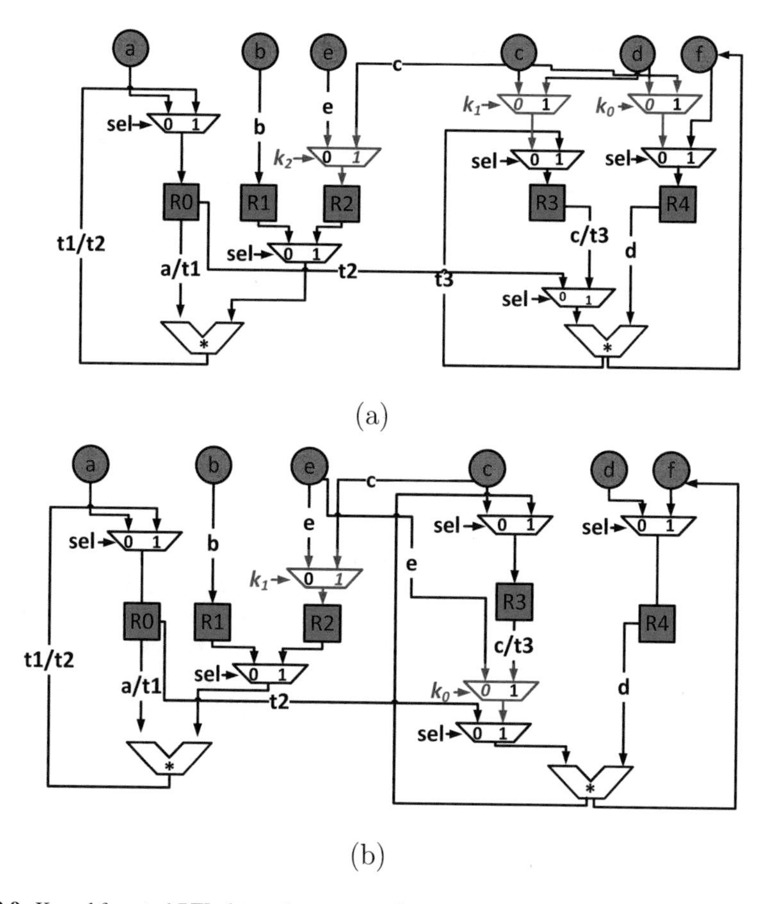

Fig. 2.9 Key-obfuscated RTL datapath corresponding to Fig. 2.8

Table 2.1 Obfuscated functions for original function, f

Schedule	Non-obfuscated function (f)	Obfuscated function (f_{obs})	Key value (k)
Figure 2.9a	$(a^*b)^*(c^*d)^*e$	$(\overline{k_2}e + k_2c)^*(a^*b)^*(\overline{k_1}c + k_1d)^*(\overline{k_0}c + k_0d)$	$\{k_2, k_1, k_0\} = \{0, 0, 1\}$
Figure 2.9b	$(a^*b)^*(c^*d)^*e$	$(\overline{k_1}e + k_1c)^*(a^*b)^*(\overline{k_0}e + (k_0(c^*d))$	$\{k_1, k_0\} = \{0, 1\}$

We can achieve a lower P value for larger values of R and n, leading to a brute-force attack for de-obfuscating the design. A large key search space will force an attacker to employ massive computational resources and excessive time to reverse engineer the design.

2.5 Experimental Evaluation

An in-house HLS tool, AUDI [31] (written in C programming language), is employed to automatically generate obfuscated RTL datapaths. A functional testing procedure to assess the obfuscated RTL design against reverse engineering is presented. We present a comprehensive evaluation of applying the proposed obfuscation scheme on HLS benchmarks in terms of security metrics and design overheads. We generate baseline design and obfuscated designs in the following three (3) stages:

1. *Baseline RTL Design Generation* We provide a behavioral VHDL input to AUDI. It converts the input into an internal format (AIF) followed by scheduling, resource allocation, and module binding for a given latency constraint. Then, a structural datapath and behavioral controller in VHDL are generated.
2. *Obfuscated RTL Design Generation* We carry out obfuscation according to the algorithm presented in Sect. 2.4 to generate key-obfuscated design. The obfuscation is carried out before the datapath/controller generation phase. While doing so, we preserve the obfuscation elements (multiplexer). Here, we obtain a functionally equivalent but structurally modified key-obfuscated RTL netlist.
3. *Logic Synthesis and Functional Equivalence Check* We synthesize the RTL design with Synopsys Design Compiler to generate logic netlist for a given PDK library. Following logic synthesis, we check the functional equivalence between the gate-level netlist and original RTL design.

2.5.1 *Benchmarks*

We validate the proposed approach with the following five (5) benchmarks:

- *Elliptic*: Fifth-order wave filter
- *FFT*: 8-point Fast Fourier Transform algorithm
- *FIR*: 5-point Finite Impulse Response algorithm
- *Lattice*: Single-input linear prediction filter
- *Camellia*: Block cipher performance-optimized core

2.5.2 *Results and Analysis*

We evaluated our proposed obfuscation technique on four datapath-intensive HLS benchmarks (Elliptic, FIR, FFT, and Lattice) and one crypto-core (Camellia) [32]. We obfuscate each design with 32-, 64-, and 128-bit key and then compare their power, timing, and area with that of the baseline design. We synthesize each design using Synopsys SAED 90 nm PDK library and target a clock period for 10 ns for three design corners (best, typical, and worst case). Our obfuscated RTL code is generated automatically from algorithmic description; hence, we do not compare our work with any contemporary work (e.g., [9]) where the obfuscation is performed manually.

2.5.2.1 Non-obfuscated Design Attributes

We report the design attributes for non-obfuscated designs in Table 2.2 to examine the power, performance, and area (PPA) variations. In generating RTL (structural VHDL), we have used minimum latency under Force-Directed Scheduling, and the value is reported in the second column. Note that the latency bound (λ) is different from target frequency used to synthesize the design and generate gate-level netlist. We indicate the total operations on both critical and non-critical paths in the third column. We have not found the relevant information for Camellia in [32].

2.5.2.2 Obfuscated Design Attributes

For each design, we calculate the PPA overhead under three design corners for three different key lengths. For 32-bit key, we incur, on average, 2.03% area overhead as shown in Fig. 2.10. For power and delay, the overheads are 0.55% and 2.47%, respectively. This is shown in Figs. 2.11 and 2.12. These overheads are reasonable for available obfuscation points and given resources. Similarly, for higher protection against malicious reverse engineering, the overheads indicate cost associated with incorporating security into the design. As it is intuitive that overhead will increase with increase in key length, we observe that the increase in design overheads is negligible with key length due to intelligent choice of obfuscation points in CDFG. In general, we observe an increasing trend of area overhead as we move from best case to worst case, which is opposite for power overhead. During best case, the tool moves registers along multiple paths to fix timing violations. For typical case, both power and delay are given equal weight during synthesis. Similarly, we observe an increasing trend of nets as we increase key length in Table 2.3. Hence, the key extraction probability becomes negligible, and for such extremely low values, an attacker will enumerate all possible netlists. To summarize, typical and best case strive for lower area and delay overhead, while worst case favors minimal increase in power overhead. For three key lengths, we report the number of cells in Table 2.4 without interconnect area. We observe that the number of cells does not increase drastically, leading to lower design overheads.

We evaluate the obfuscation on Xilinx Virtex-7 (XC7V585T) FPGA using Synopsys Synplify for a target clock frequency of 100 MHz. We report the resource utilization for three key lengths in Tables 2.5 and 2.6. As we want to preserve obfuscation modules, we did not apply optimization in the case where each obfuscation multiplexer is feeding as an individual clock to LUT. Hence, we see a higher resource utilization as reported in the row of LUTs.

Table 2.2 Non-obfuscated benchmark parameters under three design corners. + (Addition), ∗ (Multiplication), and − (Subtraction). Units of area, delay, and power are μm^2, ns, and μW, respectively

Design	Latency (λ)	Operations count	Best case			Typical case			Worst case		
			Area	Delay	Power	Area	Delay	Power	Area	Delay	Power
Elliptic	15	(26+, 8*)	106,435.69	7.78	1290	109,662.47	12.41	1029	110,941	28.04	536.03
FIR	5	(4+, 5*)	72,444.12	7.52	1410	75,467.61	11.6	1125	76,806	25.80	507.70
FFT	10	(20+, 16*, 4−)	64,561.7	6.66	690.45	6504.99	10	553.17	67,152	19.62	320.25
Lattice	10	(8+, 5*)	61,301.53	7.85	935.9	63,021	11.47	708.67	64,796	26.65	360.86
Camellia	10	–	149,660.56	2.45	2786.1	150,418.02	3.88	2139.3	170,780.36	18.43	1197.9

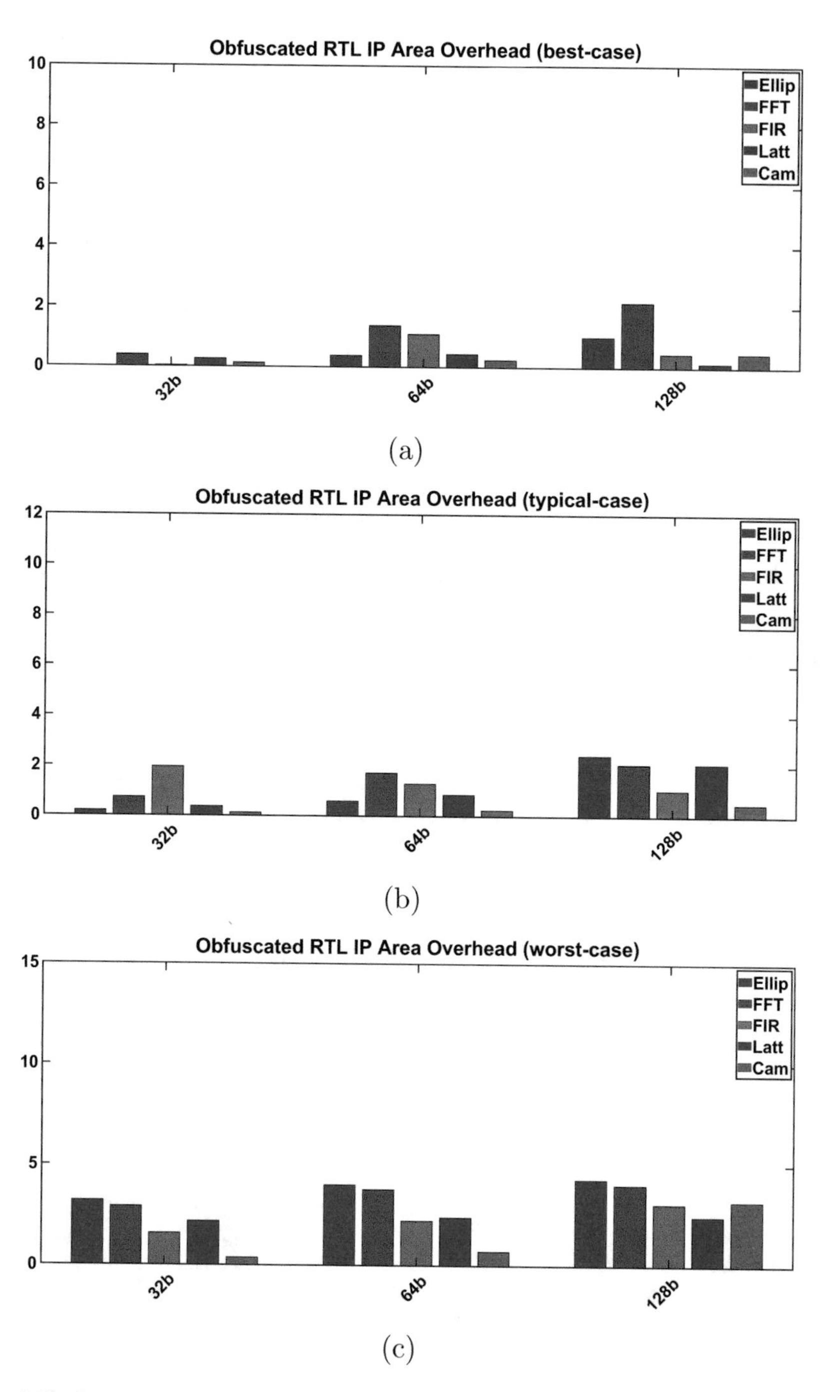

Fig. 2.10 Comparison of area overhead (%) for three different key lengths in three design corners

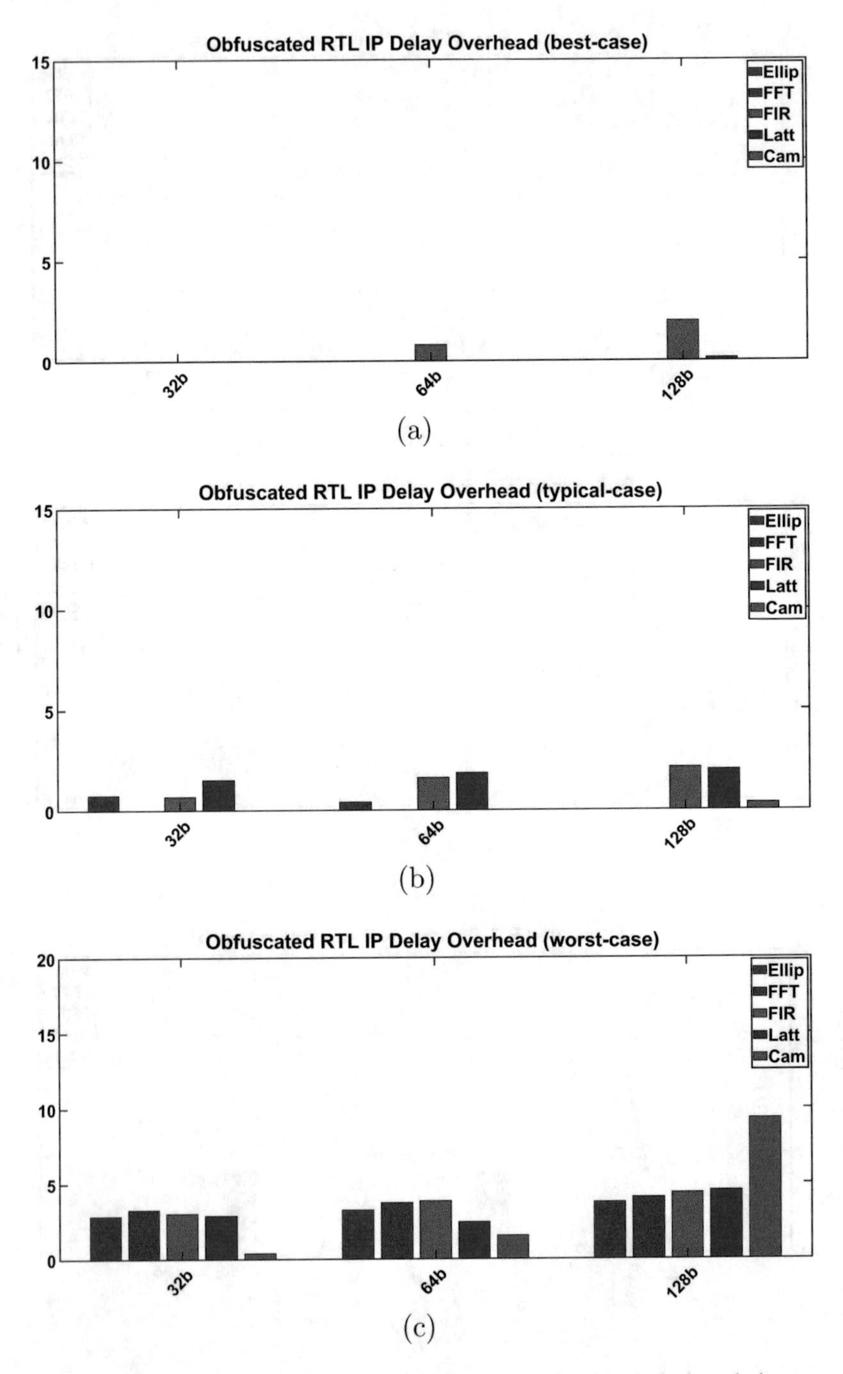

Fig. 2.11 Comparison of delay overhead (%) for three different key lengths in three design corners

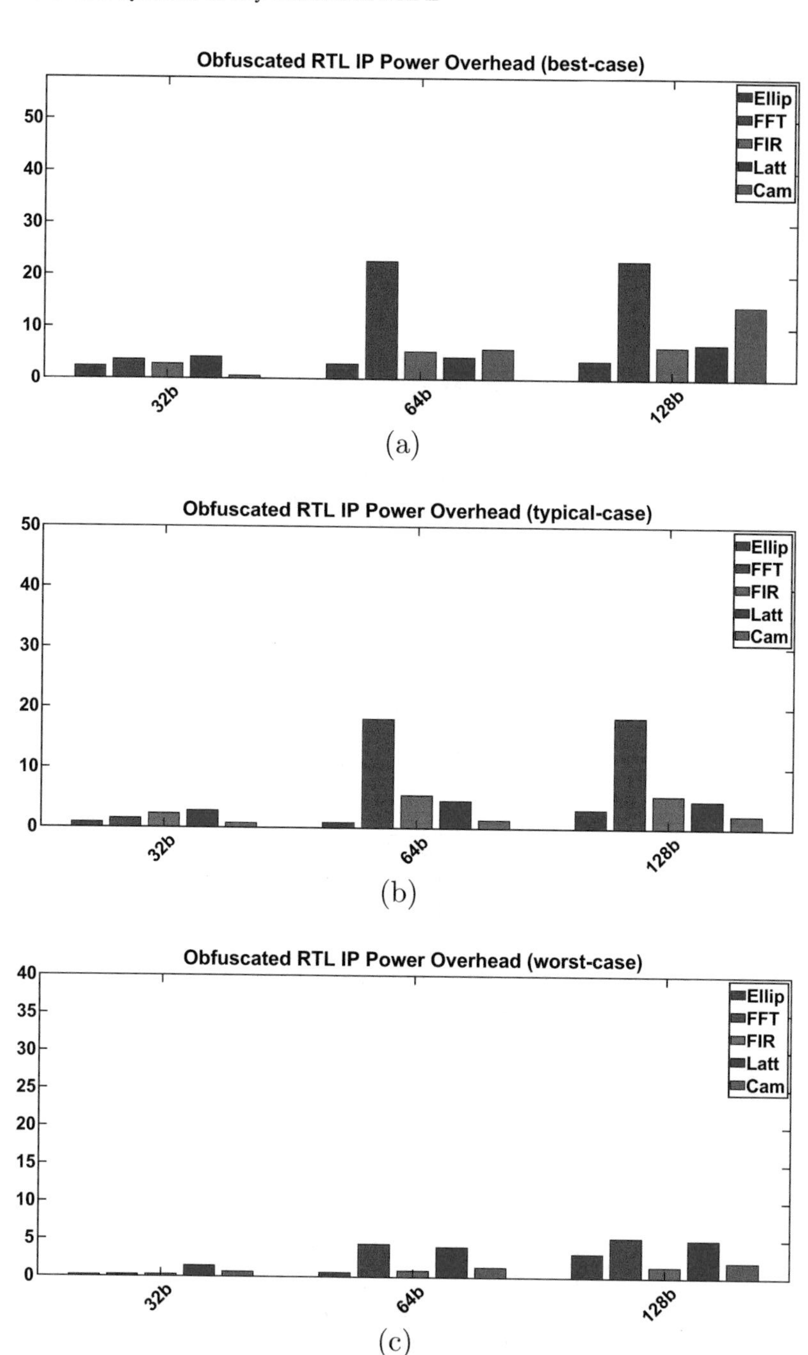

Fig. 2.12 Comparison of power overhead (%) for three different key lengths in three design corners

Table 2.3 Total number of nets under three cases for three different key lengths

Design	Obfuscation-free	Key obfuscation (worst case)			Key obfuscation (best case)			Key obfuscation (typical case)		
		32-bit	64-bit	128-bit	32-bit	64-bit	128-bit	32-bit	64-bit	128-bit
Elliptic	9147	9972	10,183	9194	13,550	9026	7925	14,082	9486	8308
FFT	8728	16,474	7186	9220	13,646	9049	8115	14,279	9292	8693
FIR	6982	17,150	10,210	19,727	13,786	9101	15,194	14,372	9495	15,183
Lattice	6087	8890	10,216	19,964	8710	9163	15,321	9117	9497	15,310
Camellia	19,520	9716	6208	20,950	8716	7891	15,577	9318	8245	15,584

Table 2.4 Cell-wise area breakdown of three protection techniques for three different key lengths. C: Number of Combinational Cells; S: Number of Sequential Cells; B/I: Number of Buffers/Inverters

Design	Obfuscation-free			Key obfuscation (32-bit)			Key obfuscation (64-bit)			Key obfuscation (128-bit)		
	C	S	B/I	C	S	B/I	C	S	B/I	C	S	B/I
Elliptic	7826	1368	826	7840	1400	833	7653	1350	1342	9007	1446	1406
FIR	5280	603	812	5317	635	822	5304	650	843	5588	643	1116
FFT	7035	932	1090	7054	959	1099	6597	1345	1207	6810	870	1512
Lattice	4461	668	548	4630	699	563	4689	701	858	4779	710	991
Camellia	13,946	1439	3814	13,971	1440	3910	14,650	1400	4001	16,264	1441	5353

Table 2.5 FPGA resources utilization (obfuscation-free and 32-bit key-obfuscated netlist)

Category	Element type	Obfuscation-free					Obfuscation (32-bit)				
		Elliptic	FIR	FFT	Lattice	Camellia	Elliptic	FIR	FFT	Lattice	Camellia
Sequential	Register	42	12	16	24	1478	42	12	16	24	1478
Combinational	LUTs	2957	984	1078	1586	3603	2971	1036	1178	1641	3771
	Shared LUTs	414	146	106	184	798	424	136	108	182	854
	MUX	458	–	325	438	–	458	–	356	436	–
	XOR	464	24	334	445	–	464	24	366	443	–
	DSP	3	6	2	3	–	3	6	2	3	–
ROM	–	–	–	–	–	64	–	–	–	–	64
IO-Pad	–	485	356	292	356	522	485	389	296	356	522

Table 2.6 FPGA resources utilization of 64- and 128-bit key for RTL obfuscation

Category	Element type	Obfuscation (64-bit)					Obfuscation (128-bit)				
		Elliptic	FIR	FFT	Lattice	Camellia	Elliptic	FIR	FFT	Lattice	Camellia
Sequential	Register	39	12	16	23	1478	32	12	16	17	1478
Combinational	LUTs	2863	986	1178	1496	3723	1918	777	1188	770	3883
	Shared LUTs	364	118	108	120	812	280	98	98	66	1248
	MUX	499	–	356	438	–	460	–	325	438	–
	XOR	506	24	366	445	–	459	24	334	445	–
	DSP	3	6	2	3	–	3	6	2	3	–
ROM	–	–	–	–	–	64	–	–	–	–	64
IO-Pad	–	485	293	296	293	522	485	199	281	262	522

Table 2.7 Obfuscation key extraction probability in obfuscated RTL

Design	(IO bit width, # registers)	Resilience (P)		
		32-bit	64-bit	128-bit
Elliptic	(32, 43)	$5.2e{-}97$	$1.77e{-}200$	$3.59e{-}398$
FIR	(32, 19)	$18.8e{-}86$	$5.38e{-}176$	$4.23e{-}349$
FFT	(32, 56)	$6.6e{-}44$	$5.44e{-}207$	$3.20e{-}413$
Lattice	(32, 21)	$7.0e{-}87$	$4.51e{-}179$	$3.85e{-}355$
Camellia	(32, 33)	$2.0e{-}95$	$2.21e{-}233$	$4.03e{-}383$

2.5.2.3 Attack Resiliency

To quantify the security of the proposed RTL obfuscation, according to Eq. (2.1) we report the obfuscation efficiency (resilience) in Table 2.7 for 32-, 64-, and 128-bit keys. The superior resilience value indicates that attacker is likely to be facing higher resistance in reverse engineering locked netlist.

2.6 Conclusion

In this chapter, we present a robust technique for obfuscating RTL design early in the design flow. During HLS, we introduce obfuscation on non-critical paths of CDFG to build high resistance against reverse engineering with minimal or no design overheads. We also provide a detailed treatment of key leakage probability for three different key lengths under three design corner cases.

References

1. Moore, G.E.: Cramming More Components onto Integrated Circuits. IEEE Solid-State Circuits Society Newsletter **38**(8), 114 (1965). https://doi.org/10.1109/N-SSC.2006.4785860
2. Stradley, J., Karraker, D.: The Electronic Part Supply Chain and Risks of Counterfeit Parts in Defense Applications. IEEE Transactions on Components and Packaging Technologies **29**(3), 703–705 (2006). https://doi.org/10.1109/TCAPT.2006.882451
3. Zhang, X., Tehranipoor, M.: Design of On-Chip Lightweight Sensors for Effective Detection of Recycled ICs. IEEE Transactions on Very Large Scale Integration (VLSI) Systems **22**(5), 1016–1029 (2014). https://doi.org/10.1109/TVLSI.2013.2264063
4. ICC. https://iccwbo.org/global-issues-trends/bascap-counterfeiting-piracy/
5. Zhang, J.: A Practical Logic Obfuscation Technique for Hardware Security. IEEE Transactions on Very Large Scale Integration (VLSI) Systems **24**(3), 1193–1197 (2016). https://doi.org/10.1109/TVLSI.2015.2437996
6. Wang, X., Jia, X., Zhou, Q., Cai, Y., Yang, J., Gao, M., Qu, G.: Secure and Low-Overhead Circuit Obfuscation Technique with Multiplexers. In: 2016 International Great Lakes Symposium on VLSI (GLSVLSI), pp. 133–136 (2016). https://doi.org/10.1145/2902961.2903000

7. Sengupta, A., Roy, D.: Protecting IP core during architectural synthesis using HLT-based obfuscation. Electronics Letters **53**(13), 849–851 (2017). https://doi.org/10.1049/el.2017.1329
8. Alkabani, Y., Koushanfar, F.: Active Hardware Metering for Intellectual Property Protection and Security. In: USENIX security symposium, pp. 291–306 (2007)
9. Chakraborty, R.S., Bhunia, S.: RTL Hardware IP Protection Using Key-Based Control and Data Flow Obfuscation. In: Proceedings of the 2010 23rd International Conference on VLSI Design, VLSID '10, pp. 405–410 (2010). https://doi.org/10.1109/VLSI.Design.2010.54
10. Li, L., Zhou, H.: Structural Transformation for Best-Possible Obfuscation of Sequential Circuits. In: 2013 IEEE International Symposium on Hardware-Oriented Security and Trust (HOST), pp. 55–60 (2013). https://doi.org/10.1109/HST.2013.6581566
11. Desai, A.R., Hsiao, M.S., Wang, C., Nazhandali, L., Hall, S.: Interlocking Obfuscation for Anti-tamper Hardware. In: Proceedings of the Eighth Annual Cyber Security and Information Intelligence Research Workshop, CSIIRW '13, pp. 8:1–8:4. ACM, New York, NY, USA (2013). https://doi.org/10.1145/2459976.2459985
12. Hansen, M.C., Yalcin, H., Hayes, J.P.: Unveiling the ISCAS-85 benchmarks: a case study in reverse engineering. IEEE Design Test of Computers **16**(3), 72–80 (1999)
13. Brglez, F., Bryan, D., Kozminski, K.: Combinational Profiles of Sequential Benchmark Circuits. In: IEEE International Symposium on Circuits and Systems, pp. 1929–1934. IEEE (1989)
14. Lao, Y., Parhi, K.K.: Obfuscating DSP Circuits via High-Level Transformations. IEEE Transactions on Very Large Scale Integration (VLSI) Systems **23**(5), 819–830 (2015). https://doi.org/10.1109/TVLSI.2014.2323976
15. Sengupta, A., Kachave, D., Roy, D.: Low Cost Functional Obfuscation of Reusable IP Cores Used in CE Hardware Through Robust Locking. IEEE Transactions on Computer-Aided Design of Integrated Circuits and Systems **38**(4), 604–616 (2019). https://doi.org/10.1109/TCAD.2018.2818720
16. Sengupta, A., Roy, D., Mohanty, S.P., Corcoran, P.: DSP design protection in CE through algorithmic transformation based structural obfuscation. IEEE Transactions on Consumer Electronics **63**(4), 467–476 (2017). https://doi.org/10.1109/TCE.2017.015072
17. Sengupta, A., Mohanty, S.P., Pescador, F., Corcoran, P.: Multi-Phase Obfuscation of Fault Secured DSP Designs With Enhanced Security Feature. IEEE Transactions on Consumer Electronics **64**(3), 356–364 (2018). https://doi.org/10.1109/TCE.2018.2852264
18. Pilato, C., Regazzoni, F., Karri, R., Garg, S.: TAO: Techniques for Algorithm-Level Obfuscation during High-Level Synthesis. In: 2018 55th ACM/ESDA/IEEE Design Automation Conference (DAC), pp. 1–6 (2018). https://doi.org/10.1109/DAC.2018.8465830
19. Pilato, C., Wu, K., Garg, S., Karri, R., Regazzoni, F.: TaintHLS: High-Level Synthesis for Dynamic Information Flow Tracking. IEEE Transactions on Computer-Aided Design of Integrated Circuits and Systems **38**(5), 798–808 (2019). https://doi.org/10.1109/TCAD.2018.2834421
20. Maes, R., Schellekens, D., Tuyls, P., Verbauwhede, I.: Analysis and Design of Active IC Metering Schemes. In: 2009 IEEE International Workshop on Hardware-Oriented Security and Trust, pp. 74–81 (2009). https://doi.org/10.1109/HST.2009.5224964
21. Koushanfar, F.: Integrated Circuits Metering for Piracy Protection and Digital Rights Management: An Overview. In: Proceedings of the GLSVLSI, pp. 449–454. ACM (2011)
22. Koushanfar, F., Hong, I., Potkonjak, M.: Behavioral Synthesis Techniques for Intellectual Property Protection. ACM Trans. Des. Autom. Electron. Syst. **10**(3), 523–545 (2005). https://doi.org/10.1145/1080334.1080338
23. Hong, I., Potkonjak, M.: Behavioral Synthesis Techniques for Intellectual Property Protection. In: Proceedings of the 36th Annual ACM/IEEE Design Automation Conference, DAC '99, pp. 849–854. ACM, New York, NY, USA (1999). https://doi.org/10.1145/309847.310085
24. Wendt, J.B., Koushanfar, F., Potkonjak, M.: Techniques for Foundry Identification. In: 2014 51st ACM/EDAC/IEEE Design Automation Conference (DAC), pp. 1–6 (2014). https://doi.org/10.1145/2593069.2593228

25. Caldwell, A.E., Hyun-Jin Choi, Kahng, A.B., Mantik, S., Potkonjak, M., Gang Qu, Wong, J.L.: Effective Iterative Techniques for Fingerprinting Design IP. IEEE Transactions on Computer-Aided Design of Integrated Circuits and Systems **23**(2), 208–215 (2004). https://doi.org/10.1109/TCAD.2003.822126
26. Chakraborty, R.S., Bhunia, S.: Hardware Protection and Authentication Through Netlist Level Obfuscation. In: 2008 IEEE/ACM International Conference on Computer-Aided Design, pp. 674–677 (2008). https://doi.org/10.1109/ICCAD.2008.4681649
27. Chakraborty, R.S., Bhunia, S.: Security Through Obscurity: An Approach for Protecting Register Transfer Level Hardware IP. In: 2009 IEEE International Workshop on Hardware-Oriented Security and Trust, pp. 96–99 (2009). https://doi.org/10.1109/HST.2009.5224963
28. Chakraborty, R.S., Bhunia, S.: HARPOON: An Obfuscation-Based SoC Design Methodology for Hardware Protection. IEEE Transactions on Computer-Aided Design of Integrated Circuits and Systems **28**(10), 1493–1502 (2009). https://doi.org/10.1109/TCAD.2009.2028166
29. Paulin, P.G., Knight, J.P.: Force-Directed Scheduling for the Behavioral Synthesis of ASIC's. IEEE Transactions on Computer-Aided Design of Integrated Circuits and Systems **8**(6), 661–679 (1989). https://doi.org/10.1109/43.31522
30. Synopsys PDK. https://www.synopsys.com/community/university-program/teaching-resources.html
31. Gopalakrishnan, C., Katkoori, S.: Behavioral Synthesis of Datapaths with Low Leakage Power. In: 2002 IEEE International Symposium on Circuits and Systems. Proceedings (Cat. No.02CH37353), vol. 4, pp. IV–IV (2002)
32. OpenCores. https://opencores.org/

Chapter 3
Source Code Obfuscation of Behavioral IPs: Challenges and Solutions

Nandeesh Veeranna and Benjamin Carrion Schafer

3.1 Introduction

The globalization of Integrated Circuit(IC) design and manufacturing process poses serious concerns about their trustworthiness and security. It is nowadays virtually impossible to fully design and manufacture an IC fully in-house using only in-house tools. To make things worse, with the increased time-to-market pressure, companies rely heavily on third parties to develop their ICs.

Most ICs are now heterogeneous system on chips (SoCs). These contain multiple in-house developed IPs and third-party IPs (3PIPs) integrated onto the same chip. IC vendors have also started to embrace High-Level Synthesis (HLS) in order to further reduce the time to market. HLS is a process that converts untimed behavioral descriptions such as ANSI-C or C++ into a Register-Transfer Level (RTL) description (e.g., Verilog or VHDL) that can efficiently execute it. This new design methodology has paved the way to the market of third-party behavioral IPs (3PBIPs). The market for 3PBIPs is theoretically larger than the traditional RT-Level IP because it includes a larger user base ranging from algorithm developers to traditional VLSI designers.

One of the main challenges for 3PBIP providers that is hampering a larger offering of these BIPs is how to protect the IP from being re-used illegally. A study conducted by SEMI [1] in 2008 estimated that the semiconductor industry loses up to $4 billion annually because of IP infringement.

N. Veeranna
ST Micro, Singapore, Singapore
e-mail: nandeesha.veeranna@st.com

B. Carrion Schafer (✉)
Department of Electrical and Computer Engineering, The University of Texas at Dallas, Richardson, TX, USA
e-mail: schaferb@utdallas.edu

S. Katkoori, S. A. Islam (eds.), *Behavioral Synthesis for Hardware Security*,
https://doi.org/10.1007/978-3-030-78841-4_3

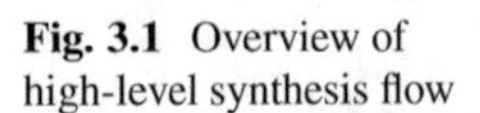

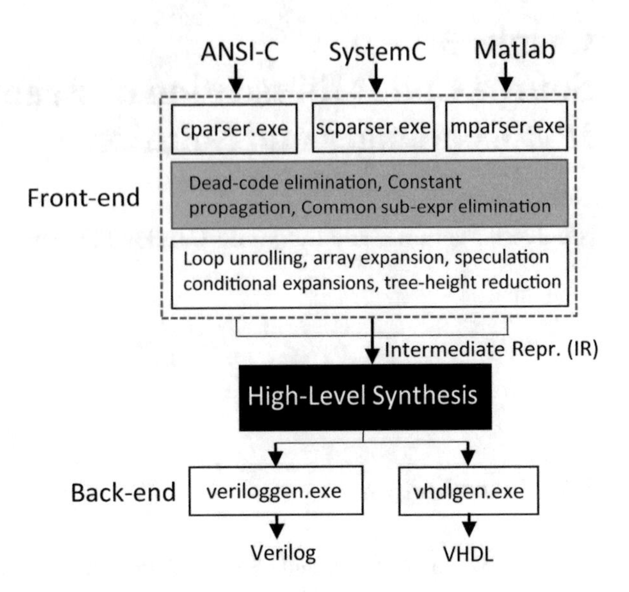

Fig. 3.1 Overview of high-level synthesis flow

Also, BIPs normally have different price policies depending on the amount of disclosure of the IP. A BIP consumer can purchase the complete source code and would not need the services of the BIP provider anymore as he would have full control and visibility of the code. An alternative service model includes the encryption of the BIP with a predefined set of constraints (e.g., IO bit widths, synthesis pragmas), which the BIP consumer cannot modify. Thus, any future alterations would require a new license agreement and a new purchase. The first type of service is obviously much more expensive than the second, typically 10–100×. This was shown by the authors in [2], where the authors showed how obfuscating/encrypting different parts of the BIP led to the control of the BIPs' search space.

Another issue that needs to be addressed is when a company wants to evaluate a 3PBIP. During the evaluation process, the BIP cannot be made visible to the company evaluating it as the BIP consumer would have full access to the IP, and thus, would not need to purchase the BIP after the evaluation process. Thus, mechanisms to protect BIPs are required.

Source code obfuscation is an easy and inexpensive form to protect IPs. In obfuscation, a functional equivalent source code description is generated that is virtually impossible for humans to understand and very difficult to reverse engineer. A typical obfuscation process removes comments and spaces, renames variables, and adds redundant expressions. There are a multitude of free or inexpensive source code obfuscators, making this a very simple and effective way to protect BIPs. The main problem when using these obfuscators is that commercial HLS tools often make use of in-house customized parsers, as front ends. These are needed as commercial HLS tools make extensive use of C/C++ extensions to allow to generate more efficient RTL (e.g., allow for custom bit widths). Figure 3.1 shows an example of a typical HLS tool structure. Different parsers parse different input languages (e.g., ANSI-C, SystemC/C++ or MATLAB) into a common intermediate internal representation (IIR). The first step in this process performs traditional technology independent compiler optimizations, e.g., constant propagation, dead-code elimination, and common sub-expression eliminations. Previous work has shown that the quality of the parsers significantly affects the quality of the synthesized circuit [3]. Parsers also perform synthesis transformations like loop unrolling, array expansion, and tree-height reductions on this simplified data structure in order to generate the IIR. This IIR is then passed to the main synthesis step to generate optimized RTL.

The main HLS synthesizer then reads this IIR and performs the three main HLS steps on it. In particular: (1) resource allocation, (2) scheduling, and (3) binding. The final step (back end) is responsible for generating the RTL code (i.e., Verilog or VHDL). This methodology has some unique advantages. One major advantage is that it is able to take as input the IIR instead of directly the source code making the entire HLS framework extensible to new languages by simply adding a new front-end parser. One obvious disadvantage is that each parser needs to be often developed in-house by the Electronic Design Automation (EDA) company that is probably not be an expert in parsers, and thus leading to sub-optimal designs due to not being able to fully optimize the input description. This is particularly important in the case of obfuscation as this process typically inserts multiple redundant operations to make the code less readable. This leads in turn to an increase in area, delay, and latency of the synthesized circuit, which is unacceptable for any BIP consumer.

The work presented in this chapter analyzes the impact of obfuscation on the quality of results (QoR) of synthesized (HLS) circuits and proposes effective obfuscation methods that do minimize the area and performance degradation, due to the obfuscation process. For this, we propose a fast and efficient heuristic obfuscation method.

3.2 Motivational Example

Figure 3.2 shows three examples of how the level of obfuscation affects the area of the synthesized circuit from a commercial HLS tool.[1] The results shown correspond to the Adaptive Differential Pulse-Code Modulation (ADPCM) benchmark, Fast-Fourier Transform (FFT), and AES encryption algorithm taken from the open-source S2Cbench [4] benchmark suite re-written to ANSI-C.

As shown in the figure, the increase in the level of obfuscation leads to the monotonically increase in the area in all of the three benchmarks. This implies that the parser cannot fully re-optimize the redundant operations added by the obfuscator, leading to circuits that are much larger than the un-obfuscated version of the same description. It is therefore important to develop techniques to maximize

[1] Tool name cannot be disclosed due to confidentiality agreement.

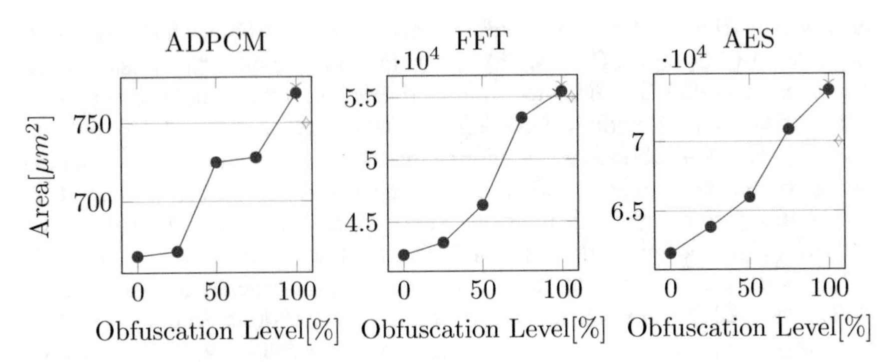

Fig. 3.2 Area degradation for three different examples from BIP [4] with the increase in level of obfuscation

the obfuscation of behavioral descriptions for HLS while maintaining the QoR as compared to the un-obfuscated version.

3.3 Previous Work

Most previous works in the field of hardware obfuscation deal with making it harder to reverse engineer a logic circuit. This chapter is mainly focused on the efficient obfuscation of the source code, which is typical in the software domain.

Previous works in this domain include [5] and [6]. This chapter represents RTL code as a data flow [5] or state transition matrix [6] graph. The graph is then modified with additional states, also called as key states, in the finite-state machine representation of the code that should be passed through with the aid of a key sequence [5] or a code word [6]. The IP will start functioning only after the application of correct keys. Otherwise, the IP will be stuck in a futile, obfuscated state. This is the main idea behind functional locking.

Similar to the software case, soft IPs in hardware can also be obfuscated in terms of readability and intelligibility. The authors in [7] incorporated the techniques such as loop transformation (loop unrolling), statement reordering, conversion of parallel processing to sequential, etc., to make the VHDL source code nearly impossible to read and yet functionally identical to the original source code. Kainth et al. [8] in their work took the leverage of control flow flattening [9, 10] approach to break the function and loop into blocks and convert into "switch" statements, which implies the control flow of the program becomes invincible to an attack. The authors in [11] introduce a technique based on a family of transformations applied at behavioral descriptions for HLS. The main goal is to obfuscate the generated hardware and not the behavioral description itself.

Although the behavioral IP descriptions for HLS are similar to the software descriptions, the software obfuscation techniques cannot be ported directed to obfuscation BIPs as it does not take into account the degradation introduced by the *unoptimized* parsers, which in turn lead to unoptimized hardware circuits.

3.4 Obfuscation

Obfuscation can be defined as the intentional act of obscuring the functionality of a program to secure the intellectual property innate in it. Formally, it can be defined as:

Definition O is an obfuscator that can transform a program P into its obfuscated version $O(P)$ that has the same functionality F as P, such that the $F\{P\} = F\{O(P)\}$ and such that it is unintelligible for an adversary who is trying to recover P from $O(P)$.

Software obfuscation is very different from hardware obfuscation. In the software obfuscation case, the source code changes its structures with the intention to make it more difficult to understand. The idea is to make it impossible for attackers to reverse engineer the code yet fully preserving the functionality. Source code obfuscation is the simplest alternative to source code encryption as it does not require to encrypt the code into a cipher text to be layer decrypted by the HLS tool.

In the hardware domain, obfuscation is mainly concerned with hiding the functionality of a logic circuit by changing making it difficult to reverse engineer. A full taxonomy of the hardware obfuscation techniques at different abstraction levels (RTL, gate-level, layout) and also some emerging techniques can be found in [12]. This taxonomy does not include the behavioral level, mainly because at this level no circuit is generated yet and thus, obfuscating at this level is equivalent to traditional software obfuscation.

The obfuscation of BIPs for HLS is similar to the obfuscation of software programs. Most of the BIP vendors supply IPs in either ANSI-C or SystemC/C++. This allows to use any commercially available software obfuscators to obfuscate these IPs.

Figure 3.3 shows a simple example of the obfuscation of a BIP program that continuously computes the moving average of eight numbers using a commercially available C/C++ obfuscator [13]. As shown in the figure, the obfuscated version of the original code snippet is extremely difficult to understand and therefore hard to reverse engineer.

3.4.1 Obfuscation Primitives

This section describes how software obfuscators work and which techniques they use. The list of obfuscation techniques includes:

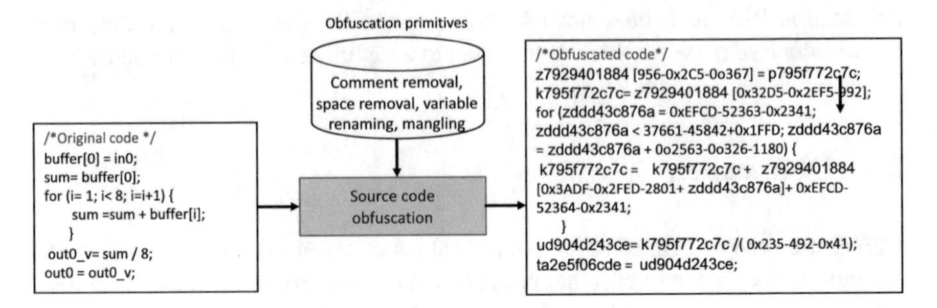

Fig. 3.3 Example of behavioral IP obfuscation process using a library of obfuscation primitives

1. *Mangle mathematical expressions and constants*: Mangling mathematical expressions and variables create extra redundancy in the code but has a zero net effect on the final functionality. For example, reverting back to Fig. 3.3, the integer "8" in the *for* loop has been replaced by "37661-45842+0x1FFD." Also, the "sum" expression, "sum =sum + buffer[i];" inside the *for* loop was replaced by "k795f772c7c = k795f772c7c + z7929401884 [0x3ADF-0x2FED-2801+ zddd43c876a]+ 0xEFCD-52364-0x2341;". Although both expressions evaluate to the same original value and expression, the mangled version is much harder to understand by anyone trying to reverse engineer the code.
2. *Comments*: Deleting comments or inserting nonsensical comments makes the source code harder to read.
3. *Stripping new lines*: This is similar to the strip spaces that remove the extra lines in a program, which leads to code congestion and poor readability.
4. *Replacing identifiers and variables*: This is one of the important techniques used by most of the obfuscators. There are different ways of replacing identifiers and variables in a program.

   (i) *Use of prefixes*: This is one of the simplest ways of replacing identifiers. In this case, every identifier is prefixed with a certain word. For example: "sum" in Fig. 3.3 is converted into "_confidential_sum." Due to its simplicity, it is also not the most effective method.
   (ii) *Using a set of similar characters*: such as I and l or O and 0 and replace them all with the combination of either I and l or O and 0.
   (iii) *Using upper and lower cases*: Since many programming languages are case sensitive, e.g., C and C++, identifiers can be declared with a mixture of lower and upper case letters.
   (iv) *Use of hash values*: Identifiers are replaced with hash values. For example, "buffer" in Fig. 3.3 is replaced by "z7929401884" generated as a hash value significantly affecting the readability of the code. The Message Digest Algorithm (md5) is one of the most common hash value generators used for this purpose.

5. *Removing spaces*: A good programmer will always make extensive use of spaces and indentations to make the source code easier to read. The main goal is to maximize its readability making the program also easier to debug. Trimming the extra spaces inserted in the code is one of the techniques used in obfuscation to reduce the code's readability.

Obfuscators typically mix all of these techniques to make the final obfuscated code as difficult to understand as possible. These techniques work extremely well with mature compilers, i.e., gcc and g++, as these are based on very efficient and robust parsers that have been developed and tested for a long period of time. Hence, the binary code ($.exe/.o$) of the obfuscated software version and the un-obfuscated versions are identical. Thus, no penalties are observed between the two versions.

As discussed in the motivational example, the parsers of different EDA tools are not as mature and can lead to different results for the obfuscated version of the code. Although the functionality is preserved, the obfuscated version incurs overheads in terms of area and delay because these parsers are not as robust as the software compiler ones, as often the input language is a subset of software languages and, hence, dedicated parsers are needed.

3.4.2 Investigating the Overheads Due to Source Code Obfuscation

Although the functionality of the obfuscated BIP is identical to that of the un-obfuscated version, the parser might fail to optimize some of the mangled mathematical expressions. This leads to additional circuitry generated by the synthesis process. For example, in Fig. 3.3, "sum =sum + buffer[i];" has been obfuscated to "k795f772c7c = k795f772c7c + z7929401884 [0x3ADF-0x2FED-2801+ zddd43c876a]+ 0xEFCD-52364-0x2341;". The obfuscation process adds the following dead code "0xEFCD-52364-0x2341." The compiler typically optimizes the code using dead-code elimination techniques. This is a very important optimization step because failure to optimize this leads to the generation of extra circuitry that affects the area, power, and delay of the final circuit. For the above example, the synthesizer may generate some extra adders for the unoptimized expression.

3.5 Proposed Optimization Method to Minimize the Impact of Obfuscation on the Quality of Synthesization of the Circuit

The motivational example showed that there is a trade-off between QoR and the level of obfuscation. In this section, we discuss two methods to minimize the QoR degradation due to the obfuscation process. The first method is a meta-heuristic-based method based on Genetic Algorithm (GA), which has been shown to lead to very good results for multi-objective optimization problems like this one where QoR is traded with amount of obfuscation. The second method is based on a fast iterative greedy method.

3.5.1 GA-Based Obfuscation

GAs were first proposed by John Holland in [14] to find solutions to problems that are computationally intractable. It is a naturally inspired meta-heuristics, which implies that it can be used to solve different types of optimization problems. Other types of these meta-heuristics include Simulated Annealing (SA) and Ant-Colony Optimization (ACO).

For this particular case, the objective is to find the source code with the highest level of obfuscation while leading to the smallest amount of degradation in terms of area, delay, and power overhead. Another option could be to generate a trade-off curve of Pareto-optimal configurations with unique degradation vs. obfuscation levels. This could be easily achieved by changing the cost function of the heuristic. Figure 3.4 shows the main steps involved in our proposed GA-based obfuscation process, where every gene corresponds to a single line in the source code to be obfuscated (BIP). Each line can be either obfuscated or left un-obfuscated.

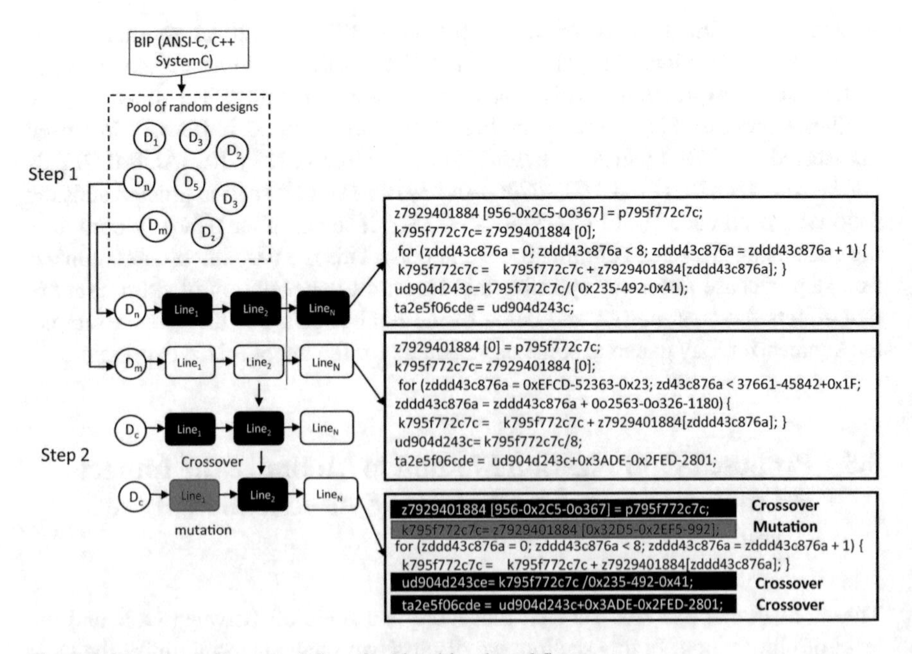

Fig. 3.4 Overview of proposed Genetic-Algorithm-based flow

The inputs to the GA-based method are the original un-obfuscated source code P, the completely obfuscated code $O(P)$ using the default settings of the obfuscator used, and a *lightly* obfuscated version $LO(P)$, which does not include mangling, as this obfuscation technique has been shown to lead to the most degradation. Thus, $LO(P)(QoR) = P(QoR)$. This *lightly* obfuscated version is used to guide the obfuscation process and is important. It is important to avoid any syntax errors when individual lines of code are fully obfuscated. In particular, the process is divided into two steps as follows:

Step 1: Initial Population Generation In this first step, an initial random population is generated, where a population is a randomly obfuscated version of the source code. Each new configuration has a random obfuscation level O_L. This is done by parsing the source code line by line and deciding if the given line should be obfuscated or not. The obfuscation probability is an input parameter to the GA. In this chapter, we set this to 30%. This will generate a random pool of parents to start generating different offsprings.

Each newly generated configuration is synthesized (HLS), and the QoR is extracted. This includes QoR = {Area, Delay} extracted and the cost C that in turn allows to measure the cost of this new configuration using $C = \alpha \times Area + \beta \times O_L$. Because of the scale difference between the cost function parameters, these have to be scaled. The values for α and β can be either fixed to find a particular design or can be dynamically updated to explore the entire search space, e.g., $\alpha = 1$ and $\beta = 0$ for the area dominant designs and $\alpha = 0$ and $\beta = 1$ to find obfuscation dominant designs, and $\alpha = 0.5$ and $\beta = 0.5$ when a compromise between QoR degradation and level of obfuscation is sought. The process is repeated for N times.

Step 2: Offspring Generation Two main functions are responsible for generating an offspring: (1) mutation and (2) crossover. First, two parents are randomly selected from the pool generated in Step 1. Then, a crossover is performed between the two parents chosen randomly from within the population. Although there are many techniques to choose parents, thischapter makes use of *adjacent* designs technique. A random cut-off point is chosen between the two parents, and the lines of code obfuscated in each parent are used to generate the offspring.

Then, individual genes are mutated. In this case, every line of code is considered a gene. The mutation is performed randomly selecting some of the lines of code and deciding if this line should be obfuscated or not. This method traverses the entire description from the first line to the last line and selects if the line should be mutated or not. A random probability P_{rand} is used for this purpose. If P_{rand} is less than a given threshold (10% in this case), then the corresponding line of the BIP is obfuscated. Finally, the offspring is synthesized and the cost for this configuration is computed. These steps are repeated until the given exit condition is met. In this case, if after N new offsprings the cost function could not be improved, where in this chapter $N = 10$.

3.5.2 Fast Iterative Greedy Method

The GA-based method can find a good result, although the running time of the method is not negligible. Thus, a faster method is desirable. In this section, we propose a fast heuristic method that generates a single design with better QoR and obfuscation level. Algorithm 3.1 shows the pseudo-code of the proposed method that is based on a fast iterative greedy heuristic. A detailed description of the algorithm is given below.

The inputs to the proposed method are the original code P, the completely obfuscated code $O(P)$, the **lightly** obfuscated code $LO(P)$(explained in the GA-based obfuscation), and the QoR synthesis results of the different variants (area and delay). The output of this method is an obfuscated source code with the highest amount of obfuscation and lowest QoR degradation (best obfuscated design $BO(P)$). The method is based on three main steps, as follows:

Step 1: Identify Mangled Expressions As already explained in the previous section, the mangled constant and expressions are mainly responsible for the QoR degradation seen, as the ANSI-C parser cannot optimize these redundant operations efficiently. Thus, in this step, the completely obfuscated code $O(P)$ is parsed and the lines of code that contain mangled constants and expressions are identified and extracted. The line numbers, where these operations take place, are stored in an array $array_1$ (lines 1–2).

Algorithm 3.1 Summary of proposed fast heuristic

Input: Original_code → P, Completely_obfuscate_code → O(P), Lightly_obfuscated_ code → LO(P), QoR of P, QoR of O(P)
Output: The best obfuscated design with minimal degradation BO(P)
Procedure:
1: Search for the lines in $O(P)$ which contains mangled constants and expressions;
2: $array_1 = search\_lines(O(P))$;
3: Identify Real Sources of Degradation;
4: **while** $array_1$ **do**
5:     Read the file O(P);
6:     **if** Line number of O(P) == $array_1$[i] **then**
7:         $O(P_{single}) = Unobfuscate\_line(i, O(P))$;
8:         $QoR(O(P_{single})) = HLS(O(P_{single}))$;
9:         **if** $QoR(O(P_{single})) < QoR(O(P))$ **then**
10:             $array_2 \leftarrow i$;
11:             break;
12:         **end if**
13:     **end if**
14:     i++;
15: **end while**
16: Generate new obfuscated code which leads to smallest QoR degradation;
17: $BO(P) = replace\_obf\_lines(array_2, O(P))$;
18: $QoR(area, delay) = HLS(BO(P))$;

Step 2: Identify Real Sources of Degradation Not all of the mangled expressions contribute toward QoR degradation, thus the ones that do need to be identified. For this, the proposed method analyzes the previously extracted lines stored in $array_1$ and determines which lines actually contribute toward any QoR degradation. For this, each line in $array_1$ is un-obfuscated separately, and a new behavioral description $O(P_{single})$ is generated and synthesized (HLS). The QoR is in turn compared with the un-obfuscated original description ($O(P)$) to determine if the particular obfuscated line leads to any degradation or not. This is done by simply comparing the area and delay of the two versions as follows: $\Delta_{QoR(O(P)-QoR(O(P_{single})} = \{Area_{O(P)} - Area_{O(P_{single})}, Delay_{O(P)} - Delay_{O(P_{single})}\}$ (lines 8–11). The result of this step is a new $array_2$ containing only the obfuscated lines that do contribute to QoR degradation ($\Delta_{QoR(O(P)-QoR(O(P_{single})} > 1$)(line 10). The advantage of this method is that the order of complexity only grows linearly with the number of lines of code (l), and hence $O(n)$. In the worst case, this step will require l synthesis, where l is the total number of lines of code of the BIP.

Step 3: Replace Obfuscation Type That Degrades QoR The $array_2$ obtained in step 2 contains the line numbers of the obfuscated code that contribute toward the QoR degradation. A new obfuscated version of the BIP is in turn generated by substituting the complex expressions in the $array_2$, by simple obfuscation techniques that are known not to result any QoR degradation. These include: instead of mangling an integer, e.g., "8" in the for loop of Fig. 3.3 with "$37661 - 45842 + 0x1FFD$," it can be obfuscated as "$5 - 6 + 9$" or any other simple expression that can be easily optimized by the parser. The same technique applies for the mangled expressions. The obfuscated revision obtained is the best obfuscated code $BO(P)$ because in this heuristic, instead of un-obfuscating every mangled integer and expression (line 17), we only obfuscate the lines that actually degrade the QoR value. Finally, the QoR of the newly generated obfuscated code $BO(P)$ is compared with the QoR of the original code P by performing a last HLS (line 18). This obfuscated code is then returned as the final solution.

3.6 Experimental Results and Discussions

Seven synthesizable SystemC benchmarks from the open-source synthesizble SystemC benchmark suite (S2CBench) [4] are used to evaluate our proposed methods. These benchmarks were manually translated into ANSI-C, since the commercial HLS tool used in this chapter uses ANSI-C as input. The software obfuscator used is Stunnix C/C++ [13]. Table 3.1 shows the qualitative results (area and maximum delay) for the original (un-obfuscated) code, the default 100% obfuscated code, and the best results found by the GA and iterative greedy method. Table 3.2 compares the running times of the two introduced methods (Genetic Algorithm vs. iterative greedy).

Table 3.1 Experimental results—Quality of Results (QoR) comparison

	Original		Fully obfuscated		Genetic algorithm			Iterative greedy		
Benchmark	Area [μm^2]	Delay [ns]	Area [μm^2]	Delay [ns]	Area [μm^2]	Delay [ns]	O_L [%]	Area [μm^2]	Delay [ns]	O_L [%]
Ave8	710	1.33	1722	2.68	710	1.33	88.88	710	1.33	94.73
ADPCM	665	5.72	669	5.89	666	5.72	99.07	665	5.72	99.07
AES	17,376	2.39	18,329	2.39	17,554	2.39	99.83	17376	2.39	99.83
Bsort	17,800	1.68	18,046	2.57	17,917	2.51	98.46	17,800	1.68	98.46
Disparity	1879	3.09	2571	4.41	1981	3.29	98.99	1879	3.09	99.61
Filter	2367	4.05	3523	4.67	2473	4.32	96.67	2363	4.05	96.67
Sobel	1173	2.94	4363	15.28	1301	4.04	97.00	1173	2.94	98.55
Avg.							96.98			98.13

Table 3.2 Experimental results—runtime comparison

Benchmark	Genetic algorithm T_{run} [s]	Iterative greedy T_{run} [s]	Speedup
Ave8	854	22	38.8
ADPCM	1062	60	17.7
AES	1256	334	3.8
Bsort	1042	26	40.1
Disparity	2256	319	7.1
Filter	926	29	31.9
Sobel	959	27	35.5
Avg.			25.0

Some important observations can be made from these results. First, by fully obfuscating the BIP leads to an area and delay degradation on average of 35% and 37.7%, respectively. This is obviously not acceptable as no BIP customer would ever accept this overhead. Second, the iterative greedy method is much faster than the GA, on average 18×. Finally, both GA and iterative greedy lead to very good results, with both of the methods allowing to fully obfuscate on average over 96% of the BIP, without degrading the area nor delay of the synthesized circuit. In this context, full obfuscation implies using complex mangling operations.

Both of the proposed methods completely obfuscated the entire BIP (all of the lines). The missing % refers to complex obfuscation not allowed. For these remaining lines of code, simple obfuscation techniques used in the *light* obfuscation version are used.

In summary, it can be concluded that the iterative greedy method is very efficient (requiring on average 60 s to execute) while leading to BIP descriptions with over 98% of obfuscated lines using complex obfuscation techniques with no penalties.

3.7 Conclusions

In this chapter, we have analyzed the impact of code obfuscation of BIPs for HLS on the quality of results (QoR) when the HLS parser cannot fully optimize complex obfuscated expressions. As shown, two functionally identical BIP, one obfuscated and the other not, can lead to significant area and delay differences when the parser, which is responsible for the technology independent optimizations, does not do a good job. Thus, two methods were proposed to maximize the amount of obfuscated source code while minimizing the amount of degradation introduced due to the obfuscation process. The first method is based on a GA meta-heuristic and is used as the baseline method against which we compare a faster heuristic based on an iterative greedy algorithm. Both methods work well, while the iterative greedy algorithm is much faster.

References

1. SEMI: Innovation Is at Risk as Semiconductor Equipment and Materials Industry Loses up to $4 Billion Annually Due to IP Infringement (2008). http://www1.semi.org/en/white-paper-ip-infringement-causes-4-billion-loss-industry-annually
2. Wang, Z., Carrion Schafer, B.: Partial encryption of behavioral IPs to selectively control the design space in high-level synthesis. In: 2019 Design, Automation Test in Europe Conference Exhibition (DATE) (2019)
3. Veeranna, N., Carrion Schafer, B.: Efficient behavioral intellectual properties source code obfuscation for high-level synthesis. In: 2017 18th IEEE Latin American Test Symposium (LATS), pp. 1–6 (2017)
4. Schafer, B.C., Mahapatra, A.: S2CBench: synthesizable SystemC benchmark suite for high-level synthesis. IEEE Embed. Syst. Lett. **6**(3), 53–56 (2014)
5. Chakraborty, R.S., Bhunia, S.: RTL hardware IP protection using key-based control and data flow obfuscation. In: 2010 23rd International Conference on VLSI Design, pp. 405–410 (2010). https://doi.org/10.1109/VLSI.Design.2010.54
6. Desai, A., Hsiao, M., Wang, C., Nazhandali, C.: Interlocking Obfuscation for Anti-Tamper Hardware. In: In: Proceedings of the Eighth Annual Cyber Security and Information Intelligence Research Workshop, pp. 1–4. ACM, New York (2013)
7. Brzozowski, M., Yarmolik, V.: Obfuscation as intellectual rights protection in VHDL language. In: 6th International Conference on Computer Information Systems and Industrial Management Applications, pp. 337–340 (2007)
8. Kainth, M., Krishnan, L., Narayana, C., Virupaksha, S.G., Tessier, R.: Hardware-assisted code obfuscation for FPGA soft microprocessors. In: 2015 Design, Automation Test in Europe Conference Exhibition (DATE), pp. 127–132 (2015)
9. Wang, C.: A security architecture for survivability mechanisms. Ph.D. thesis, University of Virginia (2000)
10. Laszlo, T., Kiss, A.: Obfuscating C++ programs via control flow flattening. In: Sectio Computatorica (2009)
11. Pilato, C., Regazzoni, F., Karri, R., Garg, S.: TAO: techniques for algorithm-level obfuscation during high-level synthesis. In: 2018 55th ACM/ESDA/IEEE Design Automation Conference (DAC), pp. 1–6 (2018). https://doi.org/10.1109/DAC.2018.8465830
12. Forte, D., Bhunia, S., Tehranipoor, M.: Hardware Protection Through Obfuscation. Springer, Berlin (2017)
13. Stunix C/C++ Obfuscator. http://stunnix.com/. Accessed 15 Jan 2017
14. Holland, J.: Adaptation in Natural and Artificial Systems. The University of Michigan Press, Ann Arbor (1975)

Chapter 4
Hardware IP Protection Using Register Transfer Level Locking and Obfuscation of Control and Data Flow

Pranesh Santikellur, Rajat Subhra Chakraborty, and Swarup Bhunia

4.1 Introduction

In the recent era of electronic design automation, system-on-chip (SoC) has evolved to meet the complex design requirements for applications. An SoC typically contains reusable intellectual property (IP) cores from various vendors. There is also a rising threat to IP vendors from reverse engineering efforts and IP piracy [1–3]. Soft IP cores are offered as synthesizable register transfer level (RTL) code and they are also popular for their flexibility and portability. Protection of IP cores against reverse engineering and piracy is of foremost importance and highly difficult due to their better clarity and intelligibility.

Hardware IP protection has been extensively studied over the past decade. The literature works can be widely categorized into two main types: (1) *Authentication* based and (2) *Obfuscation* based. The *Authentication*-based mechanisms focus on effective embedding of "digital watermark" or *authentication signature* in IPs has been broadly examined [2]. The watermark is well-hidden, difficult to remove and does not affect the normal functioning of IPs. The watermark generally exists in the form of one or more input–output response pairs.

In spite of the fact that signature mechanisms are efficient, the main disadvantage of these approaches is being *passive* methods, which only assists for evidencing ownership of IP during litigation but does not shield over piracy in the first place.

P. Santikellur · R. S. Chakraborty (✉)
Department of Computer Science and Engineering, Indian Institute of Technology, Kharagpur, West Bengal, India
e-mail: pranesh.sklr@iitkgp.ac.in; rschakraborty@cse.iitkgp.ac.in

S. Bhunia
Department of Electrical and Computer Engineering, University of Florida, Gainesville, FL, USA
e-mail: swarup@ece.ufl.edu

S. Katkoori, S. A. Islam (eds.), *Behavioral Synthesis for Hardware Security*,
https://doi.org/10.1007/978-3-030-78841-4_4

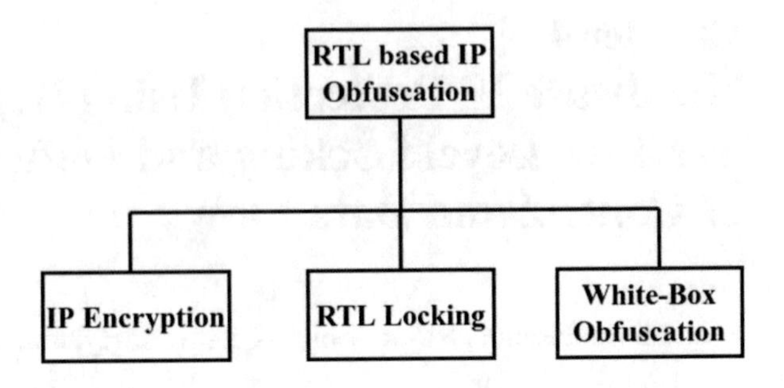

Fig. 4.1 RTL-based IP obfuscation techniques, following [12]

Hardware obfuscation is a technique to change the design in a way that it becomes hard to reverse engineer or copy. Hardware obfuscation can be achieved at various levels of design flow. Readers are requested to visit the recent literature survey [12] that describes the details of the various methodologies to achieve hardware obfuscation. Among the various types, this chapter will cover the most commonly used techniques implemented either at the register transfer level (RTL) or at the gate level. The different methods to employ RTL-based obfuscation are shown in Fig. 4.1, as described in [12]. The encryption and locking strategies have also been applied to gate-level obfuscation techniques. In *IP obfuscation through encryption*, the source code is encrypted using well-known encryption algorithms such as AES or RSA. The encryption makes IP significantly difficult to reverse engineer [1].

A major shortcoming of the above techniques is that they do not modify the functionality of an IP core and thus cannot prevent it from being stolen by an adversary and used as a "black-box" RTL module. In [1, 8], the HDL source code is encrypted and the IP vendor provides the key to decrypt the code to only its customers. But such a technique forces the use of a particular design platform, a situation that may not be acceptable to many SoC designers who often deploy multiple tools from different vendors in the design flow. Besides, IP encryption causes hardship to SoC designers during the integration and debugging of SoCs, who might have procured different IP modules from different vendors. The recent on-going research works [9, 11] focus on employing encryption with logic, fault and delay simulations enabled for SoC designers.

White-box-based obfuscation techniques focus on employing the software obfuscation methods on RTL code to prevent reverse engineering. Such techniques have also been studied extensively [5, 13]. The main drawbacks of these techniques are related to performance, as software obfuscation approaches keep different constraints (e.g. code size, run time) for their obfuscation method. These techniques also cannot shield over piracy in the first place.

RTL locking-based obfuscation technique aims to obfuscate the RTL code and allows it to function normally only after unlocking. The locking is typically achieved using a key. An effective technique was proposed in [3], which obfuscates the functionality of gate-level *firm* IP cores through modification of the *State Transition Graph* (STG) that requires a specific initialization input sequence to enable normal functional behaviour. This scheme simultaneously prevents reverse engineering as well as unauthorized use of an IP. However, extension of this approach to RTL involves major challenges, primarily due to the difficulty of obfuscating the STG modification in a highly comprehensible, industry standard RTL descriptions.

In this chapter, we present in detail an efficient and low-overhead technique for implementing RTL locking-based obfuscation [4]. The main idea of this technique is to perform judicious modification of control and data flow structures to efficiently incorporate *Mode-Control FSM* into the design in a way that the design functions in two modes: (1) *obfuscated mode* and (2) *normal mode*. The normal functionality is enabled during normal mode on successful application of predefined key, whereas the obfuscated mode generates the unexpected output without any indication of failure. The *Mode-Control FSM* is merged with the RTL of the IP, which makes it difficult to isolate and remove from the original IP. To incorporate the FSM into RTL, a few registers are judiciously selected, which we term *host registers*, and their widths are expanded if necessary. Also, their assignment conditions and values are modified. Next, the control and data flow code are enabled based on the mode *modification signal* obtained from the FSM.

We also describe a simple metric to calculate the quality of obfuscation, to estimate the hardness of reverse engineering the obfuscated RTL as experienced by an adversary. Finally, we present the results of applying the proposed obfuscation methodology on two open-source IP cores. We use Verilog HDL to describe the methodology, but the approach is equally applicable to RTL written using other HDLs, for both SoC and FPGA platforms. The results obtained from this technique indicate that effective obfuscation can be achieved at low resource and computational overhead.

4.2 Methodology

Figure 4.2 presents the general methodology adapted for this technique [3]. The incorporated Mode-Control FSM in Fig. 4.2 controls the circuit course of operation, mainly in two different modes (a) the *normal mode* when the circuit performs

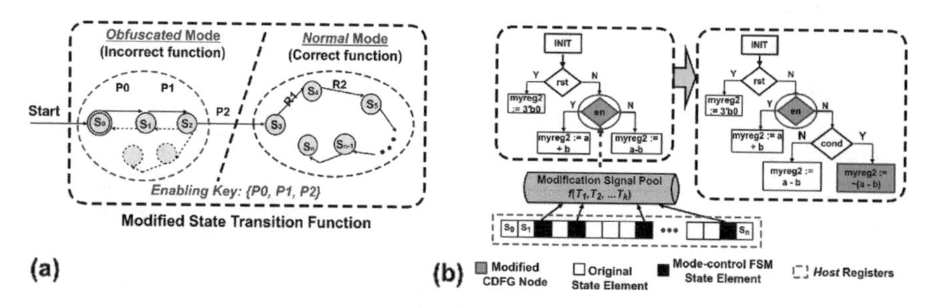

Fig. 4.2 The typical functional obfuscation scheme by state transition function modification [3]: (**a**) modified state transition function and (**b**) change in internal node structure using modification cells

predefined normal functionality and (b) the *obfuscated mode* when the circuit operates on undesired functionality. The default mode chosen on power-up is obfuscation mode. During obfuscation mode, the specific key sequence called *initialization key sequence* enables to switch to normal functioning mode by making appropriate state transitions within model-control FSM. In Fig. 4.2a, the set of transitions $P0 \rightarrow P1 \rightarrow P2$ makes switch to *normal mode* from *obfuscation mode*.

During the incorporation of FSM into RTL, modification is done at a few selected control and data flow nodes using *modification signals*. These *modification signals* are responsible to make the obfuscation mode operate in an undesired way. This is mainly done by changing the logic values at selected nodes in obfuscation mode. The approach sometimes operates on *dummy state transitions* in both modes to ensure that the mode-control FSM does not get halted at any particular state in both modes, making it hard for an adversary to distinguish the current mode and its state elements of the incorporated FSM.

This design achieves the security against piracy and reverse engineering from the below-mentioned points: (a) Evidencing ownership of IP from design features that are enabled only after correct *initialization key sequence*. (b) As explained in [3], it is very challenging to extract the key from functional simulations. However, the key issue lies in keeping mode-control FSM and RTL design modifications safe from adversary. To address this, the RTL obfuscation method uses four important steps.

1. **Build CDFG from concurrent blocks and combine them:** As a first step, the RTL code is scanned to identify parallel blocks and such blocks are transformed into CDFG data structure. A transformation of "always @()" block to a corresponding CDFG is shown in Fig. 4.3. Next, the CDFGs are integrated (whenever possible) to create combined CDFGs. A combined CDFG can be complex with a significant number of nodes involved, which improves quality of obfuscation. To illustrate a combined CDFG, all CDFGs transformed from non-blocking assignment statements to clocked registers can be merged together, which creates larger CDFG with no alteration in its functional behaviour.

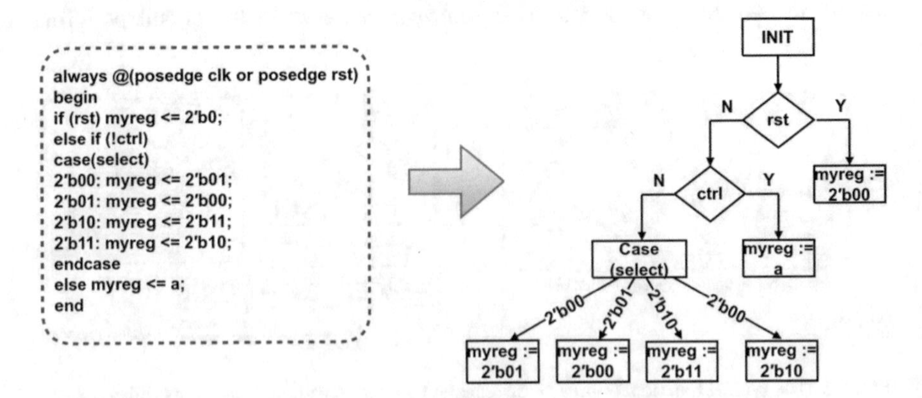

Fig. 4.3 Example of transformation of an "always @()" block to CDFG

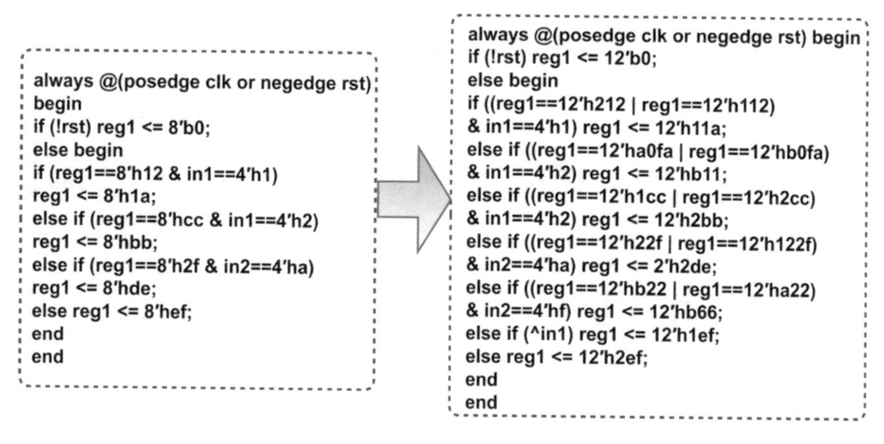

Fig. 4.4 Example of hosting mode-control FSM using registers

2. **"Hosting" the mode-control FSM in the "host registers":** Hiding the mode-control FSM in an intelligent way can improve the level of obfuscation. This technique utilizes existing registers to host the state elements of mode-control FSM. Fig.4.4 presents an example that describes the usage of host registers. In Fig.4.4, the bit width of register *reg1* is expanded from 8-bits to 12-bits, where the left 4 bits have been chosen to host the mode-control FSM. The right side RTL shown in Fig.4.4 has two operating modes controlled by register *reg1*. The circuit operates in obfuscated mode when the 4-bit values of *reg1* is either 4'ha or 4'hb and in normal mode when the values are set at 4'h1 or 4'h2. Due to bit width modification, necessary changes have been added to keep the functionality same as normal mode. Non-contiguous usage of mode-control FSM state elements can effectively improve the level of obfuscation.
3. **Modifying CDFG branches:** Once FSM "hosting" is carried out, numerous CDFG nodes are modified according to the control signal produced by FSM. In order to maximize the deviation from normal behaviour with minimum changes, larger CDFGs are good choice for modification compared to small CDFGs. As an example, Fig.4.6 shows three such different modification examples along with original RTL (a) and obfuscated RTL (b). The example considers "case()", "if()" and "assign" as three different RTL statements and *reg1*, *reg2* and *reg3* as *host registers*, respectively. The mode signals selected are *cond1*,*cond2* and *cond3* and these evaluate to logic-1 for obfuscated mode and logic-0 for normal mode. Note that mode-selection mechanism has been followed as described in step 2.
The technique focuses not only on changing the control path but also on the data path functionality. To achieve this, additional datapath elements are introduced in a way that maximum resources are shared during synthesis. However, it is important to note that datapath modifications often lead to substantial increase in hardware overhead. Figure 4.5 presents an example of datapath modification

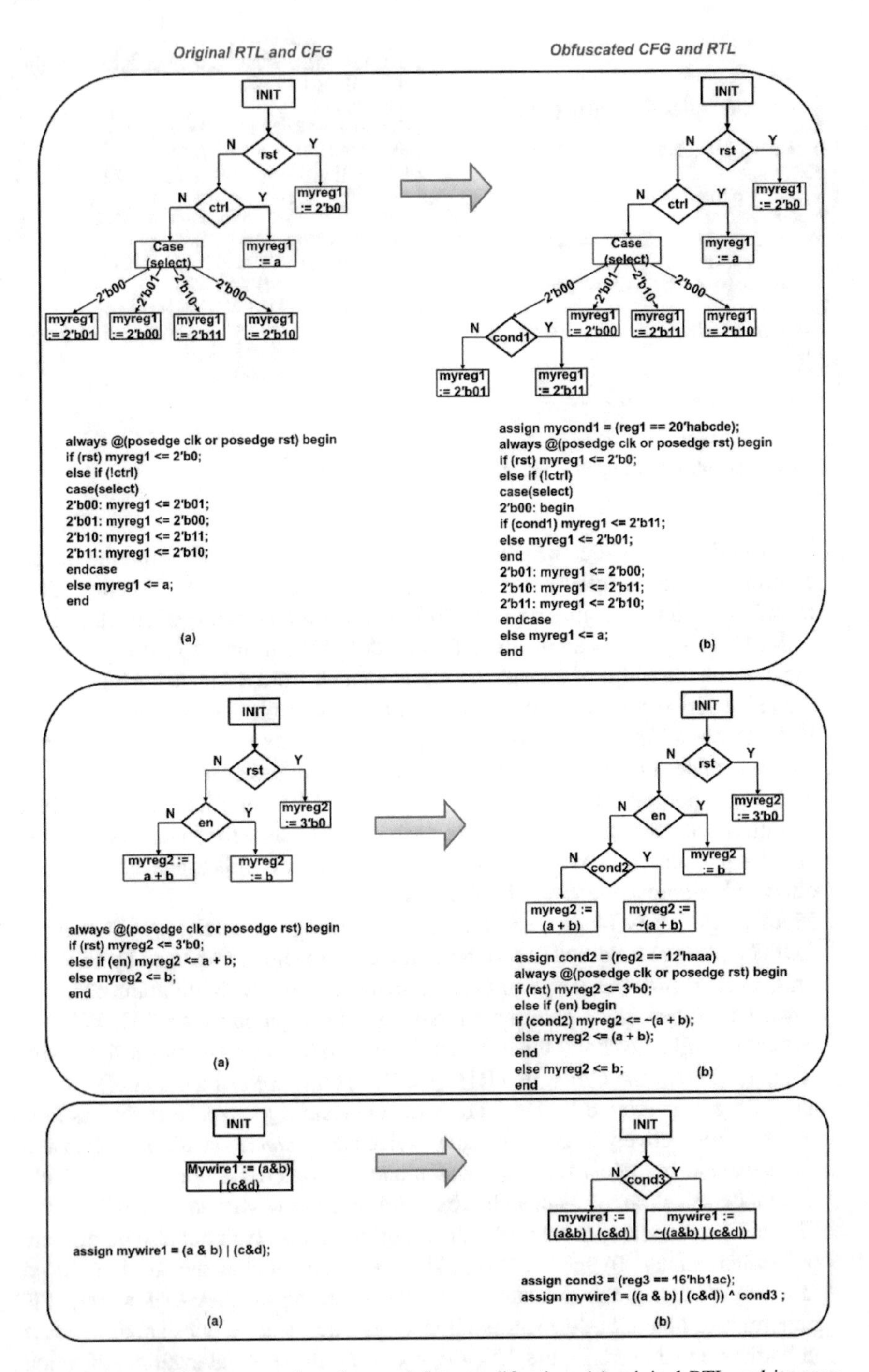

Fig. 4.5 Three different examples of control-flow modification: (**a**) original RTL and its corresponding CDFG and (**b**) obfuscated RTL and its corresponding CDFG

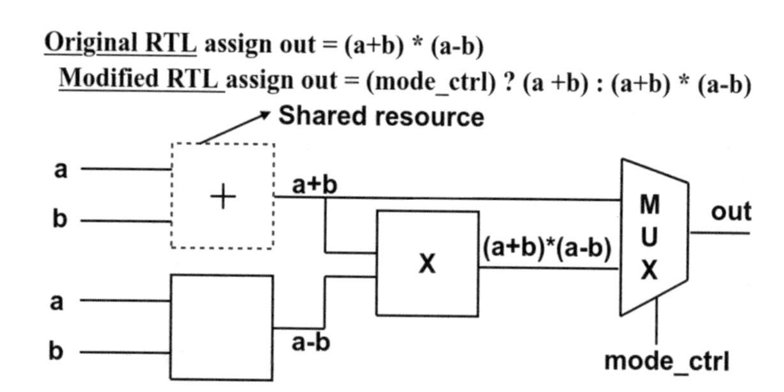

Fig. 4.6 Example of resource sharing for datapath obfuscation

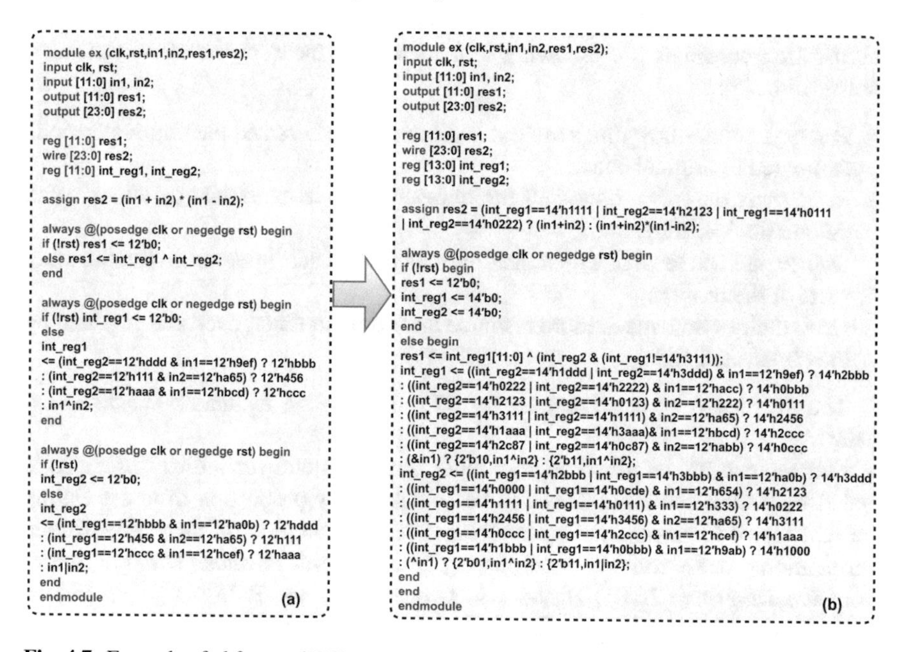

```
module ex (clk,rst,in1,in2,res1,res2);
input clk, rst;
input [11:0] in1, in2;
output [11:0] res1;
output [23:0] res2;

reg [11:0] res1;
wire [23:0] res2;
reg [11:0] int_reg1, int_reg2;

assign res2 = (in1 + in2) * (in1 - in2);

always @(posedge clk or negedge rst) begin
if (!rst) res1 <= 12'b0;
else res1 <= int_reg1 ^ int_reg2;
end

always @(posedge clk or negedge rst) begin
if (!rst) int_reg1 <= 12'b0;
else
int_reg1
<= (int_reg2==12'hddd & in1==12'h9ef) ? 12'hbbb
: (int_reg2==12'h111 & in2==12'ha65) ? 12'h456
: (int_reg2==12'haaa & in1==12'hbcd) ? 12'hccc
: in1^in2;
end

always @(posedge clk or negedge rst) begin
if (!rst)
int_reg2 <= 12'b0;
else
int_reg2
<= (int_reg1==12'hbbb & in1==12'ha0b) ? 12'hddd
: (int_reg1==12'h456 & in2==12'ha65) ? 12'h111
: (int_reg1==12'hccc & in1==12'hcef) ? 12'haaa
: in1|in2;
end
endmodule
```
(a)

```
module ex (clk,rst,in1,in2,res1,res2);
input clk, rst;
input [11:0] in1, in2;
output [11:0] res1;
output [23:0] res2;

reg [11:0] res1;
wire [23:0] res2;
reg [13:0] int_reg1;
reg [13:0] int_reg2;

assign res2 = (int_reg1==14'h1111 | int_reg2==14'h2123 | int_reg1==14'h0111
| int_reg2==14'h0222) ? (in1+in2) : (in1+in2)*(in1-in2);

always @(posedge clk or negedge rst) begin
if (!rst) begin
res1 <= 12'b0;
int_reg1 <= 14'b0;
int_reg2 <= 14'b0;
end
else begin
res1 <= int_reg1[11:0] ^ (int_reg2 & (int_reg1!=14'h3111));
int_reg1 <= ((int_reg2==14'h1ddd | int_reg2==14'h3ddd) & in1==12'h9ef) ? 14'h2bbb
: ((int_reg2==14'h0222 | int_reg2==14'h2222) & in1==12'hacc) ? 14'h0bbb
: ((int_reg2==14'h2123 | int_reg2==14'h0123) & in2==12'h222) ? 14'h0111
: ((int_reg2==14'h3111 | int_reg2==14'h1111) & in2==12'ha65) ? 14'h2456
: ((int_reg2==14'h1aaa | int_reg2==14'h3aaa)& in1==12'hbcd) ? 14'h2ccc
: ((int_reg2==14'h2c87 | int_reg2==14'h0c87) & in2==12'habb) ? 14'h0ccc
: (&in1) ? {2'b10,in1^in2} : {2'b11,in1^in2};
int_reg2 <= ((int_reg1==14'h2bbb | int_reg1==14'h3bbb) & in1==12'ha0b) ? 14'h3ddd
: ((int_reg1==14'h0000 | int_reg1==14'h0cde) & in1==12'h654) ? 14'h2123
: ((int_reg1==14'h1111 | int_reg1==14'h0111) & in1==12'h333) ? 14'h0222
: ((int_reg1==14'h2456 | int_reg1==14'h3456) & in2==12'ha65) ? 14'h3111
: ((int_reg1==14'h0ccc | int_reg1==14'h2ccc) & in1==12'hcef) ? 14'h1aaa
: ((int_reg1==14'h1bbb | int_reg1==14'h0bbb) & in1==12'h9ab) ? 14'h1000
: (^in1) ? {2'b01,in1^in2} : {2'b11,in1|in2};
end
end
endmodule
```
(b)

Fig. 4.7 Example of obfuscated RTL generation: (**a**) original RTL and (**b**) obfuscated RTL

with adder being shared across two modes. The normal mode computes $(a+b) \star (a-b)$, whereas the *obfuscated mode* outputs $(a+b)$ (Fig.4.6).

4. **Generating obfuscated RTL:** As a last step, the obfuscated RTL has to be generated from the modified CDFG. *Depth first search* is employed for efficient generation of the code. An example is shown in Fig. 4.7 to demonstrate the obfuscated RTL generation (b) from original RTL (a). The *host registers* used are *int_reg1* and *int_reg2* and a 4-bit FSM is employed. When the circuit is in obfuscation mode *int_reg1*[13:12]=2'b00, *int_reg1*[13:12]=2'b01, *int_reg2*[13:12]=2'b00 and *int_reg1*[13:12]=2'b10 holds true. The circuit

initialization sequence is *in1*=12'h654 → *in2*=12'h222 → *in1*=12'h333 → *in2*=12'hacc → *in1*=12'9ab. The datapath modifications are carried out for the outputs *res1* and *res2* using two different *modification signals*. Please note the use of *dummy state transitions* and out-of-order state transition RTL statements.

4.3 Obfuscation Efficiency

In this section, we will discuss the metric that can quantify the level of obfuscation and also a method to strengthen the quality of obfuscation achieved with a reasonable overhead. The idea of defining metrics for RTL obfuscation is derived from software obfuscation metrics as both aim to achieve the same target. Software obfuscation considers the following four metrics to measure the effectiveness of obfuscation [5]:

- *Potency*: defines how much more it is difficult to understand the obfuscated code compared to original code.
- *Resilience*: describes how well the obfuscated code can resist an attack of an automatic de-obfuscator.
- *Stealth*: defines to what extent the obfuscated code gets integrated with the other parts of the program.
- *Cost*: defines how much more resource and computational overhead is added by the obfuscated code.

The definitions of these metrics for RTL obfuscation technique are discussed below.

Potency can be described as how much structural and functional difference is added by obfuscated RTL. This can be calculated by proportion of nodes failing for obfuscated to original during the formal verification process. Generally, formal verification process consists of comparing corresponding *Reduced Ordered Binary Decision Diagram* (ROBDD) nodes in two designs, and ROBDD nodes change for each modification in control and data flow of the circuit. When the potency is high, it indicates that many nodes are failing in obfuscated circuit compared to the original, which implies that the obfuscated circuit differs significantly from its original circuit.

The next two metrics *resilience* and *stealth* for the RTL obfuscation can be defined as how difficult for the adversary to locate the *hosted* mode-control FSM and associated modification signals. For example, consider a circuit with N blocking or non-blocking assignment statements and n mode-control FSM transition statements. As the adversary does not know the number of *host registers* employed a priori, the probability to deduce it from FSM state transitions is given by: $\frac{1}{\sum_{k=1}^{n}(\binom{N}{k})}$. To apply the *initialization key sequence* in the correct order, we need $k!$ ways from k choices. Also, to estimate the accurate mode-control signal, let m be the size of modification signal pool and M be the total number of blocking, non-blocking and dataflow assignment statements present in RTL, then one from $\binom{M}{m}$ choices to choose proper modification signal. Applying these choices, the combined formula to calculate the *resilience* and *stealth* of obfuscated circuit is

$$M_{obf} = \frac{1}{\sum_{k=1}^{n} \left(\binom{N}{k} \cdot k! \right) \cdot \binom{M}{m}}. \tag{4.1}$$

The obfuscation efficiency is better if the M_{obf} value calculated from the circuit is low. For example, assume the set of values $M = 100$, $N = 30$ and $n = 3$ for a RTL code, then $M_{obf} = 7.39 \times 10^{-26}$. Practically, the values of n and N will be much larger than the considered, making the M_{obf} much lower. It is clear that brute force attack possibility remains tough for the adversary.

The *cost* metric describes the design overheads and extra time taken to compile for the obfuscated design. Section 4.4 provides details about the cost incurred.

4.4 Results

4.4.1 Implementation

Complete steps followed by obfuscation technique are shown in Fig. 4.8. The inputs to the design flow are original RTL, maximum allowable area overhead and expected level of obfuscation. The required level of obfuscation is specified using M_{obf} value. The flow starts with the design of *mode-control FSM* based on the M_{obf} value. The outputs of this step are specifications of FSM that include the *initialization key sequence* and its length, the pool of modification signals, the state encoding and state transition graph. To increase the security, *initialization key sequence* and state encoding are randomly generated.

As described in Sect. 4.2, the steps are followed accordingly to synthesize obfuscated RTL. In case, area overhead exceeds its constraint, the number of modifications (N_{mod}) is subtracted by a predefined step and the complete process is reiterated until the overhead constraints are met.

4.4.2 Implementation Results

The above-discussed flow (see Fig. 4.8) was implemented using Tcl. Two Verilog IP cores—IEEE standard 754 single precision floating-point unit ("FPU"), and a basic 12-bit RISC CPU ("TCPU"), both source codes obtained from [10], were used to

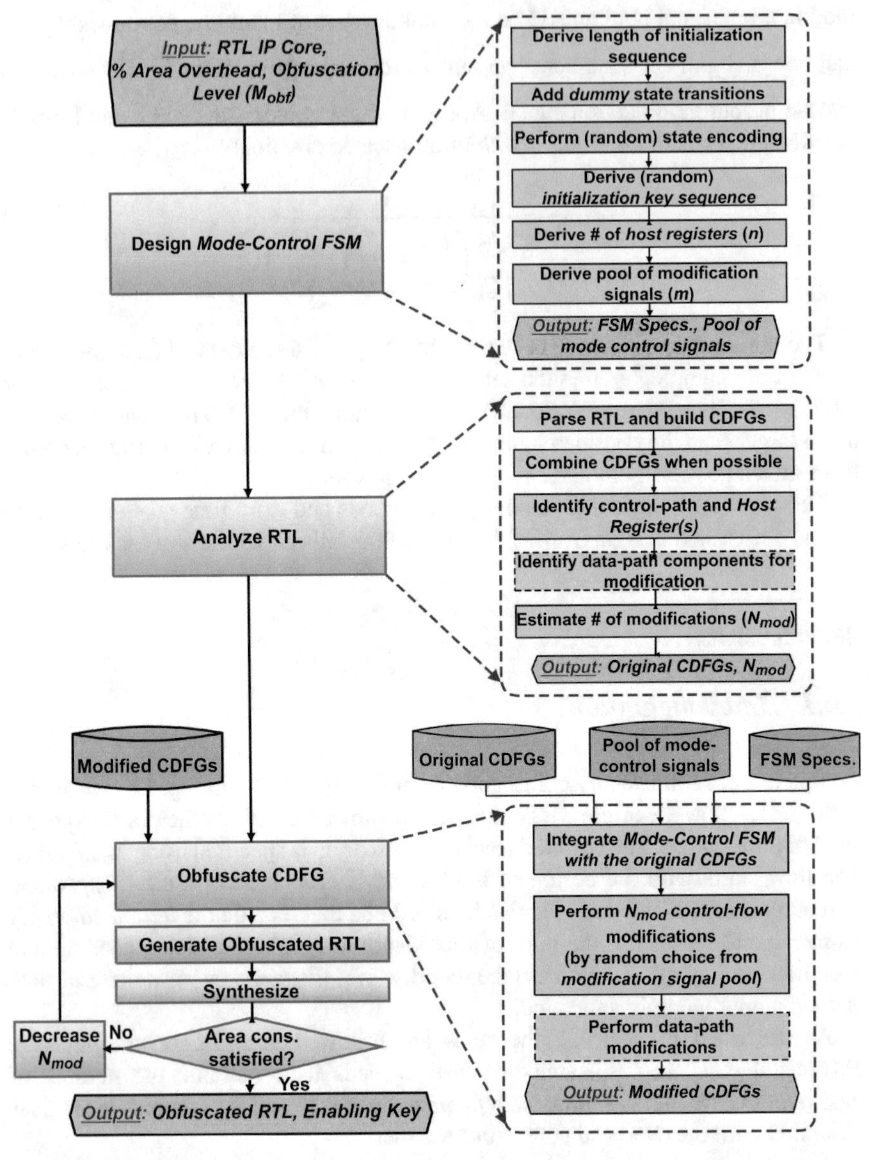

Fig. 4.8 Detailed steps of automated design flow for the discussed RTL obfuscation methodology

find the effectiveness of the obfuscation technique. For both IP cores, following parameters were used during the course of experiments:

- The *mode-control FSM* was implemented using 4 state elements.
- The length of a *initialization key sequence* was set to 3.
- $n = 3$: The *mode-control FSM* was hosted in three RTL statements.
- $m = 10$: size of the modification signal pool was set to 10.

The simulations were carried out on a Linux workstation with 1.5 GHz processor and 2 GB main memory. Synopsis *Design Compiler* tool was used to synthesize the circuit using an ARM 90-nm standard cell library. The difference between obfuscated and original design was formally verified using Synopsys *Formality* tool.

The obfuscation efficiency and related design overheads along with run time for each IP core is shown in Table 4.1. The number of modifications are set to 10% target area overhead, and all the results are reported at iso-delay. While the functional obfuscation efficiency is calculated with difference in verification failures of obfuscated and original designs, the semantic obfuscation efficiency was calculated using M_{obf} value (see Sect.4.3). It is also clear from Table 4.1 that run-time results achieved by obfuscated design are efficient. Also, the difference of compilation time between original and obfuscated flow is found to be minimal.

4.4.3 Discussions and Future Research Directions

Note that obfuscation techniques such as [3] clearly differentiate between the normal and obfuscated functional modes of a circuit, and in the obfuscated mode, it is possible that the circuit FSM get stuck at a state. This lets an adversary clearly distinguish between the obfuscated and functional modes of operation. In contrast, one of the major advantages of the approach proposed in this chapter is that when an incorrect sequence is received as input, the circuit produces incorrect output without getting stuck in any state. This helps to conceal the obfuscated functional mode to an extent, as differentiating the obfuscated mode of operation from a circuit with faults becomes somewhat difficult for an adversary. A similar approach was advocated in [6], where the correct functional mode can be entered irrespective of starting in the normal or obfuscated mode, and "Code-Words" interlocked with the state transition function allowed entry to the functional mode from obfuscation mode. The Code-Word is integrated with the with the transition function and is not explicitly stored anywhere on the chip. The advantage of this method is that the functional mode of the circuit behaves differently with different Code-Words, making it respond more dynamically and thus making reverse engineering more challenging.

There have been very few works reported to focus on attacks of FSM-based obfuscation methods. A recent work [7] provides comprehensive analysis of the difficulty of reverse engineering FSM-based hardware obfuscation methods and proposes an attack. The attack relies on the observation that among the possible states, not all of them are reachable from the starting state of the FSM. Once an adversary deduces the state transition graph of the FSM, this can help to break the path to get into functional mode. More details on the extensive study of breaking the various FSM obfuscation schemes can be found in this literature [7]. In future, we plan to make our proposed technique robust against these type of cutting-edge attacks.

Table 4.1 Obfuscation overhead (at iso-delay) and obfuscation efficiency (for 5% target area overhead)

IP cores	Sub-modules	Number of mods.	Obfuscation efficiency		Design overhead		Run time
			Failing verif. pts. (%)	M_{obf}	Area (%)	Power (%)	(s)
FPU	Post_norm	20	98.16	1.56e–37	8.22	9.14	27
	Pre_norm	20	94.16	1.24e–33	9.39	9.79	25
	Pre_norm_fmul	20	90.00	7.36e–24	8.30	9.69	20
	Except	10	100.00	6.85e–24	7.56	8.73	14
TCPU	Control_wopc	20	92.79	7.53e–43	8.74	8.97	29
	Mem	10	97.62	2.06e–20	8.29	9.76	15
	Alu	10	97.62	9.82e–16	9.59	9.88	15

4.5 Conclusion

We have described a key-based RTL obfuscation technique for hardware IP protection, applicable to RTL written in any HDL. The main idea is to incorporate the FSM that modifies the control and data flow of the original design. We have also evaluated the effectiveness of obfuscation method using appropriate metrics. The implementation results also show that the technique incurs low design and computational overhead.

References

1. Batra, T.: Methodology for protection and licensing of HDL IP. http://www.us.design-reuse.com/news/?id=12745&print=yes (2020)
2. Castillo, E., Meyer-Baese, U., Garcia, A., Parrilla, L., Lloris, A.: IPP@HDL: Efficient intellectual property protection scheme for ip cores. IEEE Trans. Very Large Scale Integr. Syst. **15**(5), 578–591 (2007)
3. Chakraborty, R.S., Bhunia, S.: Hardware protection and authentication through netlist level obfuscation. In: 2008 IEEE/ACM International Conference on Computer-Aided Design, pp. 674–677 (2008)
4. Chakraborty, R.S., Bhunia, S.: RTL hardware IP protection using key-based control and data flow obfuscation. In: 2010 23rd International Conference on VLSI Design, pp. 405–410 (2010)
5. Collberg, C., Thomborson, C., Low, D.: Manufacturing cheap, resilient, and stealthy opaque constructs. In: Proceedings of the 25th ACM SIGPLAN-SIGACT Symposium on Principles of Programming Languages, POPL '98, p. 184–96. Association for Computing Machinery, New York (1998). 10.1145/268946.268962
6. Desai, A.R., Hsiao, M.S., Wang, C., Nazhandali, L., Hall, S.: Interlocking obfuscation for anti-tamper hardware. In: Proceedings of the Eighth Annual Cyber Security and Information Intelligence Research Workshop, CSIIRW '13, pp. 1–4. Association for Computing Machinery, New York (2013). 10.1145/2459976.2459985
7. Fyrbiak, M., Wallat, S., Déchelotte, J., Albartus, N., Böcker, S., Tessier, R., Paar, C.: On the difficulty of FSM-based hardware obfuscation. IACR Trans. Cryptograph. Hardware Embed. Syst. **2018**(3), 293–330 (2018)
8. Goering, R.: Synplicity Initiative Eases IP evaluation for FPGAs. http://www.scdsource.com/article.php?id=170 (2020)
9. Millican, S., Ramanathan, P., Saluja, K.: CryptIP: An approach for encrypting intellectual property cores with simulation capabilities. In: 2014 27th International Conference on VLSI Design and 2014 13th International Conference on Embedded Systems, pp. 92–97 (2014)
10. OpenCores. http://www.opencores.org (2020)
11. Ramanathan, P., Saluja, K.K.: Crypt-delay: Encrypting IP cores with capabilities for gate-level logic and delay simulations. In: 2016 IEEE 25th Asian Test Symposium (ATS), pp. 7–12 (2016)
12. Shakya, B., Tehranipoor, M.M., Bhunia, S., Forte, D.: Introduction to hardware obfuscation: Motivation, methods and evaluation. In: Forte D., Bhunia S., Tehranipoor M.M. (eds.) Hardware Protection through Obfuscation, pp. 3–32. Springer International Publishing, Cham (2017)
13. Zhuang, X., Zhang, T., Lee, H.H.S., Pande, S.: Hardware assisted control flow obfuscation for embedded processors. In: Proceedings of the 2004 International Conference on Compilers, Architecture, and Synthesis for Embedded Systems, CASES '04, pp. 292–302. Association for Computing Machinery, New York (2004). 10.1145/1023833.1023873

Chapter 5
Protecting Behavioral IPs During Design Time: Key-Based Obfuscation Techniques for HLS in the Cloud

Hannah Badier, Jean-Christophe Le Lann, Philippe Coussy, and Guy Gogniat

5.1 Introduction

Modern hardware designs have reached a fantastic degree of complexity. To sustain this industrial challenge, design flows are increasingly distributed, with companies relying on third parties for IP development, manufacturing, and testing. This leads to substantial cost reductions, but also a growing amount of security issues, such as IP theft and counterfeit or hardware Trojan insertion. In particular, threats at early design stages are coming more and more into focus: the widespread use of third-party Behavioral IPs (BIPs), as well as external Computer-Aided Design (CAD) tools, has created new attack surfaces [1]. For example, in [2], the authors demonstrate how a high-level synthesis (HLS) tool can be manipulated to insert malicious code into designs.

In this chapter, we aim at ensuring the security of BIPs and in particular at protecting them against theft *during* HLS. We focus on the case of a cloud-based HLS service. With the recent surge in cloud computing capabilities, and the growing trend of complex computations being outsourced to dedicated Software-as-a-Service (SaaS) platforms, an HLS-as-a-Service scenario becomes more likely [3]. However, broad adoption of such a service is slowed down by legitimate security concerns. While encryption is widely used to ensure data security in the cloud, a complex operation such as HLS cannot be performed on encrypted code without decrypting the code at some point during the operation.

H. Badier (✉) · J. C. Le Lann
Lab-STICC, ENSTA Bretagne, Brest, France
e-mail: hannah.badier@ensta-bretagne.org; jean-christophe.le_lann@ensta-bretagne.fr

P. Coussy · G. Gogniat
Lab-STICC, Universite de Bretagne Sud, Lorient, France
e-mail: philippe.coussy@univ-ubs.fr; guy.gogniat@univ-ubs.fr

S. Katkoori, S. A. Islam (eds.), *Behavioral Synthesis for Hardware Security*,
https://doi.org/10.1007/978-3-030-78841-4_5

Our goal is to enable design houses to safely use cloud-based or untrusted HLS tools while mitigating the risk of BIP theft and reuse. This can be achieved by applying security measures at algorithmic level while keeping in mind that these measures should not affect design functionality, disturb the HLS process, or negatively impact design performances.

Obfuscation has been widely used as an intellectual property protection method in software and hardware design. Our work focuses on how to adapt software obfuscation techniques to BIPs while also taking into account hardware-specific constraints: directly applying traditional software obfuscation methods to high-level IPs usually causes a significant design overhead, not only by increasing the size of the original design but also by affecting the HLS quality of results [4]. By adding a de-obfuscation step at a trusted point later in the design flow, the previous overhead is avoided and the obfuscation process becomes *transient*. Furthermore, traditional software obfuscation does not modify code functionality. This means that a stolen BIP could still be re-used and studied as a black box by observing its inputs and outputs. By using key-based obfuscation techniques, inspired by logic locking [5] at lower design levels, the design functionality is locked: only the owner of the correct keys can unlock the original functionality of the circuit.

In [6], we presented a first example of how a transient, key-based obfuscation technique can be used to protect designs during cloud-based HLS. In this chapter, we extend the work by showing how several classic software obfuscation techniques can be modified to become transient, enhanced by using keys, and adapted to hardware-specific constraints. Following techniques are presented in this chapter:

- Bogus expression insertion
- Bogus basic block insertion
- Key-based control flow flattening
- Bogus CFG transition insertion
- Literal replacement

This chapter is organized as follows. Section 6.2 explains the threat model, background on software obfuscation, and related work on IP protection. In Sect. 6.3, our proposed secured design flow, as well as the obfuscation and de-obfuscation steps, is presented. Section 6.4 gives a more detailed insight into the different proposed obfuscation techniques, while Sect. 6.5 contains the experimental setup and results. Finally, conclusions and a discussion of future work are given in Sect. 6.6.

5.2 Background and Related Work

5.2.1 Threat Model

How to protect Behavioral IPs against theft has been a growing concern. In this chapter, we focus on the scenario of a design house using an external, cloud-based

HLS tool for their BIPs. We assume that the attacker is either an insider of the HLS service or an outsider who has been able to gain access to the service and is able to steal the BIP code. We envision three types of risks in this scenario:

1. *Espionage*: The attacker, working for example for a competing company or an adverse government, gains valuable information about what type of algorithms or applications the design house is working on.
2. *BIP theft*: The attacker is able to understand the stolen BIP and can modify or directly counterfeit it.
3. *Black-box usage*: Even without full insight into how the stolen BIP works, the attacker can still reuse it as a black box.

Because we cannot prevent theft on the HLS provider's side, our approach aims at minimizing the previously enumerated risks. Using software obfuscation can reduce the ability of an attacker to understand and modify a stolen design. However, on its own, software obfuscation does not modify functionality and thus does not block black-box usage. By adding hardware IP protection principles such as logic locking, we can provide a transient, key-based obfuscation scheme that protects BIPs against all three risks *during* HLS (Fig. 5.1).

5.2.2 Software Obfuscation Principles

Software obfuscation relies on several transformations applied to source code, with the goal of producing a program that is functionally equivalent but harder to understand and/or reverse engineer [7]. Works about software obfuscation can be divided into two categories.

Model-oriented, or cryptographic, obfuscation focuses on protecting code in a formally verifiable way through cryptography operations. In particular, black-box obfuscation aims at creating obfuscated code that does not provide any more information about its internal constitution than a black box would. The only information

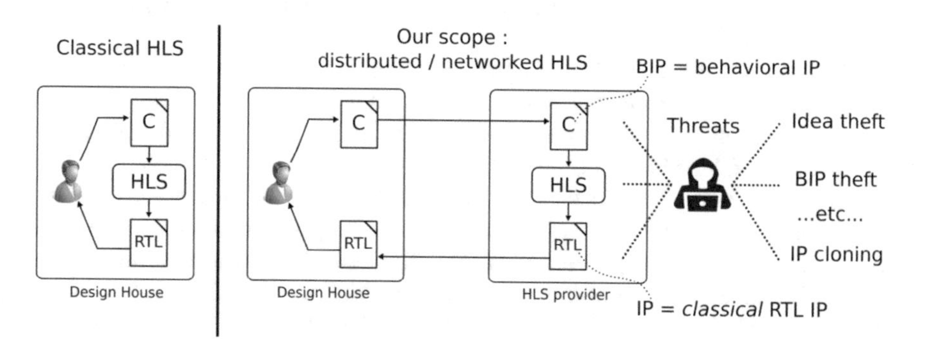

Fig. 5.1 Vulnerable HLS in the cloud design flow

that is allowed to be accessible can be gained by observing the inputs and outputs of this obfuscated code. In [8], the authors have demonstrated the impossibility of such a black-box obfuscation. While slightly weaker obfuscation schemes have been proven feasible, such as indistinguishability obfuscation presented in [9], they are not usable in a day-to-day context yet due to the complexity of the techniques and the immense size of the resulting obfuscated code [10, 11].

On the other hand, code-oriented obfuscation methods are more practical but do not give any formal proof of security. The taxonomy presented in [12] distinguishes three types of code-oriented software obfuscation techniques:

- *Layout* obfuscation aims at lowering the human readability of the code, for example by removing indentation, spaces and comments, or by renaming all variables.
- *Control Flow* obfuscation relies on obscuring the control flow of a program to make it harder to reverse engineer with automated tools or by hand. It usually involves removing structures like loops and splitting or aggregating functions.
- *Data* obfuscation modifies and hides information about data in the program, for example, by changing encoding or data representation.

5.2.3 IP Protection

IP infringement has become a critical issue for design houses. This has led to an increasing focus on IP protection techniques, at all design stages and abstraction levels. Many methods focus primarily on preventing IP theft during manufacturing, which is often outsourced to third-party foundries. Among them, hardware watermarking, which allows legal identification of the design owner, and split manufacturing, which relies on partitioning the design for fabrication in different foundries, have been widely studied [13–15]. Low-level obfuscation-based approaches, where the goal is to prevent netlist extraction, have also been proposed, for example, using Physical Unclonable Functions (PUFs) or camouflage gates. At gate level, hardware obfuscation techniques based on adding extra gates to the circuit have been studied.

However, these low-level techniques only focus on protection during manufacturing and do not protect *soft* IPs at register transfer level or Behavioral IPs at algorithmic level. Several methods have been proposed for key-based obfuscation applied to VHDL or Verilog code, either directly at RTL or during HLS. In [16], a mode-control Finite-State Machine (FSM) is added to the IP. This ensures that the circuit falls by default into an *obfuscated mode* and only reaches normal mode if the correct key sequence is applied. In [17, 18], high-level transformations are used to protect Digital Signal Processing (DSP) circuits. Finally, in [19, 20], HLS tools are extended to add key-based obfuscation during the HLS process, in order to produce obfuscated RTL. The goal in these works is still to protect IPs during manufacturing or during sales by third-party IP vendors. They give no protection to BIPs before or during HLS, which is considered a trusted design step.

A first method for BIP protection using obfuscation is presented in [4]. This heuristic approach is based on a study of the impact of commercial and free software obfuscators on HLS quality of results. Their method focuses on how best to apply software layout obfuscation techniques to BIPs before HLS and can thus increase difficulty of an attacker to understand stolen BIPs. However, the proposed techniques do not modify functionality of the code and therefore do not hinder "black-box" usage of the BIP.

In this chapter, we aim at preventing both BIP theft and black-box reuse, which is why we propose to use *key-based* obfuscation techniques. However, the methods presented in [4] are compatible with our approach and could be combined to increase the protection level.

5.3 Transient Obfuscation: Proposed Approach

5.3.1 Secured HLS Design Flow

The goal of this chapter is to protect a design house against BIP theft and black-box usage during obfuscation. We propose a transient protection method by adding two steps to the conventional design flow, before and after HLS:

1. *Obfuscation*: The original C code is obfuscated, making it harder to understand for an attacker. We use key-based obfuscation that modifies the design's functionality by tying the correct behavior to a set of obfuscation keys. This prevents black-box usage of the design.
2. *De-obfuscation*: After HLS, the resulting RTL code still relies on keys to function correctly and contains all the obfuscated code. Directly synthesizing the IP at this point would result in a malfunctioning, oversized design. By de-obfuscating with the correct keys, the original functionality is recovered and the design overhead is strongly reduced.

By combining these two steps, as shown in Fig. 5.2, our protection method becomes transient and has no significant lasting impact on the design. Both steps have been implemented in a fully automated way in an open-source tool, KaOTHIC (Key-based Obfuscating Tool for HLS in the Cloud), and are detailed in the next subsections.

5.3.2 Obfuscation Flow

All of our proposed obfuscation techniques have been implemented as code transformations in an in-house source-to-source compiler. Since this tool is open-source, we assume that an attacker also has access to it and thus cannot rely on the

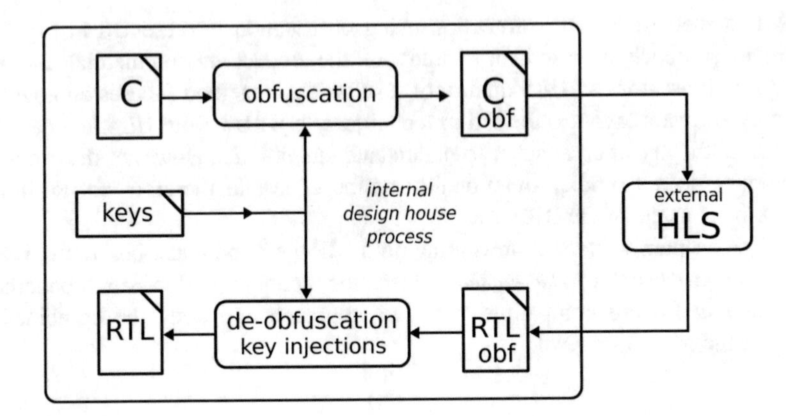

Fig. 5.2 HLS in the cloud design flow secured by KaOTHIC, where the external HLS is considered untrusted

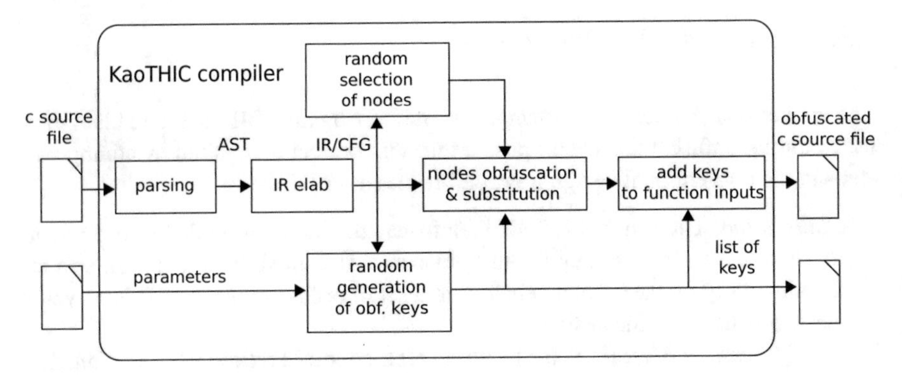

Fig. 5.3 Obfuscation flow in KaOTHIC compiler

secrecy of the process to ensure security. To prevent an attacker from reproducing obfuscation results, randomness is introduced wherever possible in the obfuscation flow.

The complete obfuscation flow is illustrated in Fig. 5.3. Our tool takes as input the original BIP in the form of C source code. It can then be guided with a series of parameters: obfuscation level, applied transformations, and order of transformations. The tool outputs the obfuscated BIP again as C code, as well as the list of obfuscation keys. In detail, the following steps are performed:

1. The C files are parsed, and an Abstract Syntax Tree (AST) is generated.
2. The AST is traversed to build the Control Flow Graph (CFG) of the program.
3. From the CFG, a list of "candidate" basic blocks that can be obfuscated is built.
4. Based on the desired obfuscation level and/or detailed user instructions, basic blocks are randomly selected among the candidates.
5. Each selected candidate is obfuscated and replaced in the original code.

6. For each selected candidate, a key variable is added as input to the obfuscated function and a random value is generated as key.
7. The AST is rebuilt from the modified CFG.
8. From the modified AST, the final, obfuscated C code is recreated.

This generic flow can be adapted to different code transformations. A detailed presentation of all implemented obfuscation techniques can be found in Sect. 5.4.

5.3.3 De-obfuscation

To protect BIPs during cloud-based HLS, we aim at proposing a *transient* protection scheme. To achieve this, we have added a de-obfuscation step after HLS. When the correct obfuscation keys are provided, this step performs two functions: it recovers correct functionality of the IP and limits design overhead. Most implemented obfuscation techniques rely on adding bogus code and then using key-based predicates to choose what code to execute, see Sect. 5.4.1. In that case, de-obfuscation removes all bogus code.

De-obfuscation is performed on the obfuscated RTL code resulting from HLS, usually VHDL or Verilog code. At RT level, the keys that were added as arguments to the main obfuscated function in C are now inputs to the component. For all implemented obfuscating transformations, the resulting RTL code contains the same artifacts: comparisons between key signal and an expected value, and *obfuscation multiplexers* that select what code should be executed based on the results of these comparisons.

5.3.3.1 Naïve Key Injection

This first approach consists of injecting the correct obfuscation keys at the circuit interfaces in the RTL code. During logic synthesis, the following behavior is then expected: if the correct keys are injected, through constant propagation the predicates should evaluate correctly, and the bogus code branches should be removed by logic simplification. The circuit would then function correctly, and the design overhead compared to the original, unobfuscated circuit should be close to null.

In practice however, this technique does not perform well: the sequential datapath created by the HLS tool scheduling prevents full de-obfuscation. Experiments performed on several benchmarks have shown that the obfuscation multiplexer is often scheduled in a different clock cycle than the corresponding comparison between key input and value, as shown in Fig. 5.4b. This means that a dedicated register is used to store the result of the key comparison ("reg1" in Fig. 5.4). Injecting the correct key value as a constant input results in a constant value in the comparison register. However, logic synthesis tools have no way of knowing

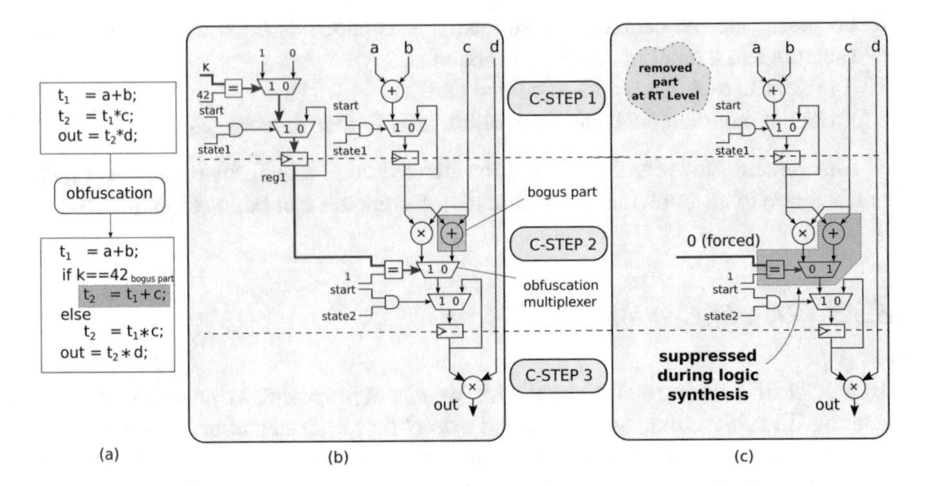

Fig. 5.4 Example RTL datapath before and after de-obfuscation. (**a**) obfuscation example. (**b**) obfuscated RTL datapath generated by Vivado HLS. (**c**) RTL datapath after full de-obfuscation

that the register always contains the same value and could thus be removed. As a consequence, the obfuscation multiplexer is not removed either, and most of the bogus code added by obfuscation remains.

While the final circuit is functionally correct, the remaining bogus code causes a significant design overhead.

5.3.3.2 Targeted RTL Modification

We propose to improve upon this first, *naïve* approach, by using a more targeted and in-depth de-obfuscation method, illustrated in Fig. 5.4c. Instead of injecting the keys externally and relying purely on logic collapsing, we locally modify the RTL code itself by performing the following operations for each obfuscation key:

- The given key and its expected value in the code are compared.
- The result of this comparison is stored as "0" or "1." This is the value that would during execution be in the aforementioned comparison register.
- The key comparison and the comparison register are completely removed from the RTL code.
- The stored result, "0" or "1," is directly injected into the obfuscation multiplexer.
- Logic synthesis is performed as for a normal RTL design.

This more targeted approach enables logic synthesis tools to completely collapse and remove any remaining obfuscation artifacts, resulting in close to no overhead. This guarantees that our obfuscation process is transient.

This second approach was fully automated for VHDL code and could easily be extended for Verilog as well.

5.4 Obfuscation Techniques

5.4.1 *Bogus Code Insertion*

This set of techniques is based on control flow splitting, as presented in [12]. It relies on a simple principle: bogus code is added to the design. A set of predicates, in the form of if/else statements, is then added to split the control flow between the correct, original code, and the bogus code. The goal is to make it near impossible for the attacker to guess which part of the code is fake. This means that the bogus code should be stealthy by strongly resembling the original code but still provide enough differences to have a real impact on the circuit's behavior (Fig. 5.5).

5.4.1.1 Key-Based Predicates

In conventional control flow splitting implementations, *opaque predicates* [7] are used to split the control flow: for example, based on complex mathematical expressions, the value of opaque predicates is known at obfuscation time but hard to evaluate for an attacker. In this chapter, we propose to instead use *key-based* predicates, where an input value is tested against a constant. Similar to logic locking techniques used for hardware obfuscation, only the correct input combination of inputs will allow the code to execute correctly. A simplified example in the form of a Data Flow Graph (DFG) is given in Fig. 5.6. The input variable "key" is tested against the value "42." If the test evaluates to true, then the original code branch is executed. Otherwise, a bogus code branch is executed.

5.4.1.2 Bogus Expressions

This obfuscation technique, illustrated in Fig. 5.7, can be directly performed at AST level. Candidate AST nodes for obfuscation are binary expressions of the form: `<binaryExpression>::=<expression><op><expression>`, where *op* is a bitwise or arithmetic operator. For each selected binary expression, several similar bogus expressions are created with slight variations, see Fig. 5.5b. The order of the operands in the original expression is randomly shuffled. Furthermore, the operators are randomly replaced by other, similar operators. For example:

$$a = (b + c)^{*}2 \quad \Longrightarrow \quad a = (2 - c) + b.$$

The choice of the operators added in the bogus expressions has some importance in improving security but also in reducing overhead. The chosen operators have to be computationally similar to increase stealth (a bitwise operator cannot be added for an operation between two integers, for example) and of equal or lower complexity. Special care has to be given to the fact that the added bogus code, which is generated

```c
int gcd(int m, int n){
    while(m != n){
        if (m > n){
            m = m - n;
        } else {
            n = n - m;
        }
    }
    return m;
}
```

(a)

```c
int gcd_obf(int m, int n, int k1, int k2){
    while (m != n){
        if (m > n){
            if (k1 == 42){ // true
                m = m - n;
            } else if (k1 == 25){ // false
                m = n - m;
            } else { // false
                m = m + n;
            }
        } else {
            if (k2 == 54){ // false
                n = n + m;
            } else if (k2 == 12){ // false
                n = m - n;
            } else { // true
                n = n - m;
            }
        }
    }
    return m;
}
```

(b)

```c
int gcd_obf(int m, int n){
    int control = 1;
    while (control != 0){
        switch(control){
            case 1: {
                if (m != n){
                    control = 2;
                } else {
                    control = 0;
                }
                break;
            }
            case 2: {
                if (m > n){
                    control = 3;
                } else {
                    control = 4;
                }
                break;
            }
            case 3: {
                m = m - n;
                control = 1;
                break;
            }
            case 4: {
                n = n - m;
                control = 1;
                break;
            }
        }
    }
    return m;
}
```

(c)

```c
int gcd_obf(int m, int n, int k1, int k2, int k3,
            int k4, int k5, int k6, int k7){
    int control = k1;
    while (control != 0){
        switch(control){
            case 1: {
                if (m != n){
                    control = k2;
                } else {
                    control = k3;
                }
                break;
            }
            case 2: {
                if (m > n){
                    control = k4;
                } else {
                    control = k5;
                }
                break;
            }
            case 3: {
                m = m - n;
                control = k6;
                break;
            }
            case 4: {
                n = n - m;
                control = k7;
                break;
            }
        }
    }
    return m;
}
```

(d)

Fig. 5.5 Different obfuscation techniques applied by KaOTHIC to a function calculating the GCD of two numbers. (**a**) No obfuscation. (**b**) Bogus expression insertion. (**c**) Control flow flattening. (**d**) Key-based control flow flattening

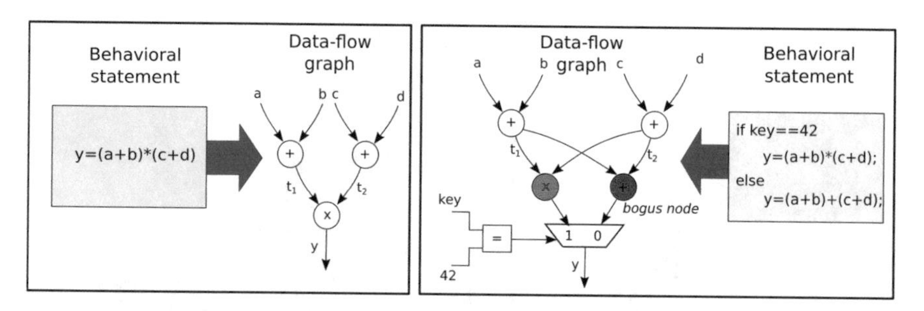

Fig. 5.6 Behavioral and dataflow representation of a code with bogus code insertion

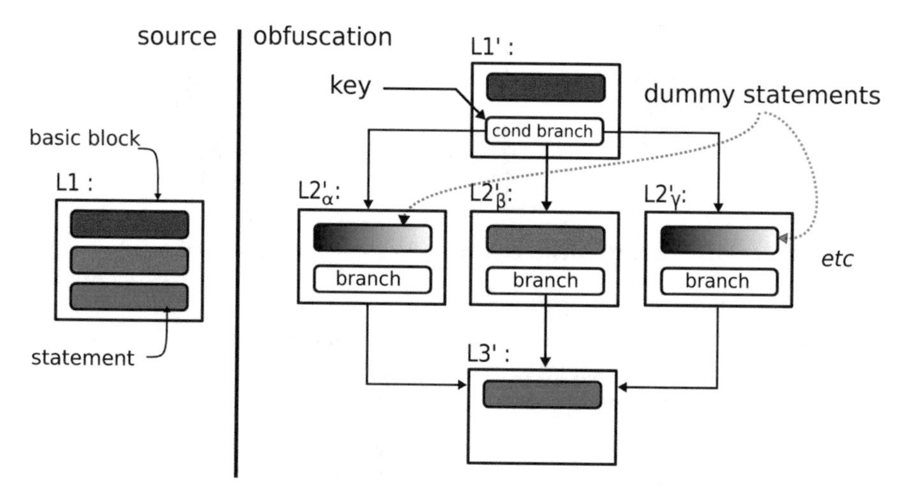

Fig. 5.7 Basic block example before and after obfuscation: insertion of two bogus expressions

in a completely automated way by our obfuscation tool, should still be compilable. Applying careful, small modifications to existing expressions significantly eases this process and decreases the risk of adding code leading to compilation errors while still maintaining a high level of stealth.

5.4.1.3 Bogus Basic Blocks

In order to diversify and improve upon the previously presented technique, we also propose to insert complete bogus basic blocks, instead of just individual bogus expressions. For this technique, the first step builds the CFG of the source program. A list of candidate basic blocks for obfuscation is then established. For each selected basic block, one or several bogus basic blocks are created, as illustrated in Fig. 5.8. The creation of these basic blocks relies on the same principles as for bogus expressions, presented in Sect. 5.4.1.2:

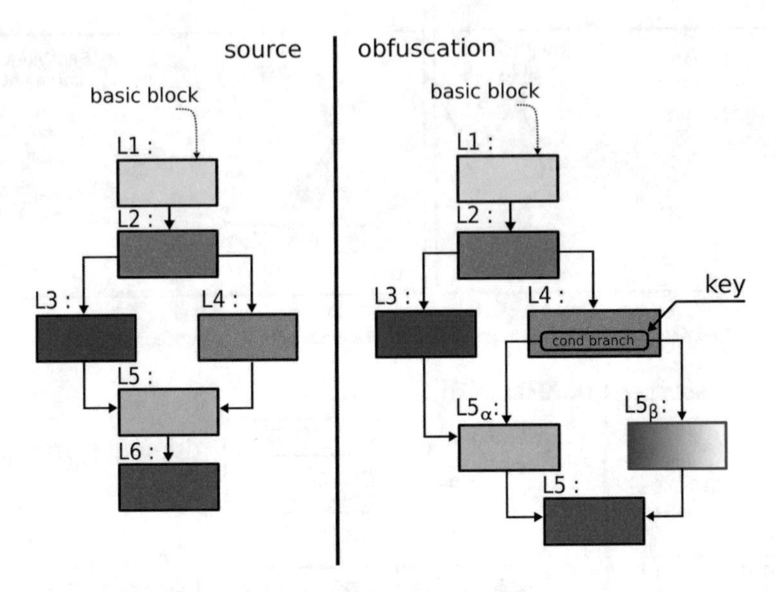

Fig. 5.8 Example CFG before and after obfuscation by adding bogus basic blocks

- In individual expressions, order of the operands and operations is shuffled.
- Operators are randomly replaced in each expression.
- The order of all the expressions in the block is shuffled.

As before, key-based predicates are then added to split the control flow. Finally, an AST is rebuilt from the modified CFG and C-level source code is returned. The success of this technique relies on correctly identifying in an automated way which basic blocks *can* be obfuscated without preventing the correct rebuilding of the AST, and carefully selecting which basic blocks *should* be obfuscated without hindering HLS optimizations. For example, the increment instruction of a for loop is usually represented as a basic block when a CFG is generated. However, duplicating and modifying this block during obfuscation would prevent correct rebuilding of the loop when printing the obfuscated C code.

5.4.2 Control Flow Flattening

Control flow flattening is a software obfuscation technique first introduced in [21]. Control flow structures that are easily identifiable such as loops or if/else statements are replaced by one global switch statement. The goal is to hinder static analysis of the program by hiding the targets of the code branches.

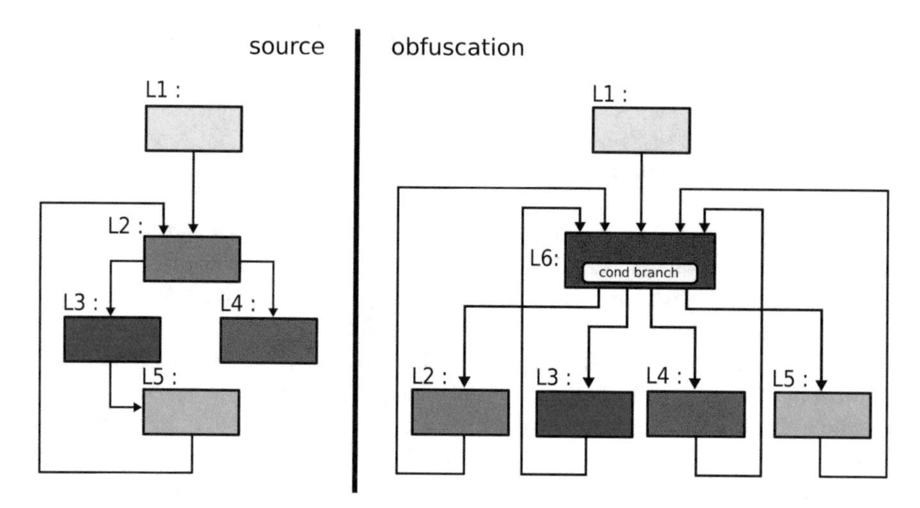

Fig. 5.9 Example CFG before and after obfuscation by control flow flattening

5.4.2.1 Key-Based Control Flow Flattening

An algorithm to flatten the control flow of a C++ program was presented in [22]. The first step generates the Control Flow Graph of the original program by separating it into basic blocks. Next, each basic block is rewritten as a case in one global switch statement. This switch statement is embedded in a loop. A routing variable is used to direct the control flow of the program. At the end of each case, a new value is attributed to the routing variable to indicate the next state. When the value 0 is given, the exit condition for the surrounding loop is reached. An illustration of control flow flattening can be found in Fig. 5.9.

One of the vulnerabilities and easiest attack vectors for de-obfuscation of classic control flow flattening is the fact that the routing variable is not hidden, as shown in the example Fig. 5.5c. An attacker can thus follow the code execution and reconstruct the original control flow. Several techniques to hide the values assigned to the routing variable have been proposed, for example, by using a computationally hard problem in the dispatcher [23], or by using a one-way function to update the routing variable [24].

In accordance with our threat model, we only want to secure one design step, HLS, which is why transient obfuscation can be used. To increase security of control flow flattening without further increasing overhead, we propose a key-based control flow flattening method.

We propose to replace the literal values assigned to the routing variable by variables that are inputs to the function, as can be seen in Fig. 5.5d. This means that if the wrong inputs, or keys, are given, the basic blocks will be executed in the wrong order. Only the exact correct sequence of inputs can ensure correct code execution. A code example with classic and key-based control flow flattening can be found in Fig. 5.5d.

5.4.2.2 Partial Control Flow Flattening

When using control flow flattening, every loop gets broken up and transformed into individual basic blocks. This means that HLS tools are no longer able to perform common optimizations on loops (unrolling, pipelining, etc.). Without these optimizations, the final overhead can be too high for some users.

An additional feature in KaOTHIC was added to enable users to fine-tune which parts of the code should be flattened. By placing pragmas in the code, a user can for example choose to flatten everything except for a particular loop. Another option is to perform what we called "coarse flattening": instead of creating a case for each basic block, several basic blocks can be grouped together in one case, for example all the blocks inside a loop. This selective control flow flattening can help improving the results in terms of design overhead but comes with a decrease in security.

5.4.3 Further Hiding of the Control Flow

To further improve the security level of the previous obfuscation techniques, several variations as well as new transformations can be proposed. For example, in Sect. 5.4.2, we used a switch statement as *dispatch*, i.e., as a method for selecting which basic block to execute next in a flattened CFG. In our environment KaOTHIC, we implemented another type of dispatch, *Goto dispatch*, where Goto statements are used to direct the control flow. This involves labeling each statement of the original code and then adding Goto statements at the end of each basic block to lead to the next block. Instead of directly pointing to the label of the next block, our Goto statements point to a key input, making it once again hard for an attacker to find the correct control flow. To further increase opaqueness for attackers, we also propose to add bogus transitions (see Fig. 5.10) between basic blocks, i.e., bogus Goto statements, which can be removed during de-obfuscation with the correct keys. It should furthermore be noted that all of the transformations here can be combined in different orders to increase robustness of the obfuscation.

5.4.4 Data Obfuscation by Literal Replacement

To enhance the previous work in KaOTHIC, mainly focused on hiding the control flow of a program, basic data obfuscation was also added. One of the first steps in data obfuscation is often to hide literals by replacing them with opaque expressions. In our case, we can take advantage of the fact that the obfuscation is temporary.

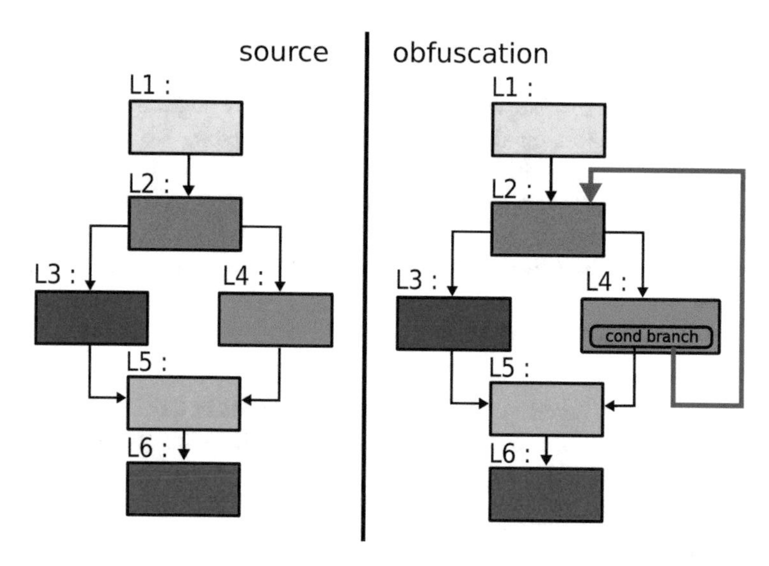

Fig. 5.10 Example CFG before and after obfuscation by adding bogus transitions

Instead of using opaque expressions, which a motivated attacker might be able to break given enough time, we simply remove literals completely from the code and replace them with placeholder variables. During de-obfuscation, the key values are added again, ensuring correct functionality of the circuit.

In practice, we implemented a basic transformation that parses the whole code at Abstract Syntax Tree (AST) level and flags all integer literals. Depending on the obfuscation level desired by the user, part or all of the integer literals are replaced by variables. These variables are added as inputs to the main function. The real integer values are stored in a file that stays on the user's machine. During de-obfuscation, the values provided by the user are hard-coded as inputs to the circuit. During logic synthesis, constant propagation will lead to the removal of these now unnecessary inputs and the recovery of the original integer literals. The same process can be applied for other literals such as string literals.

5.5 Experimental Study

5.5.1 Experimental Setup

To validate our transient obfuscation approach, we tested the different obfuscation techniques applied individually to five (5) benchmarks from MachSuite [25] and CHStone [26]: advanced encryption standard (`AES`), adaptive differential pulse code modulation encoder (`Adpcm`), optimal sequence alignment algorithm (`Needwun`),

Merge Sort algorithm (`Merge_sort`), and three-dimensional stencil computation (`Stencil3d`).

For each benchmark, we applied HLS followed by logic synthesis once to obtain a baseline report on design size and timing. Next, we obfuscated each benchmark with a single obfuscating transformation, followed by HLS with the same parameters as for the original, unobfuscated design. Then de-obfuscation was applied to the resulting obfuscated RTL IP, followed by logic synthesis. The results of these syntheses were used to calculate design overhead of the different approaches.

We also varied the two following parameters for bogus code insertion techniques:

- *Obfuscation level*: Percentage of candidate AST nodes or CFG basic blocks that are obfuscated. 100% means that all possible candidates are obfuscated, leading to the highest level of security.
- *Branching degree*: Amount of bogus expressions or blocks added for each obfuscated item. Increasing it has a strong impact on the number of guesses necessary to find the correct key combination.

We used Xilinx Vivado HLS as an HLS tool. Logic synthesis was performed with Synopsys Design Compiler, targeting the Synopsys SAED 90nm educational library for ASIC, as well as with Xilinx Vivado for a Nexys 4 DDR Artix-7 FPGA board.

5.5.2 Results and Analysis

In order to analyze the different parameters at play during obfuscation, as well as the validity of our whole process, we performed more detailed tests for one transformation in particular, bogus expression insertion.

5.5.2.1 De-obfuscation

We simulated each circuit before and after de-obfuscation using GHDL, an open-source simulator for VHDL language, as well as the co-simulation tool provided by Vivado HLS. These simulations show that before de-obfuscation, the obfuscated circuits only perform correctly when the exact right key sequence is applied as inputs. After de-obfuscation, the keys are no longer necessary and the circuit always behaves as expected. This proves that our obfuscation process is transient where functionality is concerned and that the automated de-obfuscation tool does not remove any needed logic. To prove that any additional logic added by obfuscation is removed, and to show the benefit of our de-obfuscation approach, we extended the previously presented test flow by performing in total four (4) logic syntheses per test:

Table 5.1 Average design overhead on ASIC

	No De-obfuscation		Naive De-obfuscation		Full De-obfuscation	
Benchmark	Area	Delay	Area	Delay	Area	Delay
Adpcm	2.61%	−0.06%	2.29%	−0.08%	−0.08%	−0.01%
AES	0.3%	<0.01%	0.31%	<0.01%	0.03%	2.72%
Merge sort	0.07%	<0.01%	0.07%	<0.01%	0.03%	0.01%
Mips	1.14%	0.5%	1.45%	0.5%	−1.12%	−7.47%
Needwun	9.01%	11.69	8.39%	11.69%	0.26%	<0.01%
Stencil3d	9.38%	0.73%	9.2%	0.73%	1.54%	−0.14%

- The original, never obfuscated, RTL IP is used as baseline.
- The obfuscated RTL IP, with **no** de-obfuscation.
- The obfuscated RTL IP, with **naïve** de-obfuscation.
- The obfuscated RTL IP, with **full** de-obfuscation.

The overheads in percent when compared to the original IP are presented for each benchmark and for both ASIC and FPGA in Tables 5.1 and 5.2. These tables contain the results with a branching degree of 1, meaning that for each obfuscated expression, one bogus expression was added. These results show that with no de-obfuscation, design overhead is usually high, up to 12% on ASIC and up to more than 200% on FPGA. Increasing the branching degree during obfuscation would automatically further increase the area overheads, since adding additional expressions in C code results in additional logic in the final design.

When using naïve de-obfuscation, results indicate that the overall overhead decreases slightly but is still significant. On the other hand, de-obfuscating the IPs in a targeted way with our proposed full de-obfuscation method strongly reduces overhead. This proves that we are successfully able to remove the code added during obfuscation.

In some cases, the overhead is already close to 0% (e.g., AES on ASIC) without any de-obfuscation. With de-obfuscation, overhead can even be negative. This may happen when the obfuscated code, greatly increased in size due to added bogus code, forces the HLS tool into different, sometimes better optimization, choices. In some cases, the different choices thus result in overall smaller final designs. In other cases, the small remaining overhead after full de-obfuscation (e.g., Stencil3D) is in our opinion also due to the choices made by the HLS tool: adding code at behavioral level will, for example, cause the tool to handle resource sharing differently. Furthermore, the syntactic variances caused by the obfuscation process can also have a significant impact on the HLS tool, see [27].

5.5.2.2 Obfuscation Level

We call *obfuscation level* the ratio of effectively obfuscated expressions or basic blocks with respect to the total number of such items in the design. For bogus

Table 5.2 Average design overhead on FPGA

Benchmark	No De-obfuscation			Naive De-obfuscation			Full De-obfuscation		
	LUT	Register	Delay	LUT	Register	Delay	LUT	Register	Delay
Adpcm	5.82%	28.9%	15.9%	−0.08%	−0.2%	10.85%	−0.12%	−0.2%	11.06%
AES	29.0%	20.0%	1.98%	7.05%	0.42%	3.35%	8.97%	0.42%	4.49%
Merge sort	−1.37%	18.7%	2.57%	−6.18%	4.08%	−1.27%	−5.99%	4.08%	−3.77%
Mips	45.9%	210%	7.51%	1.46%	−1.39%	4.51%	1.78%	−1.39	4.61%
Needwun	23.9%	33.5%	−4.22%	0.79%	0.01%	−2.78%	−0.6%	−0.25%	−1.98%
Stencil3d	20.7%	22.4%	−9.44%	1.38%	4.56%	−3.78%	−1.25%	5.57%	−0.86%

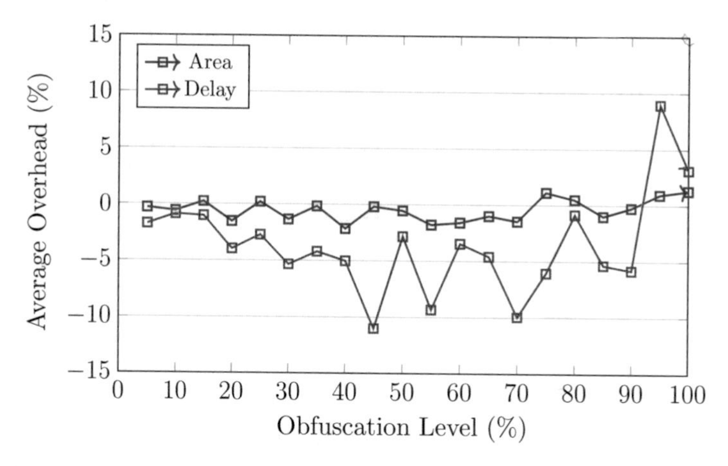

Fig. 5.11 Average design overhead per obfuscation level, for bogus expression insertion—ASIC

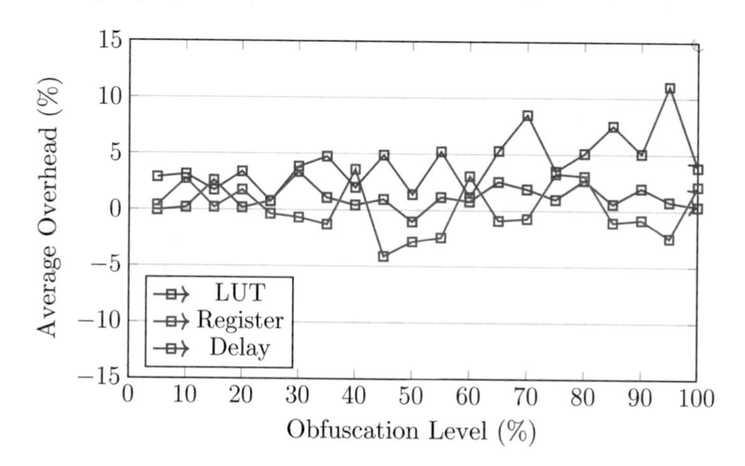

Fig. 5.12 Average design overhead per obfuscation level, for bogus expression insertion—FPGA

basic block insertion, for example, an obfuscation level of 100% means that all possible basic blocks were obfuscated. Increasing the obfuscation level results in an increased number of obfuscation keys to guess for an attacker and thus in a higher level of security. Ideally, we should always choose an obfuscation level of 100%.

We studied the impact of obfuscation level on overhead using the bogus expression insertion technique as an example. By doing a series of tests with the obfuscation level varying from 5% to 100%, we established whether there is a correlation between obfuscation level and overhead. The results, shown in Figs. 5.11 and 5.12 for ASIC and FPGA, respectively, indicate that there is no such correlation and in particular that area overhead does not increase with obfuscation level. This positive result demonstrates that our de-obfuscation process is effective

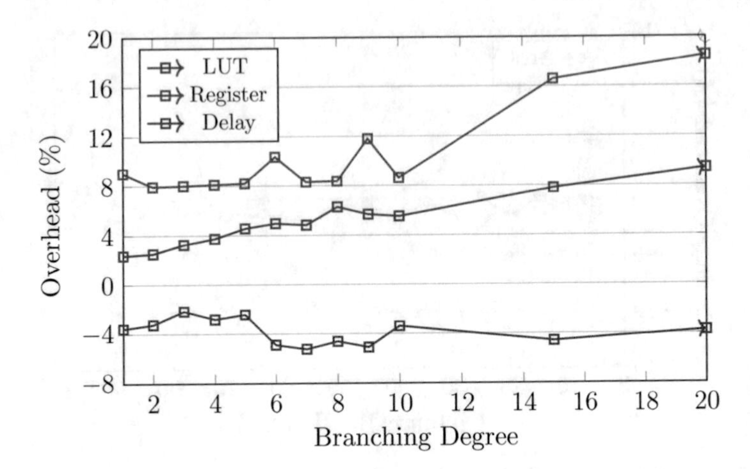

Fig. 5.13 Design overhead per branching degree, for bogus basic block insertion—FPGA—*obfuscated*

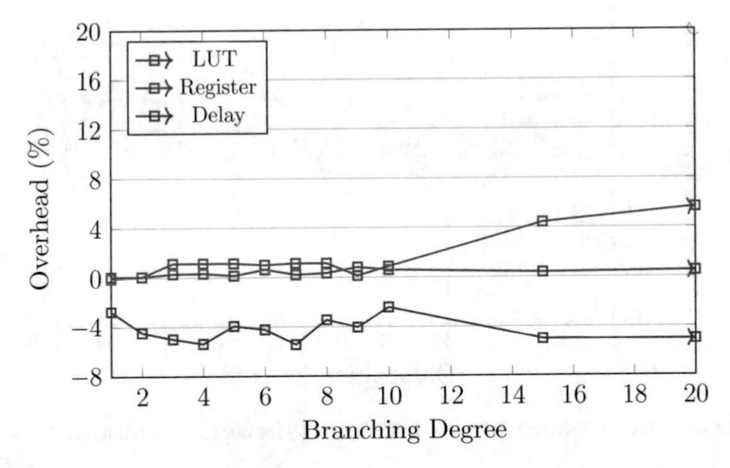

Fig. 5.14 Design overhead per branching degree, for bogus basic block insertion—FPGA—*de-obfuscated*

no matter what amount of code is obfuscated. We conclude that the maximum level of obfuscation and thus of security, i.e., 100%, can always be chosen.

5.5.2.3 Branching Degree

We call *branching degree* the number of bogus elements added for each obfuscated expression or basic block. A higher branching degree increases the amount of choices an attacker has for each obfuscated item. By combining the maximum obfuscation level, and a high branching degree, security can be significantly increased.

Preliminary tests were made for one obfuscation technique in particular, bogus basic block insertion. In Figs. 5.13 and 5.14, results for Adpcm benchmark, which are representative of results found with other designs, are shown. They show that with a higher obfuscation degree, meaning a higher amount of bogus basic blocks added, the design size increases for the obfuscated design. This is to be expected since the number of basic blocks in the obfuscated design is much higher. For the de-obfuscated design however, while the number of registers slightly increases with a high branching degree, the overall overhead remains below a threshold of around 5%. These results show that for each obfuscated basic block, up to 20 bogus blocks can be added without any significant impact on design performances and size.

5.5.2.4 Obfuscation Techniques

We have so far performed in-depth testing mainly for bogus code insertion and control flow flattening techniques. For all of these tests, average runtime overhead for HLS is negligible, at under 5%. For bogus code insertion techniques (bogus expressions and bogus basic blocks), the design overhead after logic synthesis is overall satisfactory. In the majority of tests, area and delay overheads are below 5%, which we consider an acceptable threshold. For some benchmarks, overhead can be higher than tolerable. However, this can be mitigated by choosing carefully which obfuscation technique to apply and which values to use for obfuscation level and branching degree.

For key-based control flow flattening, the results are not as encouraging. When used for software obfuscation, control flow flattening has been shown to result in a high overhead: in [22], the authors explain that on average, runtime and size of a program double. In an HLS context, this problem becomes even more predominant. Not only does control flow flattening hinder compiler front-end optimizations, but it can also severely impact later stages of the HLS process. For example, since loops are no longer easily identifiable in the code, any loop-related optimizations such as loop pipelining or unrolling cannot be performed anymore. This issue can be mitigated by using coarser control flow flattening and leaving some control flow structures intact instead of flattening everything. This does however decrease the level of security. There is thus a trade-off between overhead and security, which has to be balanced for each individual design.

We have also noticed that the dispatch structure used for control flow flattening, either in the form of a global switch-case statement, or in the form of a series of Goto and labeled statements, has a severe impact on overhead. Flattening the control flow at behavioral source code and rewriting the BIP with such a dispatch structure are effectively equivalent to creating a Finite-State Machine (FSM). However, this FSM-like structure is not recognized as such by HLS tools, which results in an additional FSM being created on top of the first one. The final overhead due to this behavior is significant (up to 35% on FPGA and 45% on ASIC) and can, in our opinion, not be avoided. It should be noted however that switching from classic control flow obfuscation to key-based control flow obfuscation, which highly improves the security level, does not have any further negative effect on overhead.

Key-based control flow flattening technique, especially when used in conjunction with bogus code insertion techniques, can strongly decrease readability of the code and thwart static code analysis tools. From a security point of view, this is a powerful technique that could be used to secure cloud-based HLS flows. However, it can lead to significant design overhead when used in a hardware context and should thus be applied only on select designs.

5.6 Conclusions and Future Work

In this chapter, we showed how software obfuscation techniques can be adapted to protect BIPs against theft and reuse. In particular, we proposed several variations of these techniques to secure a cloud-based HLS flow during design time. By using key-based methods and by de-obfuscating at RTL, we ensure that this protection is transient and does not cause too high overhead. Experimental results show that some techniques are better suited to hardware design context than others and that simply reusing known software obfuscation techniques is not enough to ensure satisfactory results. Further work will focus on more detailed analysis and test campaigns of several techniques but also on how this work can be extended for other security topics such as watermarking or prevention of hardware Trojan insertion in early design steps.

References

1. Polian, I., Becker, G.T., Regazzoni, F.: Trojans in early design steps—An emerging threat. In: Proceedings of the Conference on Trustworthy Manufacturing and Utilization of Secure Devices (TRUDEVICE) (2016)
2. Pilato, C., Basu, K., Regazzoni, F., Karri, R.: Black-Hat high-level synthesis: Myth or reality? IEEE Trans. Very Large Scale Integr. Syst. **27**(4), 913–926 (2018)
3. Dashtbani, M., Rajabzadeh, A., Asghari, M.: High level synthesis as a service. In: 2015 5th International Conference on Computer and Knowledge Engineering (ICCKE), pp. 331–336. IEEE, Piscataway (2015)
4. Veeranna, N., Schafer, B.C.: Efficient behavioral intellectual properties source code obfuscation for high-level synthesis. In: 2017 18th IEEE Latin American Test Symposium (LATS), pp. 1–6. IEEE, Piscataway (2017)
5. Baumgarten, A., Tyagi, A., Zambreno, J.: Preventing IC piracy using reconfigurable logic barriers. IEEE Design Test Comput. **27**(1), 66–75 (2010)
6. Badier, H., Le Lann, J.C., Coussy, P., Gogniat, G.: Transient key-based obfuscation for HLS in an untrusted cloud environment. In: 2019 Design, Automation & Test in Europe Conference & Exhibition (DATE), pp. 1118–1123. IEEE, Piscataway (2019)
7. Collberg, C., Thomborson, C., Low, D.: Manufacturing cheap, resilient, and stealthy opaque constructs. In: Proceedings of the 25th ACM SIGPLAN-SIGACT Symposium on Principles of Programming Languages, pp. 184–196 (1998)

8. Barak, B., Goldreich, O., Impagliazzo, R., Rudich, S., Sahai, A., Vadhan, S., Yang, K.: On the (Im)possibility of obfuscating programs. In: Annual International Cryptology Conference, pp. 1–18. Springer, Berlin (2001)
9. Garg, S., Gentry, C., Halevi, S., Raykova, M., Sahai, A., Waters, B.: Candidate indistinguishability obfuscation and functional encryption for all circuits. SIAM J. Comput. **45**(3), 882–929 (2016)
10. Xu, H., Zhou, Y., Kang, Y., Lyu, M.R.: On secure and usable program obfuscation: A survey (2017). Preprint arXiv:1710.01139
11. Apon, D., Huang, Y., Katz, J., Malozemoff, A.J.: Implementing cryptographic program obfuscation. IACR Cryptol. ePrint Archive **2014**, 779 (2014)
12. Collberg, C., Thomborson, C., Low, D.: A Taxonomy of Obfuscating Transformations. CiteSeerX (1997)
13. Kirovski, D., Hwang, Y.Y., Potkonjak, M., Cong, J.: Intellectual property protection by watermarking combinational logic synthesis solutions. In: Proceedings of the 1998 IEEE/ACM International Conference on Computer-Aided Design, pp. 194–198 (1998)
14. Lewandowski, M., Meana, R., Morrison, M., Katkoori, S.: A novel method for watermarking sequential circuits. In: 2012 IEEE International Symposium on Hardware-Oriented Security and Trust, pp. 21–24. IEEE, Piscataway (2012)
15. Imeson, F., Emtenan, A., Garg, S., Tripunitara, M.: Securing computer hardware using 3D integrated circuit IC technology and split manufacturing for obfuscation. In: Presented as Part of the 22nd {USENIX} Security Symposium ({USENIX} Security 13), pp. 495–510 (2013)
16. Chakraborty, R.S., Bhunia, S.: RTL hardware IP protection using key-based control and data flow obfuscation. In: 2010 23rd International Conference on VLSI Design, pp. 405–410. IEEE, Piscataway (2010)
17. Lao, Y., Parhi, K.K.: Obfuscating DSP circuits via high-level transformations. IEEE Trans. Very Large Scale Integr. Syst. **23**(5), 819–830 (2014)
18. Sengupta, A., Roy, D., Mohanty, S.P., Corcoran, P.: DSP design protection in CE through algorithmic transformation based structural obfuscation. IEEE Trans. Consumer Electr. **63**(4), 467–476 (2017)
19. Islam, S.A., Katkoori, S.: High-level synthesis of key based obfuscated RTL datapaths. In: 2018 19th International Symposium on Quality Electronic Design (ISQED), pp. 407–412. IEEE, Piscataway (2018)
20. Pilato, C., Regazzoni, F., Karri, R., Garg, S.: TAO: Techniques for algorithm-level obfuscation during high-level synthesis. In: Proceedings of the 55th Annual Design Automation Conference, pp. 1–6 (2018)
21. Wang, C., Hill, J., Knight, J., Davidson, J.: Software tamper resistance: Obstructing static analysis of programs. Technical Report CS-2000–12, University of Virginia, 12 2000 (2000)
22. László, T., Kiss, Á.: Obfuscating C++ programs via control flow flattening. Annales Univ. Sci. Budapest. de Rolando Eötvös Nominatae, Sect. Comput. **30**, 3–19 (2009)
23. Chow, S., Gu, Y., Johnson, H., Zakharov, V.A.: An approach to the obfuscation of control-flow of sequential computer programs. In: International Conference on Information Security, pp. 144–155. Springer, Berlin (2001)
24. Cappaert, J., Preneel, B.: A general model for hiding control flow. In: Proceedings of the Tenth Annual ACM Workshop on Digital Rights Management, pp. 35–42 (2010)
25. Reagen, B., Adolf, R., Shao, Y.S., Wei, G.Y., Brooks, D.: MachSuite: benchmarks for accelerator design and customized architectures. In: 2014 IEEE International Symposium on Workload Characterization (IISWC), pp. 110–119. IEEE, Piscataway (2014)
26. Hara, Y., Tomiyama, H., Honda, S., Takada, H., Ishii, K.: CHStone: A benchmark program suite for practical C-based high-level synthesis. In: 2008 IEEE International Symposium on Circuits and Systems, pp. 1192–1195. IEEE, Piscataway (2008)
27. Chaiyakul, V., Gajski, D.D., Ramachandran, L.: High-level transformations for minimizing syntactic variances. In: 30th ACM/IEEE Design Automation Conference, pp. 413–418. IEEE, Piscataway (1993)

Chapter 6
Protecting Hardware IP Cores During High-Level Synthesis

Christian Pilato, Donatella Sciuto, Francesco Regazzoni, Siddharth Garg, and Ramesh Karri

6.1 Introduction

The design flow for producing an Integrated Circuit (IC) is shown in Fig. 6.1. It is composed of several phases, ranging from the design of the components to the physical implementation in the target technology and the fabrication of the device on silicon. While the design phases only require access to commercial CAD tools, fabrication requires expensive infrastructure to function. For example, TSMC expects to invest more than 20 billions of US dollars for 3 nm foundries [1]. Many semiconductor companies cannot afford the increasing cost of IC manufacturing [2]. As the technology scales, more and more companies are becoming *fabless*, outsourcing the IC fabrication to third-party foundries [3]. While this process allows for a cost reduction, it creates security threats in the semiconductor supply chain: an attacker that has access the design files can reverse engineer the functionality and steal the intellectual property (IP) [4]. Since these ICs often implement proprietary optimized designs, this malicious process can cause significant economic harm to the design houses. While IC watermarking is able to determine the real ownership of an IC during litigation, this is a passive method that requires to identify the illegal copy and enter into a legal dispute [5].

C. Pilato (✉) · D. Sciuto
Politecnico di Milano, Milan, Italy
e-mail: christian.pilato@polimi.it; donatella.sciuto@polimi.it

F. Regazzoni
ALaRI Institute, Università della Svizzera italiana, Lugano, Switzerland
e-mail: regazzoni@alari.ch

S. Garg · R. Karri
New York University, New York, NY, USA
e-mail: siddharth.garg@nyu.edu; rkarry@nyu.edu

S. Katkoori, S. A. Islam (eds.), *Behavioral Synthesis for Hardware Security*,
https://doi.org/10.1007/978-3-030-78841-4_6

Semiconductor design houses are thus showing an increasing interest for anti-reverse engineering techniques for untrusted foundries. For example, *split manufacturing* divides computing resources and interconnections, with the two parts fabricated in different foundries. This process is based on the assumption that collusion between the two foundries is unlikely. However, designing for split manufacturing is complex and expensive. *Obfuscation* and *logic locking* have been extensively investigated for this purpose as well [6]. The designer adds additional inputs and modules to the design to hide the correct functionality (obfuscation), while a *locking key* (unknown to the foundry and written later in a tamper-proof memory) activates the IC (logic locking). These methods are usually applied on the gate-level netlist [7, 8]. With the increasing complexity of ICs, designers are migrating to *high-level synthesis* (HLS) to automate the design process [9]. While security features can be applied at any design steps, more robust solutions can be applied in the early stages, i.e., during HLS [10–12]. For example, the Stripped-Functionality Logic Locking approach (SFLL) [13] has been recently extended to HLS [14]. However, a holistic solution that brings together HLS and obfuscation is still missing.

In this chapter, we discuss a possible approach to raise the abstraction level of register transfer-level (RTL) obfuscation by embracing a security-aware HLS flow to generate obfuscated designs by construction. We propose an approach based on *algorithm-level obfuscation*, which aims at developing anti-reverse engineering techniques based on the characteristics of the algorithm during the different HLS steps. The approach we present starts from a high-level description of the functionality in C language and uses HLS methods to produce the corresponding obfuscated RTL description. This is achieved by obfuscating the HLS results or the generated RTL description. For doing this, HLS algorithms are extended to obfuscate the most sensitive details of an algorithm. After compiler analysis, the information that comes from the specification (e.g., constant values, loop bounds) and the information generated by HLS (e.g., control states, used and unused datapath resources, execution latency) are obfuscated. It is possible to obfuscate complex functions as part of a comprehensive HLS-based obfuscation design flow. As a proof-of-concept, the obfuscation techniques are implemented in BAMBU [15], an open-source HLS framework, and applied the resulting flow to benchmarks that are much larger than the ones commonly used for gate-level obfuscation.

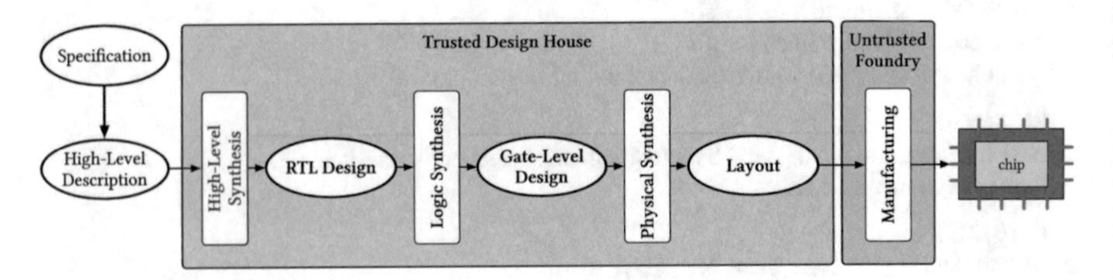

Fig. 6.1 IC design flow

6.1.1 Contributions

Complete HLS solutions for obfuscating an IC are unavailable. However, this is a promising approach to target complex designs while protecting the semantics of an IC more efficiently. The obfuscation techniques discussed in this chapter are imposing constraints on both the compiler front end and the HLS engine to expose the elements to obfuscate instead of spreading them inside the design. Working at the HLS level allows to remove the sensitive algorithmic information and combine it with the locking key. The main contributions of the proposed approach are:

- The attack scenario consists of the untrusted foundry as the adversary in an oracle-less threat model for low-volume customers (Sect. 6.4).
- The approach embraces a set of obfuscation techniques that address the different elements of an algorithm to protect (Sect. 6.5.1).
- The techniques are integrated in an HLS-based design flow for algorithmic obfuscation that starts directly from C code (Sect. 6.5.2).
- Different solutions to manage the locking key are presented (Sect. 6.5.3).

The approach is evaluated by applying it to common HLS benchmarks, showing promising results for high-level obfuscation similar to program obfuscation.

6.1.2 Roadmap

After introducing logic locking (Sect. 6.2) and presenting the model of the components that we aim at protecting (Sect. 6.3), we introduce our threat model (Sect. 6.4). Then, we present our approach for algorithm-level obfuscation, showing how it is implemented in an HLS flow (Sect. 6.5). In Sect. 6.6, we evaluate the area and performance overhead for the obfuscation techniques and present a validation of the obfuscated designs.

6.2 Background on Logic Locking

IC counterfeiting is a critical issue for *fabless* companies since they may lose billions of dollars for IP theft and overselling [4]. Addressing the problem, several IP protection techniques have been proposed at different stages of the design process [16]. To protect the intellectual property of an IC, the designer has to leverage intrinsic hardware properties of the device with Physical Unclonable Functions (PUFs) [17] or to modify the manufacturing process to separate the fabrication of the interconnections from the rest of the chip [18]. However, the

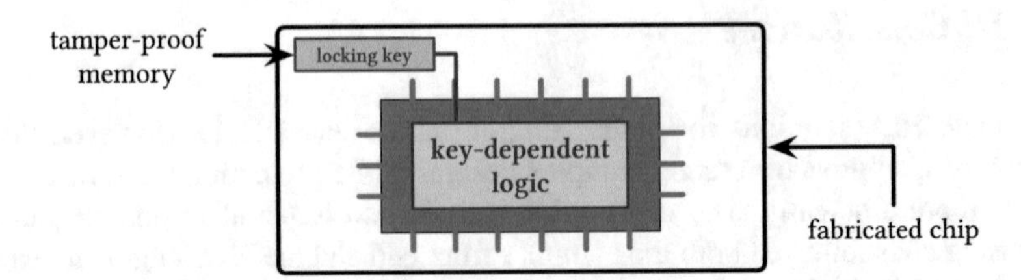

Fig. 6.2 Logic locking requires an additional input to deliver the *locking key* to the IC

process requires a 2.5D integration technology. Furthermore, these solutions require an intimate knowledge of the target technology and the back-end tool chain.

Different solutions have been proposed for adding a "signature" to verify the ownership of an IC during litigation (*watermarking*) [19, 20]. Most of these approaches operate on the gate-level netlist or during layout generation [21–23], aiming at embedding a unique "signature." Also in this case, approaches have been proposed to raise the abstraction level for IP watermarking and operate during the component generation [5, 10, 24]. However, these methods are passive and cannot be used to prevent an illegal IC copy.

Logic locking is a well-known technique to thwart a potential attacker who wants to reverse engineer and copy the IC design. To do so, it hides the IC functionality against reverse engineering by using extra logic controlled by a key known only to the designers (see Fig. 6.2) [6, 7, 25]. High-level transformations have been already proposed but only to obfuscate DSP circuits [26]. To thwart attacks aiming at recovering the key, several methods have been proposed to improve logic locking at the gate level [8, 13] or to raise the abstraction level and perform obfuscation during HLS [14], aiming at removing semantic information like in *program obfuscation* [27].

SAT-based attacks can extract these keys [8, 25]. In [12], the authors propose RTL hardening techniques by adding extra connections among the functional units. While this approach is more powerful than gate-level methods, constant values and branches are challenging to obfuscate since the design is already optimized. For instance, interconnections between resources and multiplexers have been sized based on the given precision. However, this reveals information on their range. Since we operate at a higher level of abstraction, it masks sensitive details of the algorithm by hiding sensitive constants and encrypting them during the front end with a limited overhead. Key management is another aspect of algorithmic obfuscation. Many companies are proposing solutions to store keys in off-chip tamper-proof memories (e.g., one-time-programmable memories). These approaches are complementing the approach we present here wherein the keys are stored in on-chip tamper-proof and non-volatile memories.

6.3 Design Model and High-Level Synthesis

The approach we present generates RTL components with logic locking using HLS. The accelerator model couples a *controller* and a *datapath*, as in the *finite-state machine with datapath* (FSMD) model [28]. The *controller* is a finite-state machine (FSM) that determines which operations execute in each clock cycle based on the evaluation of certain data-dependent conditions. Based on the set of operations to execute in each given clock cycle, the controller sends the proper control signals to trigger the functional units, the registers, and the interconnections in the *datapath* to drive the data values and perform the computation. Both parts are required to replicate the IC's functionality.

High-level synthesis (HLS) is a design methodology that allows designers to automatically derive an RTL design from its high-level description. The classic HLS flow is shown in Fig. 6.3, assuming the input functionality is specified in C language. The HLS tool leverages state-of-the-art compilers (e.g., GCC or LLVM) to parse the input code, apply compiler optimizations (like loop transformations), and extract a language-independent intermediate representation (IR) [9]. The core HLS steps manipulate this IR as follows. *Scheduling* determines the operations to execute in each clock cycle based on data dependencies and the available hardware resources (e.g., functional units and memories). During *module binding*, operations scheduled in different clock cycles are analyzed for potential resource sharing to reduce area occupation. Data values crossing the clock boundaries are assigned to registers (*register binding*) [29]. In *interconnection binding*, the different resources (functional units, registers, and memories) are interconnected and multiplexers are added to correctly drive the signals when multiple data sources share the same target port. Ultimately, all resources are analyzed to derive the control signals needed in each clock cycle and the FSM controller is accordingly generated during *controller synthesis*. The output is an RTL design in Verilog or VHDL ready for logic synthesis.

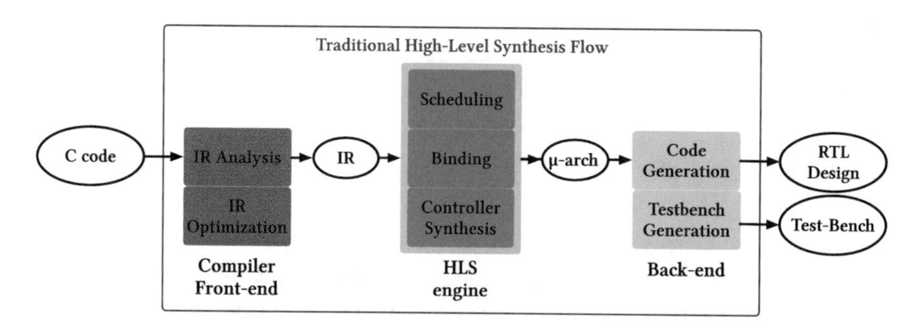

Fig. 6.3 Classic HLS flow

6.4 Threat Model: The Untrusted Foundry

6.4.1 Untrusted Foundry's Objective

The main goal of the rogue in the untrusted foundry is to reverse engineer and replicate the target IC. For doing this, adversaries aim at recovering the correct sequence of states executed by the controller (execution traces) corresponding to given input sequences, along with the corresponding signals provided to the datapath (operations to execute, registers, and interconnections) in each given clock cycle. Once the entire design is recovered, the foundry can reproduce the component, thus misappropriating the IP. In case of designs protected with logic locking, this reverse engineering process requires to identify also the correct locking key to obtain a working IC copy. So, the attacker aims at determining the design alternatives based on the values of the key bits, eventually ruling out implausible key values, i.e., values for which the design becomes clearly wrong.

6.4.2 Foundry's Capabilities

The semiconductor design houses use logic synthesis and physical design tools on the HLS results to obtain the GDSII file (i.e., the layout) of the IC to fabricate. The layout is then provided to the untrusted foundry for fabrication. However, the rogue in the untrusted foundry can also access the GDSII file to reverse engineer the functionality, attempting to break logic locking. To do so, we assume that the foundry can reverse engineer the types of modules used in the design (i.e., registers, functional units, and interconnection elements) and can identify the operations executed by each functional unit (i.e., arithmetic, relational, and logic operations). The foundry can also perform simulations with different inputs and locking key values to extract information from the circuit that can help rule out implausible key values. However, the untrusted foundry does *not* have access to the correct key or a functioning unlocked IC (*oracle-less* attacks).

6.4.3 Target of the Attacks

The oracle-less attacks considered in this chapter are common in low-volume customers who build sensitive designs (e.g., US DoD). These designs are typical targets for attacks from untrusted foundries under pressure from their government to acquire proprietary cutting-edge technology. Until recently, IBM was maintaining the trusted foundry for the US government. Once it got acquired by Global Foundries (owned by an entity outside USA), there is no trusted US foundry anymore, demanding effective methods for IC protection.

6.5 High-Level Synthesis Techniques for Algorithm Obfuscation

To protect the semantic and, in turn, the intellectual property (IP) of an algorithm via obfuscation, it is necessary to protect the following elements:

- *Constant values* contain proprietary information (e.g., coefficients) or reveal details of the algorithm (e.g., loop bounds).
- *Data flow* describes how many and which operations are executed in each clock cycle together with their dependencies (i.e., which values are elaborated). This information is represented by the scheduled data-flow graph (DFG).
- *Control flow* represents the sequence of FSM states traversed during the execution for the given inputs. It represents protocol implementations in control-dominated applications.

The elements must also be obfuscated to prevent further logic-level optimizations that can reveal proprietary information. For example, constant values are propagated to simplify the logic. Also, all elements must be obfuscated because they are connected and leaking information on one set of elements can aid recover details on the others. For example, multiplications by a constant that is power of two are often converted into shift operations that are more hardware-friendly. The optimization results can leak information both on the original operation (i.e., the multiplication) and the constant value (i.e., the power-of-two value).

The obfuscation techniques that we consider follow the same principles as in *program obfuscation* [30]. The real functionality is hidden with the creation of *opaque variables* or *opaque predicates*. A variable is opaque if it has some property (e.g., a value) that is known during obfuscation but is difficult for the attacker to deduce. Similarly, a predicate is opaque if its outcome is known only during obfuscation. To create opaque variables and predicates, expressions are combined with the locking key bit values that are known during obfuscation, but unknown to the attacker.

6.5.1 *Obfuscation Techniques*

In this section, the techniques proposed to obfuscate the elements discussed above are presented. All these elements require specific bits for obfuscation.

6.5.1.1 Constant Obfuscation

Constant values are an essential part of the IP specification and may disclose sensitive information about the implemented algorithm. Consider a component with proprietary coefficients that realizes a specific digital filter. Such coefficients include

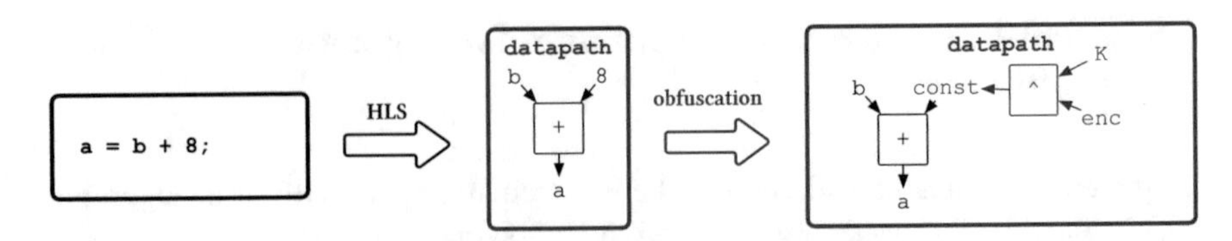

Fig. 6.4 Design modifications to implement constant obfuscation during RTL generation

also the number of taps in the filter. Without having this information, it is possible to replicate the type of component, but it is almost impossible to match the exact same filter function. Also, HLS tools optimize the datapath based on the data bit width to reduce the IC cost [9, 31]. However, using the minimum number of bits to represent a constant leaks information about its range.

To enable constant obfuscation, sensitive constants in the datapath are replaced with opaque variables as shown in Fig. 6.4. The variable associated with a constant V_i is obtained by combining a value encoded in the circuit V_i^e with the locking key bits K_i as

$$V_i^p = V_i^e \oplus K_i, \tag{6.1}$$

where V_i^e is the obfuscated value that will be stored in the RTL micro-architecture, while K_i is a C-bit signal that represents the part of the working key dedicated to obfuscating the constant V_i. The encoded value associated with each constant V_i of the input algorithm is obfuscated as

$$V_i^e = V_i \oplus K_i. \tag{6.2}$$

Clearly, the correct value is re-obtained only when the correct key is provided. Instead, if a wrong key is provided, the resulting value will be different from the one contained in the initial specification, but an attacker cannot determine this. Even when the constant represents a loop bound, the exact number of execution clock cycles for complex specifications is unknown to the attacker.

The number of bits C to implement all the constants of the function is pre-defined. The use of a pre-defined number of bits for all constants (regardless of their real size) hides the real bit width of the specific constant, thwarting the identification of the correct range. However, extracting the constants from the circuit may rule out subsequent logic optimizations (e.g., constant propagation and logic trimming), increasing the obfuscation cost.

Example Consider a constant $V_i = 10$ to be stored using 5 bits (`5'b01010`). The same value can be obfuscated as an 8-bit value as $V_i^e =$ `8'b01010111` or $V_i^e =$ `8'b11001101` based on locking keys $K_i =$ `8'b01011101` and $K_i =$ `8'b00100111`, respectively. □

This example also shows that the same constant V_i is encoded in different ways (resulting in different V_i^e values) based on the specific value of the locking key. This prevents the attacker from breaking the obfuscation and recovering the constant by comparing different versions of the design.

This step can be applied at any step of the design flow. However, it is applied at the beginning, right after the compiler phase, to avoid constant-related HLS transformations and optimizations.

6.5.1.2 Data-Flow Obfuscation

Algorithm 6.1 Algorithm for data-flow obfuscation

Procedure: `CreateDFGvariant`(DFG_i, k_i)

Data: DFG_i is the DFG of the basic block BB_i; k_i represents the key bits assigned to BB_i

Result: $VDFG_i$ is the set of DFG variants associated with BB_i

```
1:  Variants ← ∅
2:  V ← ComputeKeyVariants(k_i)
3:  for v ∈ V do
4:      dist_v ← ComputeDistance(v, k_i)          //compute distance between v and k_i
5:      DFG*_i ← CopyDFG(DFG_i)                   //create a copy of the current DFG
6:      OP ← ClusterOperations(DFG*_i)
7:      for op ∈ OP do
8:          op_j ← GetOperation(op, dist_v)        //return an operation at distance dist_v
            mod clusters
9:          SwapOperationTypes(op, op_j) //statistically swap the types of the two operations
10:     end for
11:     for dep ∈ DFG*_i do
12:         dep_j ← GetDependence(d, dist_v)       //return a dependence at distance dist_v
13:         RearrangeDependence(dep, dep_j)        //statistically reorganize the dependences
14:     end for
15:     Variants ← Variants ∪ DFG*_i
16: end for
    return Allocation
```

To hide the arithmetic operations performed in the datapath, several DFG variations for each basic block are created. The DFG variations are based on the results of the HLS engine. Indeed, each basic block is scheduled to determine the minimal number of functional units and registers to perform the computation, along with the latency (i.e., number of clock cycles), to perform the corresponding computation. Algorithm 6.1 shows how this information is used as constraints for creating the set $Variants$ of DFG variations starting from a valid schedule DFG_i and the k_i key bits assigned to the basic block b_i. The number of key bits assigned to the basic block b_i is proportional to the number of operations in b_i. In this way, a large number of variants are created only for more data-intensive basic blocks, while small basic blocks (with less sensitive operations) are left untouched. First, the 2^{B_i-1} key variants are computed, beginning from the given key bits k_i. These key values will be associated with each DFG variant (ComputeKeyVariants), and values have the same number of bits as k_i but different values to distinguish the variants from the correct functionality. Then, for each variant, the Hamming distance between the corresponding key value and the obfuscation key bits k_i (ComputeDistance) is computed. A copy of the current schedule (CopyDFG) is produced, and the operations are topologically ordered and then clustered based on the operation types. For each operation, the distance values are used as parameters for selecting the operations to alter and determining a reciprocal one in an alternative cluster (GetOperation). Once the operations are selected, the two types are swapped with probability 0.5 (SwapOperationTypes). The algorithm proceeds to change DFG dependences in a similar way (RearrangeDependences). The set of resulting DFGs are merged together to create a single datapath micro-architecture, where multiplexers are inserted to drive the signals and implement one of the variants based on the values of the key bits.

Figure 6.5 shows the application of this algorithm to a simple example. Starting from a DFG, pairs of different operations are selected for swapping their operand (step ① in Fig. 6.5). For every DFG edge, an alternative edge is selected and the dependencies to return a credible DFG (step ② in Fig. 6.5) are restructured. Finally,

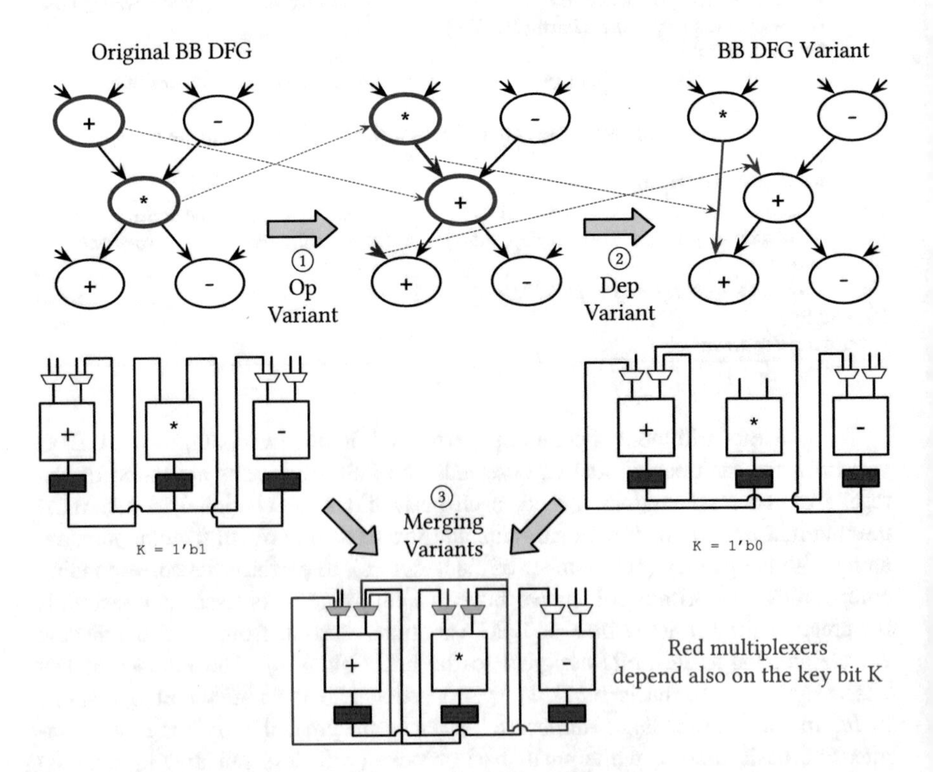

Fig. 6.5 Generation of DFG variants for operation obfuscation

the architectures corresponding to each variant are recombined into a single datapath micro-architecture, restructuring the interconnections using extra multiplexers and control signals (step ③ in Fig. 6.5). In each clock cycle, the functionality to execute is selected through a combination of key bits (to select the variant) and scheduling information (to select the operations).

6.5.1.3 Control-Flow Obfuscation

Each branch in the CFG corresponds to a branch also in the corresponding controller FSM to determine the next state to execute. The target state depends upon the outcome (either true or false) of a predicate evaluation. The predicate is computed in the datapath (e.g., an arithmetic comparison or a Boolean operation) but is evaluated in next-state function of the controller. The identification of the correct condition (i.e., `true` and `false`) is thwarted, and, in turn, the corresponding target state is also thwarted by assigning a key bit K_j to each branch j and changing the corresponding test in the controller to be of the form:

$$\texttt{test} \oplus K_j == 1'b1. \tag{6.3}$$

To maintain the semantic equivalence of the branch, the two branches are reordered based on the value of the key bit K_j. For instance, the `true` and `false` blocks are swapped when $K_j = 1$ because the `xor` operation inverts the value of the variable `test`. This transformation corresponds to the creation of an opaque predicate because the result of the `xor` is known during obfuscation because the value of the key bit is known. On the contrary, the attacker cannot determine which is the actual `true` (`false`) block without knowing the value of the key bit. Figure 6.6 shows this transformation on a simple example.

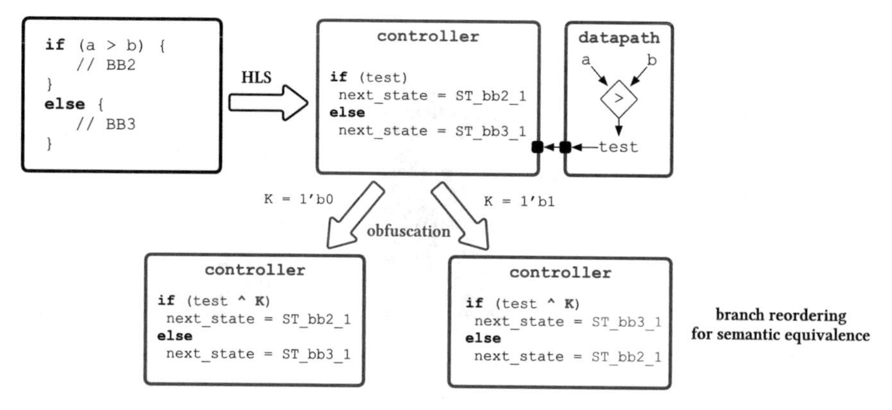

Fig. 6.6 Control-flow obfuscation. The combination with the assigned key bit can create different versions with branch reordering to thwart the identification of the correct `true` and `false` blocks

Example Consider the `if-then` statement in the black box shown in Fig. 6.6. When `a` is greater `b`, the control transfers to `BB2`; otherwise, it transfers to `BB3`. After performing the traditional HLS, we obtain the controller and datapath shown in the red boxes of Fig. 6.6. Based on the results of the test, the next state is the first state of `BB2` or `BB3`. An attacker can determine which part of the algorithm executes when the condition is `true`. Conversely, our obfuscation technique can yield alternative versions of the controller (shown in the blue boxes in Fig. 6.6). The two resulting tests are perfectly equivalent, but the target state in case of `true` (`false`) result is different based on the key bit. So, the attacker cannot determine which is the real `true` block without knowing the correct value of the key bit. □

The same transformation applies to the test conditions of the `for`/`while` loops because the front-end compiler translates them into an identical form. One can obfuscate also complex branch constructs (e.g., `switch`) by using more key bits.

6.5.2 *Obfuscation Approach*

Since the components generated with HLS require a strict interaction between the datapath and the controller, an implementation of the approach should necessarily be comprehensive, embracing all HLS steps to automatically implement the obfuscation transformations presented in Sect. 6.5. Such a comprehensive solution has been implemented extending the traditional HLS flow (see Sect. 6.3), making reverse engineering and hence the IP theft more difficult. This enhanced flow is shown in Fig. 6.7 and starts from the C code of the component to generate and the *locking key* K, which is generated by the designer to activate the IC after fabrication. The final output is a locked RTL design (with an extra input for providing the key) ready for logic synthesis and physical design.

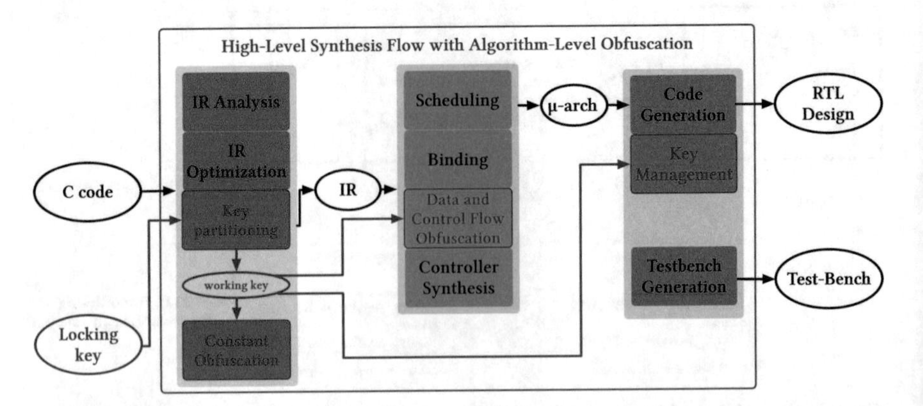

Fig. 6.7 Proposed HLS flow. Each phase is extended with additional phases to implement key-based obfuscation

6.5.2.1 Compiler Front End

The input C code is processed with the compiler front end to generate the internal intermediate representation (IR). Then, common compiler transformations, including function inlining and loop optimizations, are applied to the IR to prepare it for HLS. Constant propagation is instead disabled so that the information can be obfuscated.

The IR generated and optimized during the HLSfront end is processed to determine the number of key bits needed to obfuscate the different elements of the algorithm. For this, the call graph is extracted to figure out the list and hierarchy of functions to synthesize [9]. Other information consists of the number of basic blocks[1] and the resulting control-flow graph (CFG). A fixed number C of key bits are assigned to obfuscate each constant, B_i key bits to obfuscate the scheduled DFG of each basic block b_i, and one bit to obfuscate each control branch. By combining this information with the IR details, the number of bits of the internal obfuscation key W, called *working key*, needed to obfuscate the algorithm is determined:

$$W = Num_{if} + Num_{const} * C + \sum_{i=0}^{BB} B_i, \qquad (6.4)$$

where Num_{if} and Num_{const} are the number of branches and constants, respectively. C is the number of key bits assigned to implement each constant and B_i is the number of key bits assigned to the basic block BB_i. So, the size of the working key W depends on the complexity of the algorithm to protect and is usually larger than the locking key K provided by the designer. Section 6.5.3 describes how to generate the working key K starting from the input locking key K.

6.5.2.2 HLS Engine

In the mid-level phase, we perform the traditional HLS steps, extended with the obfuscation techniques. First, the constants are extracted and obfuscated (see Sect. 6.5.1.1) to prevent HLS transformations and optimizations based on their bit widths and values. For example, multiplications by constants are usually simplified to obtain more efficient hardware [32]. However, this may reveal sensitive information that cannot be removed after HLS. The resulting IR is input to the HLS steps to create the datapath and the controller of each sub-function.

For creating the datapath, after scheduling each basic block, several variants are created with the goal of thwarting the identification of the arithmetic operations and dependencies (see Sect. 6.5.1.2). Since B_i obfuscation bits are used for the basic block b_i, the corresponding key value is assigned to the correct version and the

[1] A basic block is a sequence of instructions with a single entry point and a single exit point.

other $2^{B_i} - 1$ values to the variants. In the resulting datapath micro-architecture, extra connections between functional units and registers are added to implement the functionality of the different variants, and multiplexers to activate the execution of the variant associated with the value of the corresponding key portion.

For creating the controller, the FSM associated with each scheduled module is determined and each control branch is obfuscated (see Sect. 6.5.1.3). In case of a conditional jump, the result of the condition evaluation performed in the datapath is masked with a key bit. The next-state functions of the controller are thus masked with key bits to obfuscate the correct transitions while maintaining logical but incorrect execution flows in case of wrong locking keys.

At the end of the HLS steps, the module of each function is created by combining the corresponding datapath and controller. The hierarchy of the modules as in the traditional HLSflow is also created.

6.5.2.3 Back End

This step generates the RTL description and the logic for key management of the obfuscated design. The component will feature an additional input port to load the locking key from the system, while the working key used for obfuscation is stored internally and derived from the input locking key. The strategy to manage locking and working keys is described in Sect. 6.5.3.

6.5.3 *Key Management*

6.5.3.1 Storing the Locking Key

The *locking key* is the only extra input used to lock the circuit, as shown in Fig. 6.2. This key is given to our obfuscation approach for applying the obfuscation techniques but not to the foundry. Instead, it is stored in a tamper-proof memory (e.g., EEPROM, eFuses, or non-volatile memory [33]) after IC fabrication [7, 12]. The number of *locking key* bits that one can deliver to the IC may be fixed and limited by the technology. On the contrary, the number of key bits needed by an algorithm for obfuscation (*working key*) depends on the number and size of the basic blocks, number of control branches, and number of constants and the obfuscation techniques that are applied.

6.5.3.2 Generating the Working Key

When the number of working key bits is smaller than the number of available locking key bits, there is a one-to-one correspondence between the working and locking key bits, which are thus directly connected. This situation is ideal because

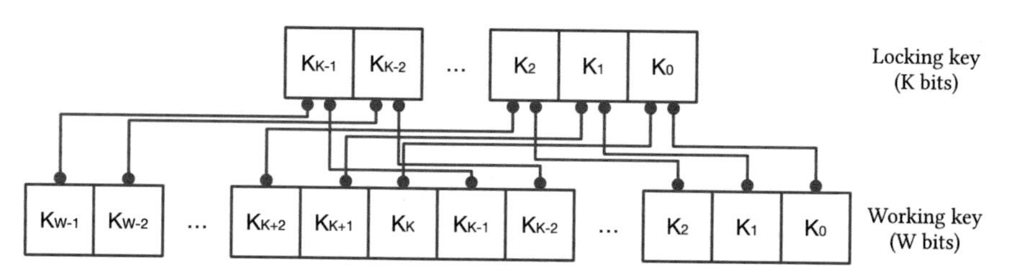

Fig. 6.8 When working key requires more bits than the locking key ($W > K$), the working key is obtained by replicating the locking key as many times as needed

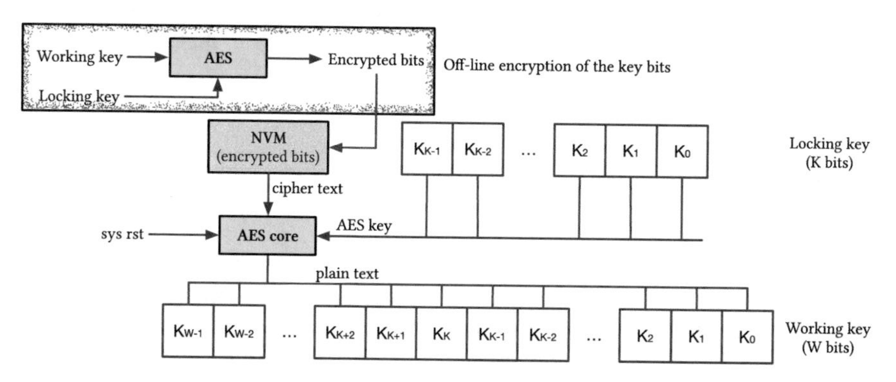

Fig. 6.9 When working key requires much more bits than the locking key ($W > K$), the working key is obtained with AES decryption of the values stored in the non-volatile memory at power-up

each key bit is unique and there is no additional overhead for key management. However, this is not always possible, and cases exist where more working key bits than available locking key bits need to be considered. When it is needed to derive many working key bits from a smaller number of locking key bits, one solution entails reusing the locking key bits as many times as needed to generate the working key, as shown in Fig. 6.8. In this situation, each key bit has a maximum fan-out of $f = \lceil W/K \rceil$, which may leak information to break the logic locking. Indeed, the attacker can use correlation analysis to extract information about each single key bit. If the attacker can extract the value of one working key bit, the corresponding locking key bit and, in turn, all its replicas are extracted. This may compromise the security of the generated IC for large values of f, i.e., when the number of working key bits is much larger than the number of available key bits.

For this case, an alternative solution is shown in Fig. 6.9. The locking key is used by an AES cryptographic core to encrypt the working key at design time. The resulting cipher text is stored in a non-volatile memory (NVM) that is added to the IC. At power-up, the NVM values are decrypted using the given locking key and placed into the working key registers. So, the correct working key is loaded only when the correct locking key is placed into the IC and used for decryption. This solution does not introduce any performance overhead for deriving the working key since this process is performed only at power-up, when the system is not operational.

Also, since this solution leverages the security guarantees of AES, we can use a 256-bit locking key, which is a reasonable size for existing technologies, to secure a large number of working key bits.

6.6 Experimental Evaluation

The obfuscation approach is validated by extending BAMBU, an open-source HLS framework [15]. Since BAMBU has a modular organization, the obfuscation techniques are implemented as additional steps in the HLS flow. The resulting version of the tool is called LOCKBAMBU.

6.6.1 *Experimental Setup*

LOCKBAMBU is used to generate obfuscated circuits on five HLS benchmarks from a range of application domains: GSM is a linear predictive coding analysis for telecommunication. ADPCM is an algorithm for adaptive differential pulse code modulation, SOBEL is an image-processing algorithm. BACKPROP is a method for training neural networks, and VITERBI is a dynamic programming method for computing probabilities on a hidden Markov model. These algorithms represent applications that a designer may want to obfuscate because of proprietary implementations. Table 6.1 shows the characteristics of the benchmarks.

For each benchmark, Table 6.1 reports the number of constants (`# Const`), basic blocks (`# BB`), and control branches (`# CJMP`) following the compiler optimizations. Together with the number of lines of C code (`# C lines`), they capture the algorithm complexity. The bit width of each obfuscated constant is set to 32 bits (i.e., $C = 32$), while the original constants range between 8 (`char` values) and 32 bits (`int` values). One bit is assigned to each control branch. Finally, four bits are assigned to each basic block to generate up to 16 DFG variants (i.e., $B_i = 4$ for all basic blocks). Table 6.1 reports the working key bits required for each algorithm (`W`). The resulting number of working key bits shows that constant obfuscation requires a large number of bits. For example, VITERBI and GSM have

Table 6.1 Characteristics of the benchmarks

Benchmark	`# C lines`	`# Const`	`# BB`	`# CJMP`	`W (bits)`
GSM	110	4	88	4	484
ADPCM	412	5	100	5	565
SOBEL	65	2	11	2	110
BACKPROP	264	12	123	11	887
VITERBI	144	117	98	9	4145

more or less the same number of basic blocks and control branches, but the former has many more constants, requiring 10× more key bits than the latter. These benchmarks are bigger than those commonly used for logic obfuscation. Working at a higher abstraction allows us to obfuscate larger circuits.

For each benchmark, 256-bit locking keys are generated and the effects of the obfuscation technique are evaluated in terms of *obfuscation potency* (how much is the attacker confused?) and *obfuscation cost* (what is the obfuscation overhead?). To evaluate the obfuscation techniques, the baseline designs (generated with BAMBU) are compared with the corresponding obfuscated ones (generated with LOCKBAMBU). BAMBU generates RTL testbenches to validate the circuit for a series of input values through RTL simulations. These executions are compared against the respective executions of the input specification in software. In LOCKBAMBU, these testbenches are extended to specify different locking keys as input to verify the execution for each of them. Simulations are performed with Mentor ModelSim SE 10.3 and are instrumented to report if the execution provides the same results as the baseline design (to evaluate *obfuscation potency*) and the number of cycles (to evaluate *obfuscation cost* in terms of performance overhead). The baseline and obfuscated versions of the circuits are synthesized using Synopsys Design Compiler J-2014.09-SP2 targeting the Synopsys SAED 32nm Generic Library at 500 MHz (to evaluate *obfuscation cost* in terms of area overhead).

6.6.2 Evaluation of Obfuscation Potency

For each benchmark, 100 random keys are generated. The first key is used as input for LOCKBAMBU (locking key), while the others are used to test for security evaluation. First, the generated circuits are simulated with the correct locking key corroborating that the circuits produce the same results as in the baseline version. All other keys result in wrong results, and this assures that the attacker cannot turn on the circuit with another key. The obfuscation potency (i.e., how much attacker is confused) is quantified using the "output corruptibility" of each locked circuit, computed as the Hamming distance with respect to the output of the baseline circuit [8]. The ideal obfuscation procedure should provide an output corruptibility of 50% to avoid any bias in the output bits that can leak information on the key values. When combined, the three obfuscation techniques produce an average HD of 62.2% over the five benchmarks, which is a good result. Also, designs can leak information through variations in the execution latency (*timing channels*). However, incorrect locking keys impact the performance only when they modify the loop bounds. Other constants have no effect, while datapath obfuscation works on a valid schedule without altering the total number of cycles. It is difficult for an attacker to tell whether a circuit is behaving properly or not. While the alternative DFGs are conceptually similar to the creation of the Super CDFG [12], SAT-based attacks are much harder to apply because the oracle chip is unavailable in the untrusted foundry threat model and the complexity of the circuits demands novel methods to apply these attacks on large sequential circuits composed of datapath and controller. Moreover, in case of constants, the information is fully cut out from the circuit, and one cannot recover it without the correct locking key. In conclusion, the circuits generated by LockBambu have a higher security level than previous obfuscation techniques operating at the logic level.

6.6.3 Evaluation of Obfuscation Cost

To evaluate the cost of each obfuscation technique, LockBambu is modified to selectively apply the methods through command-line options. Since these transformations are orthogonal, different versions of the circuits are generated for each of them.

RTL simulations are performed to check the *performance overhead* concerning the circuit latency (in clock cycles). When the correct key is applied, there is no performance overhead on the generated designs concerning the *baseline* versions. However, the target frequency is decreased by 8% on average when we apply data-flow obfuscation because of the more additional multiplexers. Also, the drop-off in frequency is proportional to the number of key bits assigned to each basic block because creating more variants requires more multiplexers. Obfuscating the control branches has a negligible impact on the frequency (less than 1%). Representing the constants by a pre-defined number of bits C increases the size of multiplexers and reduces logic optimizations. However, the impact on the critical path is minimal (around 4%). When all obfuscations are applied, the target frequency is decreased by less than 10% on average that can be reduced to 6% on average with more aggressive logic synthesis optimizations.

Logic synthesis is carried out to evaluate the area overhead of the various obfuscation techniques. Figure 6.10 shows the results, where each value is normalized against the area of the respective *baseline* version (obtained with the original version of Bambu). The results indicate that obfuscating the control flow has practically no area impact. This technique only adds a few exclusive-or gates to the controller. Obfuscating constants increases the area by 10% on average since it creates larger multiplexers and prevents logic-level optimizations. Data-flow obfuscation has the most impact, increasing the area by around 21% on average. This area overhead is mainly due to the additional multiplexers to connect functional units and registers. This obfuscation is appropriate for benchmarks where the computational part has simple functional units (e.g., shifters and Boolean operations) or has many basic blocks. BACKPROP is the benchmark with more basic blocks and has the largest overhead (>30%). Similarly to the frequency, the area overhead is proportional to the number of key bits assigned to the basic blocks. When all obfuscation techniques are applied together, the overhead adds up, resulting in a total overhead between 20 and 45%. It is worth noting that memory controllers to access the external memory are responsible for a significant portion of the circuit area, but they are not obfuscated because their implementation is generic and does not contain any

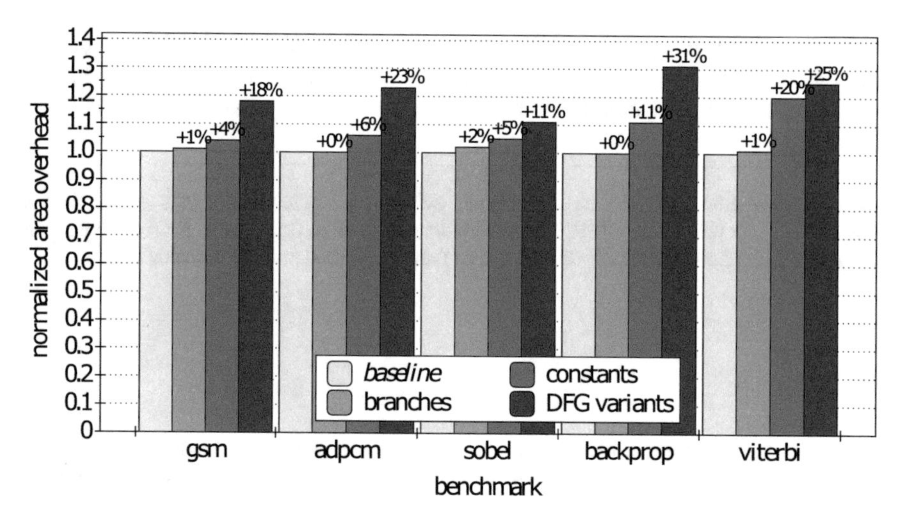

Fig. 6.10 Area overhead of the proposed obfuscation techniques

algorithm-dependent part. This significantly reduces the impact of obfuscation on the final design.

In the basic approach of replicating the key bits, there is no performance or area overhead. The signals are coming from the tamper-proof memory where the locking key is stored and directly connects to the points where one uses the working key. For the AES-based solution, there are two contributions to the area overhead: one part is the AES decryption module, and the other one is the NVM used to store the encrypted key bits and the flip-flops to save the decrypted values. The first contribution is fixed and depends on the AES implementation. The second contribution is proportional to the number of working key bits. The key decryption is performed only once at power-up, and there is no performance overhead once the chip is ready to use.

6.7 Conclusion

In this chapter, we present an approach for implementing obfuscation during high-level synthesis that is able to hide the algorithm semantics to the attacker and thwart reverse engineering of the corresponding physical design. The presented approach starts from a C-level description of the algorithm and creates a version of the corresponding RTL component by masking all relevant algorithm portions through opaque predicates that are based on an input locking key. In particular, techniques for obfuscating constant values, arithmetic operations, and control branches have been presented. These techniques have been implemented within Bambu, a state-of-the-art, open-source HLS tool. This combination allowed us to obtain a comprehensive solution for obfuscation during HLS that has been validated on a set of representative benchmarks. These techniques do not incur performance overhead, and each of them has a maximum area overhead of around 30% (20% on average).

Acknowledgments R. Karri is supported in part by NSF (A#: 1526405) and CCS-AD. S. Garg is supported in part by an NSF CAREER Award (A#: 1553419). S. Garg and R. Karri are both with the NYU Center for Cybersecurity (cyber.nyu.edu) and supported in part by Boeing Corp.

References

1. Reuters Semiconductors: TSMC says latest chip plant will cost around $20 bln (2017). https://www.reuters.com/article/tsmc-investment/tsmc-says-latest-chip-plant-will-cost-around-20-bln-idUSL3N1O737Z
2. Heck, S., Kaza, S., Pinner, D.: Creating value in the semiconductor industry. McKinsey on Semiconductors pp. 5–144 (2011)
3. Hurtarte, J., Wolsheimer, E., Tafoya, L.: Understanding Fabless IC Technology. Elsevier, Amsterdam (2007)
4. Guin, U., Huang, K., DiMase, D., Carulli, J.M., Tehranipoor, M., Makris, Y.: Counterfeit integrated circuits: a rising threat in the global semiconductor supply chain. Proc. IEEE **102**(8), 1207–1228 (2014)
5. Abdel-Hamid, A.T., Tahar, S., Aboulhamid, E.M.: IP watermarking techniques: survey and comparison. In: Proceedings of the IEEE International Workshop on System-on-Chip for Real-Time Applications (IWSOC), pp. 60–65 (2003)
6. Vijayakumar, A., Patil, V.C., Holcomb, D.E., Paar, C., Kundu, S.: Physical design obfuscation of hardware: a comprehensive investigation of device and logic-level techniques. IEEE Trans. Inf. Forensics Secur. **12**(1), 64–77 (2017)
7. Roy, J.A., Koushanfar, F., Markov, I.L.: Ending piracy of integrated circuits. Computer **43**(10), 30–38 (2010)
8. Xie, Y., Srivastava, A.: Anti-SAT: mitigating SAT attack on logic locking. IEEE Trans. Comput.-Aid. Des. Integr. Circuits Syst. **38**(2), 199–207 (2019)
9. Nane, R., Sima, V., Pilato, C., Choi, J., Fort, B., Canis, A., Chen, Y.T., Hsiao, H., Brown, S., Ferrandi, F., Anderson, J., Bertels, K.: A survey and evaluation of FPGA high-level synthesis tools. IEEE Trans. Comput.-Aid. Des. Integr. Circuits Syst. **35**(10), 1591–1604 (2016)
10. Pilato, C., Basu, K., Shayan, M., Regazzoni, F., Karri, R.: High-level synthesis of benevolent trojans. In: Proceedings of the Design, Automation & Test in Europe Conference (DATE), pp. 1124–1129 (2019)
11. Pilato, C., Garg, S., Wu, K., Karri, R., Regazzoni, F.: Securing hardware accelerators: a new challenge for high-level synthesis **10**(3), 77–80 (2018)
12. Rajendran, J., Ali, A., Sinanoglu, O., Karri, R.: Belling the CAD: toward security-centric electronic system design. IEEE Trans. Comput.-Aid. Des. Integr. Circuits Syst. **34**(11), 1756–1769 (2015)
13. Yang, F., Tang, M., Sinanoglu, O.: Stripped functionality logic locking with hamming distance-based restore unit (SFLL-hd) – unlocked. IEEE Trans. Inf. Forensics Secur. **14**, 2778–2786 (2019)
14. Yasin, M., Zhao, C., Rajendran, J.J.: SFLL-HLS: stripped-functionality logic locking meets high-level synthesis. In: Proceeding of the IEEE/ACM International Conference on Computer-Aided Design (ICCAD), pp. 1–4 (2019)
15. Pilato, C., Ferrandi, F.: Bambu: a modular framework for the high-level synthesis of memory-intensive applications. In: Proceedings of the International Conference on Field programmable Logic and Applications (FPL), pp. 1–4 (2013)

16. Rostami, M., Koushanfar, F., Karri, R.: A primer on hardware security: models, methods, and metrics. Proc. IEEE **102**(8), 1283–1295 (2014)
17. der Leest, V.V., Tuyls, P.: Anti-counterfeiting with hardware intrinsic security. In: Proceedings of the Design, Automation & Test in Europe Conference (DATE), pp. 1137–1142 (2013)
18. Imeson, F., Emtenan, A., Garg, S., Tripunitara, M.: Securing computer hardware using 3D integrated circuit (IC) technology and split manufacturing for obfuscation. In: Proceedings of the USENIX Conference on Security (SEC), pp. 495–510 (2013)
19. Charbon, E.: Hierarchical watermarking in IC design. In: Proceedings of the IEEE Custom Integrated Circuits Conference (CICC), pp. 295–298 (1998)
20. Qu, G., Potkonjak, M.: Intellectual Property Protection in VLSI Designs: Theory and Practice. Kluwer, Cambridge (2003)
21. Cui, A., Chang, C.H., Tahar, S., Abdel-Hamidothers, A.T.: A robust FSM watermarking scheme for IP protection of sequential circuit design. IEEE Trans. Comput.-Aid. Des. Integr. Circuits Syst. **30**(5), 678–690 (2011)
22. Kahng, A.B., Lach, J., Mangione-Smith, W., Mantik, S., Markov, I.L., Potkonjak, M., Tucker, P., Wang, H., Wolfe, G.: Constraint-based watermarking techniques for design IP protection. IEEE Trans. Comput. Aid. Des. **20**(10), 1236–1252 (2001). https://doi.org/10.1109/43.952740
23. Kahng, A.B., Mantik, S., Markov, I.L., Potkonjak, M., Tucker, P., Wang, H., Wolfe, G.: Robust IP Watermarking Methodologies for Physical Design. pp. 782–787. ACM, New York (1998)
24. Sengupta, A., Roy, D.: Antipiracy-aware IP chipset design for CE devices: a robust watermarking approach [Hardware Matters]. IEEE Consum. Electron. Mag. **6**(2), 118–124 (2017)
25. Subramanyan, P., Ray, S., Malik, S.: Evaluating the security of logic encryption algorithms. In: Proceedings of the IEEE International Symposium on Hardware Oriented Security and Trust (HOST), pp. 137–143 (2015)
26. Lao, Y., Parhi, K.K.: Obfuscating DSP Circuits via High-Level Transformations. IEEE Trans. Very Large Scale Integr. Syst. **23**(5), 819–830 (2015)
27. Xu, H., Zhou, Y., Kang, Y., Lyu, M.R.: On secure and usable program obfuscation: a survey. ArXiv (2017)
28. De Micheli, G.: Synthesis and Optimization of Digital Circuits. McGraw-Hill, New York (1994)
29. Stok, L.: Data path synthesis. Integr. VLSI J. **18**(1), 1–71 (1994)
30. Collberg, C., Thomborson, C., Low, D.: A Taxonomy of Obfuscating Transformations. Tech. Rep. 148, Department of Computer Science, The University of Auckland (1997)
31. Gal, B.L., Andriamisaina, C., Casseau, E.: Bit-width aware high-level synthesis for digital signal processing systems. In: Proceedings of the IEEE International SOC Conference (SOCC), pp. 175–178 (2006)
32. Boullis, N., Tisserand, A.: Some optimizations of hardware multiplication by constant matrices. IEEE Trans. Comput. **54**(10), 1271–1282 (2005)
33. Forte, D., Bhunia, D., Tehranipoor, M.: Hardware Protection Through Obfuscation. Springer, Berlin (2017)

Part II
Hardware IP Protection Through Watermarking and State Encoding

Chapter 7
Hardware (IP) Watermarking During Behavioral Synthesis

Anirban Sengupta and Mahendra Rathor

7.1 Introduction

Data-intensive applications, such as digital filtering, compression, decompression, etc., are important part of Consumer Electronics (CE) systems. These applications are facilitated by Digital Signal Processing (DSP) algorithms such as Finite Impulse Response (FIR) filter, Discrete Cosine Transform (DCT), etc. [1, 2]. Realizing DSP algorithms in the form of hardware accelerators is vital for achieving high performance of CE devices such as digital camera, tablet, cell phones, laptop, etc. Since the DSP algorithms have higher complexity and larger size, behavioral synthesis (or high-level synthesis) plays a crucial role in mitigating the complexity of the design process [3–5].

A System-on-Chip (SoC) employed in CE devices integrates both non-data-intensive hardware such as general purpose processor and data-intensive hardware such as DSP Intellectual Property (IP) cores. Both categories of hardware IPs are vulnerable to attacks, such as piracy, false claim of ownership, etc., because of globalization of design supply chain. Therefore, security algorithms should be integrated with the traditional design flow. However, the security paradigms for both categories of hardware are quite different. This is because the design process of data-intensive hardware starts with behavioral level, whereas the design process of non-data-intensive hardware mostly starts at register-transfer level (RTL) [6]. Therefore, in the context of data-intensive applications such as DSP/multimedia cores, security technique such as hardware watermarking during behavioral synthesis becomes essential [7].

A. Sengupta (✉) · M. Rathor
Indian Institute of Technology Indore, Indore, India
e-mail: asengupt@iiti.ac.in; mrathor@iiti.ac.in

S. Katkoori, S. A. Islam (eds.), *Behavioral Synthesis for Hardware Security*,
https://doi.org/10.1007/978-3-030-78841-4_7

In the era of multi-vendor IP integration, there may exist several hardware-related vulnerabilities including piracy, fraud claim of ownership, and reverse engineering (RE). The role of watermarking is to embed digital evidence into the design for enabling detective control and proving authentic IP ownership [8–13]. Integrating watermarking algorithms in the behavioral synthesis design flow generates a security aware high-level synthesis (HLS) process, which provides manifold advantages such as better handling design overhead, efficiently handling design complexity, and securing RTL designs of soft IP cores [14, 15].

7.2 High-Level Synthesis (HLS)-Based Watermarking

Design process of highly complex data-intensive applications such as DSP hardware accelerator and multimedia cores start with HLS (a higher abstraction level) because of their higher algorithmic complexity and size. HLS plays a very critical role in securing such designs from piracy and fraud claim of ownership threats, using watermarking, because of the following reasons:

- Complexity of employing robust watermarking increases while managing design overhead as we traverse down the design abstraction level, i.e., from behavioral level to gate level and further down to layout level.
- Employing watermarking technique during high level or behavioral level offers the flexibility of performing design space exploration (DSE), which enables design parameter trade-offs such as area–speed, area–power, and power–speed trade-offs. This helps in achieving optimal watermark.
- Embedding watermarking at high (behavioral) level also leads to secured designs at the subsequent lower abstraction levels, as RTL/gate-level design carries the implanted watermarking constraint information post-synthesis.
- Embedded watermark during HLS is highly robust as watermarking constraints get distributed throughout the design, thus protecting against tampering attack.

This section discusses some of the popular watermarking techniques employed during behavioral or high-level synthesis.

7.2.1 Single-Phase Multi-Variable Watermarking [14]

To secure highly complex design such as DSP hardware accelerator against false claim of IP, piracy, or counterfeiting and cloning threats, Sengupta and Bhadauria [14] presented a novel approach of hardware watermarking. The watermarking approach [14] is based on multi-variable author signature and is employed during register allocation phase of behavioral synthesis. Figure 7.1 provides an overview of the single-phase multi-variable watermarking approach. As shown in the figure, algorithmic description (such as C/C++, transfer function) of target application is

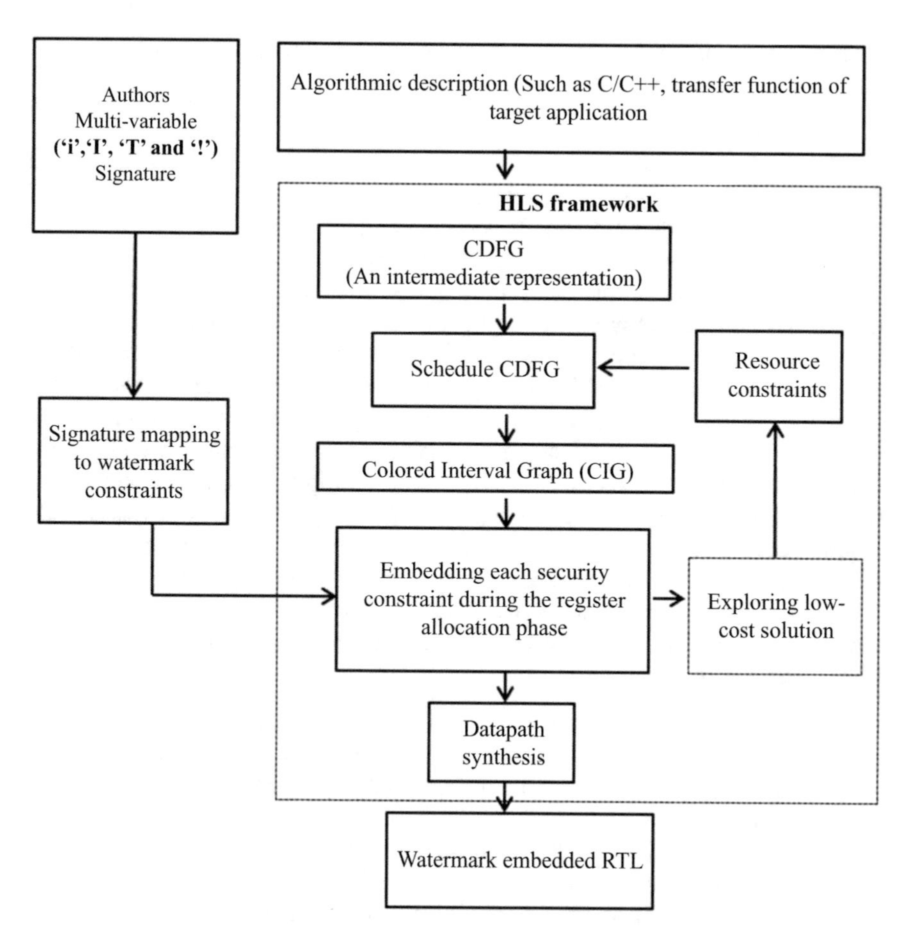

Fig. 7.1 Single-phase watermarking approach during behavioral synthesis [14]

fed to HLS process. The watermarking process is integrated with HLS process. In order to embed watermark, first, the Control Data Flow Graph (CDFG) of the target application is generated. Thereafter, CDFG is scheduled using designer's selected resource constraints. Post scheduling, the next phases of HLS are resource (i.e., functional unit and registers) allocation and binding. In this work, the security constraints corresponding to owner's signature are implanted into the design during register allocation phase of HLS. Implanting watermarking constraints is realized using colored interval graph (CIG) framework. However, since embedding watermark may result in design cost overhead, in order to minimize the impact of embedded watermark on design cost, the low-cost solution is explored during HLS. This results in a low-cost optimal watermark solution. Furthermore, datapath synthesis phase in HLS is performed, which generates a low-cost secured register-transfer-level design containing owner's secret information in the form of robust watermark.

The details of single-phase multi-variable watermarking approach [14] are discussed in the following sub-sections:

(a) Signature Encoding [14]
As discussed earlier, the register allocation phase of HLS (leveraging CIG) is used for inserting owner's watermark into the design. Before discussing the decoding of signature into secret watermark constraints, understanding of the concept of CIG is discussed. A CIG represents the allocation of storage variables of the design to minimum possible datapath registers. The storage variables include the primary/intermediate inputs and outputs of the design. All storage variables are represented by nodes in a CIG. Each node has a specific color, which represents assignment of storage variables to a specific register. The number of colors used in the CIG represents the minimum number of registers required to accommodate all storage variables. Since all storage variables do not remain alive throughout the execution, a single register can be allocated to multiple storage variables within the entire execution time. This concept of reusing registers leads to less register requirement for storing all storage variables. Furthermore, overlapping of lifetime of two storage variables is indicated by an edge between respective nodes in the CIG. Therefore, an edge between two nodes of same color cannot exist in the CIG, because same register cannot be used to store two live variables at a time (in the same control step).

Watermarking constraints are embedded into the design during register allocation phase in the form of additional secret (artificial) edges added into the CIG of the DSP design. Manifold edges can be inserted additionally into the CIG depending on the strength of the signature used. Thus, the potential edges to be added as watermarking constraints are derived using owner's signature, where each digit of the signature corresponds to a secret edge of the CIG. It is important to note that a decoded edge from a signature digit may also exist by default in the CIG. However, if it does not exist, then it is additionally added by adhering to the requirement of an edge between two nodes of a CIG. The selection of node pairs for embedding additional edges depends on the encoding rules of signature variables. Sengupta and Bhadauria [14] proposed four unique signature variables where each variable has a specific encoding. A signature constructed using four independent variables is much stronger in contrast to a binary (two-variable) signature. The author's signature used by Sengupta and Bhadauria [14] is a combination of the following four variables: i, I, T, and !, where each variable can be repeated multiple times to create a robust signature of desired strength. More the digits in the signature, the more number of additional edges is embedded into the CIG, indicating robust watermarking. The encoding of each variable of the signature, which decides the location of insertion of additional edges into the CIG, is shown in Table 7.1.

(b) Signature Embedding [14]
In order to embed watermarking constraints, first, the author's signature of desired strength is chosen. Furthermore, each digit of the signature is converted to the

Table 7.1 Signature encoding

Signature variable	Encoding
i	Encoded value of edge with node pair id as (prime, prime)
I	Encoded value of edge with node pair id as (even, even)
T	Encoded value of edge with node pair id as (odd, even)
!	Encoded value of edge with node pair id as (0, any integer)

watermarking constraints based on encoding rules shown in Table 7.1. The end-to-end process of embedding watermark is as follows [14]:

(i) The algorithmic description of the target design is first converted into an intermediate representation called CDFG.
(ii) The CDFG is scheduled based on the designer's chosen resource constraints to obtain a scheduled CDFG.
(iii) A CIG is created using scheduled CDFG, where nodes in the CIG indicate storage variables assigned in CDFG to store primary and internal inputs and outputs, and colors used in the CIG indicate the minimum number of registers required to execute all storage variables.
(iv) Subsequently, a controller for register allocation corresponding to the CIG is generated. This controller provides the information about the assignment of storage variables/nodes to registers/colors in control steps (CS).
(v) A sorted list of storage variables is created, where variables are sorted based on their number in ascending order.
(vi) Once owner's signature combination is chosen, each digit is converted to potential edge to be added into the CIG based on encoding as per Table 7.1. The node pairs in the CIG, for all potential edges to be added, are chosen by traversing the sorted list obtained in step (v) from left to right.
(vii) Constraint edges are added into the CIG one by one. However, adding some edges may violate the condition of an edge between two nodes. This is because an edge between two nodes indicates overlapping of lifetime of two storage variables, and therefore both cannot be executed through same register/color. In this case, both nodes should be assigned to different colors. In order to accommodate an edge in such a scenario, the color of one of the nodes is swapped with another node in the CIG.

Post embedding all edge constraints into the CIG, the controller timing table of the register allocation is modified. Here, it can be observed that some storage variables are deliberately forced to execute through different registers because of imposing of owner's secret constraints. Thus, the owner's chosen signature-based dynamic watermark is embedded during register allocation phase of HLS process of the design of a reusable IP core. The embedded watermark does not affect the functionality of the design. Once the owner's watermark is embedded during register allocation phase, the datapath synthesis phase of HLS is performed. Post datapath synthesis, an RTL datapath of the target design is generated. The effect of

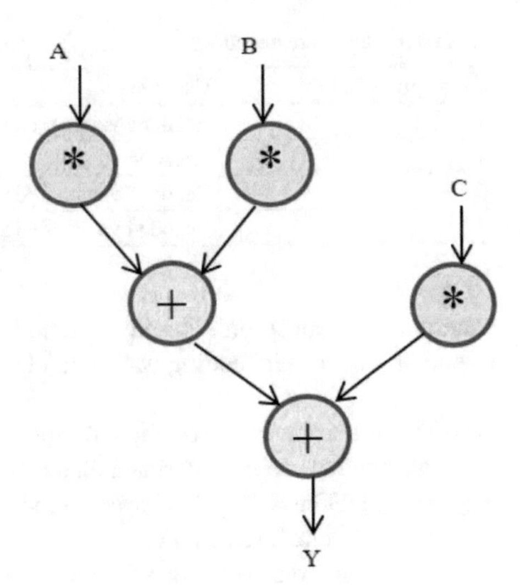

Fig. 7.2 DFG of sample application [14]

embedded watermark incurs in the RTL datapath in terms of alteration in the inputs of multiplexers associated with respective registers.

(c) Demonstration Example of Single-Phase Multi-Variable Watermarking [14]
The demonstration of single-phase multi-variable watermarking is given using a sample application. The DFG of the sample application is shown in Fig. 7.2, where A, B, and C are the inputs and Y is the output. Furthermore, list scheduling of DFG is performed based on the designer's chosen resource constraints of 3 multipliers (M) and 1 adder (A). Thus, scheduled DFG, shown in Fig. 7.3, is generated, which executes in three control steps CS1 to CS3. Primary/intermediate inputs and outputs of scheduled DFG are assigned to eight storage variables s0 to s7 as shown in the figure. These eight storage variables are executed through minimum three different registers, R1, R2, and R3. The allocation of storage variables to minimum possible registers is shown using CIG in Fig. 7.4. As shown in the figure, each storage variable has been represented using a node in the CIG, and the total three colors have been used where each color is representing a distinct register. An edge between two nodes has been drawn if lifetime of corresponding storage variables is overlapping. For example, as shown in Fig. 7.3, storage variables s0, s1, and s2 are alive in CS0, and therefore edges exist among them in the CIG. These three storage variables are executed through three registers R1 (Red), R2 (Blue), and R3 (Green) in CS0. After CS1, these three storage variables are no longer alive, whereas the three new storage variables s3, s4, and s5 are born, and therefore same registers R1 (Red), R2 (Blue), and R3 (Green) are reused to execute them. The controller table corresponding to the CIG is shown in Table 7.2.

Now, the process of generating watermarking constraints corresponding to the author's chosen signature and embedding them into the CIG is demonstrated. First of all, an n-digit signature is selected by the owner. For example, the chosen 6-digit

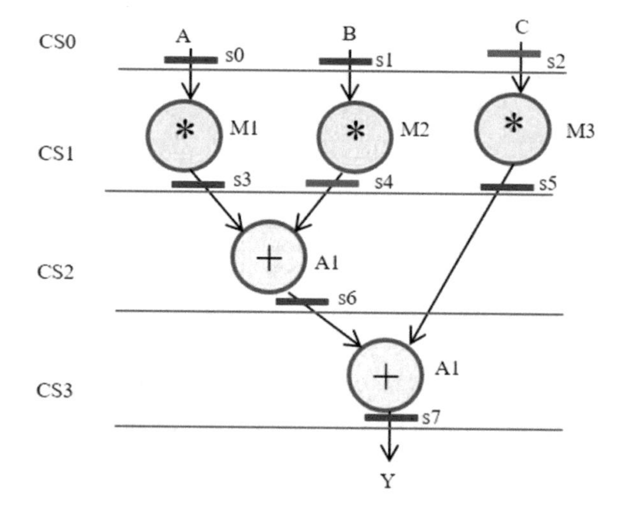

Fig. 7.3 Scheduled DFG based on 3M, 1A [14]

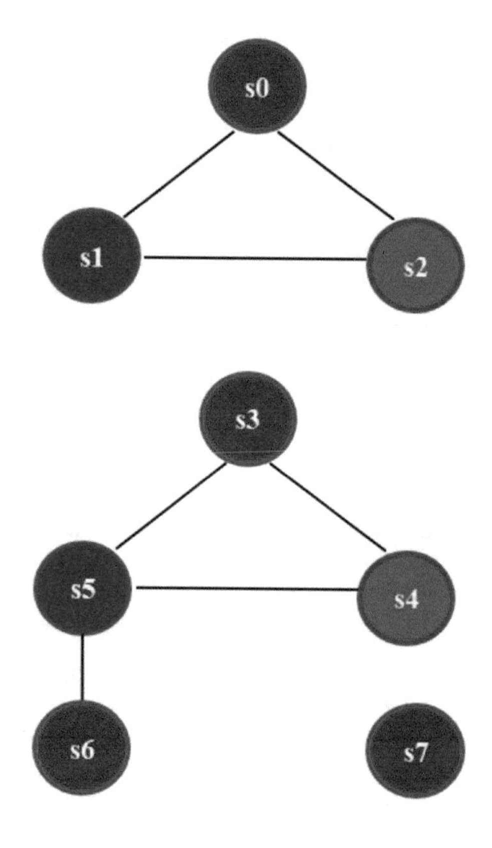

Fig. 7.4 CIG before embedding watermark [14]

Table 7.2 Controller table for register allocation before embedding watermark. "CS" indicates control step and "R" indicates register

CS	R1 (red)	R2 (blue)	R3 (green)
0	s0	s1	s2
1	s3	s5	s4
2	s6	s5	–
3	s7	–	–

Table 7.3 Mapping of digits of owner's chosen signature into edge constraints

Signature digits	i	i	I	!	T	I
Additional edges into the CIG	<s2,s3>	<s2,s5>	<s2,s4>	<s0,s1>	<s1,s2>	<s2,s6>

signature of an author is as follows: "i i I ! T I." Based on the encoding rules shown in Table 7.1, each digit is converted to an additional secret edge to be added into the CIG. For adding an edge, corresponding node pair is chosen by picking up storage variables from their sorted list. For example, the sorted list of storage variables is {s0, s1, s2, s3, s4, s5, s6, s7}. The first digit in the signature is "i," and therefore it indicates that an edge between node pair id <prime, prime> is to be added into the CIG. The first two storage variables (nodes) bearing prime number in the sorted list are s2 and s3, and therefore the additional edge to be embedded corresponding to the first signature digit "i" is <s2, s3>. Similarly, additional edges to be embedded into the CIG for remaining signature digits are also shown in Table 7.3. As shown in the table, the constraint secret edges corresponding to second and sixth signature digits are <s2, s5> and <s2, s6>, which are added additionally into the CIG as shown in Fig. 7.5. However, the constraint edges corresponding to fourth and fifth digits are <s0, s1> and <s1, s2>, which already exist in the CIG. Therefore, watermarking constraints corresponding to fourth and fifth digits satisfy by default. Furthermore, the constraint edge corresponding to the third digit in the signature is <s2, s4>, and however an edge cannot be straightly added between the corresponding nodes. This is because both nodes (storage variables) are assigned to the same color (register), i.e., Green (R3) as shown in the CIG (Fig. 7.4). Therefore, to satisfy this edge constraint, the conflict can be resolved using either of the following solutions:

1. The color of either of the nodes is swapped with another node or storage variable executing in the same CS.
2. If the first solution is incapable to resolve the conflict, then a distinct register or color is used in the CIG to accommodate the respective edge.

In this case, the edge conflict cannot be resolved using the aforementioned first solution. This is because, s4 (Green) can neither be swapped with s3 (Red) nor with s5 (Blue) in the same control step, i.e., CS1. The reason is that the swapping of s4 with s3 turns the color of s3 into Green. This will create conflict between newly added additional edge <s2, s3> as the color of s2 is already Green. Furthermore, the swapping of s4 with s5 turns the color of s5 into Green. This will also create conflict between newly added additional edge <s2, s5> as the color of s2 is already Green. Therefore, to accommodate additional edge <s2, s4> into the CIG, the

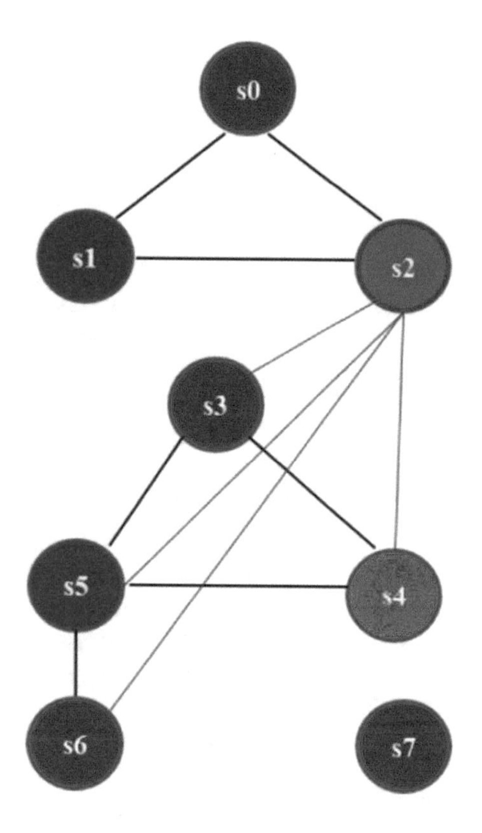

Fig. 7.5 CIG after embedding watermark [14]

second solution is adopted. Based on the aforementioned second solution, a distinct color/register, i.e., orange (R4), is used in the CIG to accommodate the edge <s2, s4>. The orange color is assigned to the node s4. Now, an additional edge is drawn between both storage variables s2 and s4 into the CIG, as they are now executing through distinct colors. The final modified CIG, post embedding all edge constraints, is shown in Fig. 7.5. As shown in the figure, four additional edges got embedded into the CIG corresponding to the six digits of the signature because two edge constraints are coincidentally satisfying by default. The modified controller table of the register allocation is shown in Table 7.4. In this particular example, an extra register is required to insert the watermarking constraints in the design. However, this is not always the case. In many cases, inserting watermarking constraints does not require any extra register, thus leading to zero overhead. This depends on the chosen signature combination, designer's encoding rule, and type/size of the target application to be watermarked.

(d) Using Encryption in Generating Digital Signature

As discussed earlier, the owner's signature is encoded in terms of secret watermarking constraints in order to embed it into the design. It is important to note that distinct encoding rules can lead to different strength or quality of watermarking.

Table 7.4 New controller table for register allocation after embedding watermark

CS	R1 (red)	R2 (blue)	R3 (green)	R4 (orange)
0	s0	s1	s2	–
1	s3	s5	–	s4
2	s6	s5	–	–
3	s7	–	–	–

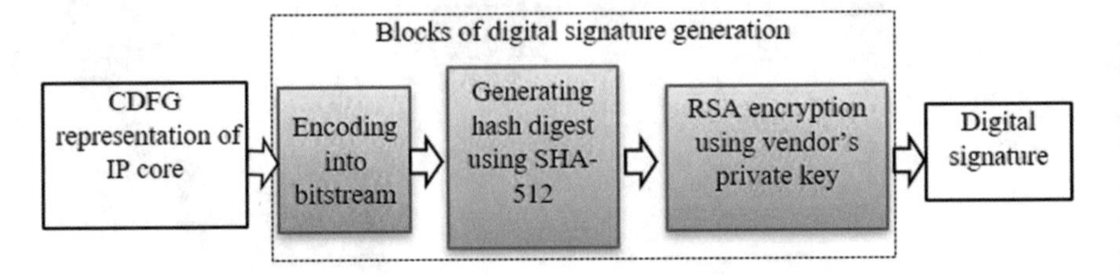

Fig. 7.6 Digital signature generation process through hashing and encryption [4]

Therefore, a strong encoding scheme is an indispensable part, which can be obtained by incorporating multiple variables in the signature like single-phase multi-variable watermarking [14]. In addition, instead of directly choosing the author's signature, a digital signature can be generated using a combination of encoding and encryption. Sengupta et al. [4] presented a method of generating digital signature corresponding to a DSP application. Here, the DSP design is first encoded, and then it is enciphered by applying cryptographic methods to generate watermarking constraints. This technique of performing encoding followed by encryption is more robust compared to regular watermarking because of the following reasons: (1) it thwarts an unauthorized user to obtain useful information of the owner's signature, (2) only the authorized user can decrypt the signature correctly as he/she is aware of the points of concatenation used for reforming the digital signature, and (3) in case the signature is leaked to adversary, he/she cannot prove ownership because he/she does not know the owner's encoding algorithm and encryption private key. In the digital signature-based watermarking approach, Sengupta et al. [4] employed secured hashing algorithm (SHA-512) followed by RSA-encryption algorithm for encrypting the owner's secret information. This approach first encodes the DSP design (CDFG) using owner's specified encoding technique, and next the encoded information is converted into a hash digest using SHA-512. Furthermore, the hashed information is encrypted using a strong private key (say 128 bits) though RSA encryption, thus generating digital signature as shown in Fig. 7.6. The bit-stream information obtained, post applying encryption, is further encoded in terms of watermarking constraints to be embedded into the design. This kind of process of generating digital signature leads to a highly robust watermark, which is arduous to be removed or tampered by an adversary and is robust against fraud claim of IP ownership. Unawareness of the owner's encoding rules and cryptographic key renders the watermark embedded into the design highly robust against standard threats such as fraud ownership and piracy.

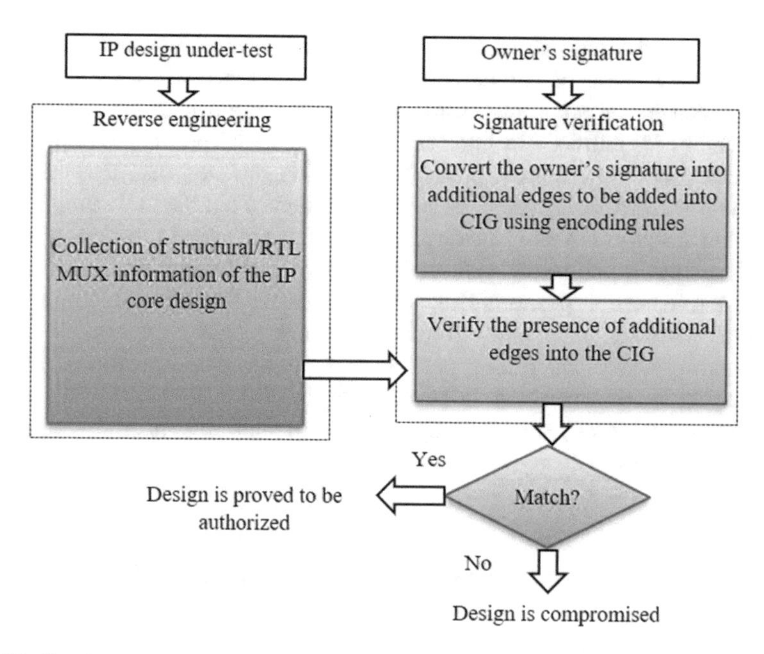

Fig. 7.7 Signature detection process [14]

(e) Signature Detection [14]

Signature detection process is an integral component of any watermarking process. To detect counterfeiting or cloning and to prove ownership of an IP, the owner's signature is detected in the design. By detecting the signature of the genuine owner in the design, false claim of IP ownership can be nullified. Similarly, by inspecting the original watermark in the design, pirated (counterfeited and cloned) designs can be detected and authenticity of the genuine design can be proved. Figure 7.7 depicts the process of detecting signature in the design of an IP core under test. As shown in the figure, the following steps are performed to detect the owner's signature embedded into the design:

1. *Reverse Engineering:* The design of an IP under test is reverse engineered to acquire structural/RTL information of the design. For example, in a single-phase multi-variable watermarking approach, the information about the inputs of those multiplexers that are associated with registers is acquired.
2. *Signature Verification:* In this step, the presence of owner's signature is verified. First, the owner's signature is converted, using encoding rules, into the corresponding additional edges to be embedded into the CIG. Next, the presence of constraint edges is verified in the register multiplexer inputs (obtained in the previous step through reverse engineering). If the register mux design satisfies the watermarking constraints (corresponding to the original signature), then it is proved to be authentic. Similarly, if the authors' signature is found to be

satisfied in the register mux design, then the original owner is awarded IP ownership in case of conflict. Based on the presence of owner's watermark in the design, ownership is awarded to the genuine owner and false claim of ownership is nullified. In addition, the presence of original watermark also indicates that the design is probably not a counterfeited one. However, if the original watermark is present in the same design of a different brand, it implies that the design is probably a cloned version. It is noteworthy that any alterations made at the lower levels such as logic, gate, or physical do not disturb the signature detection process. This is because, these alterations do not affect the digital evidence (watermark information) at high (behavioral) level. In contrast, implanting watermark during behavioral synthesis phase preserves (does not affect) the secret watermark information in the design used for detectability (in detection algorithm), even if changes (due to partial/full tampering) are made at the lower levels.

7.2.2 Other Watermarking Approaches During HLS

There are some other watermarking approaches in the literature that also exploit HLS framework to embed the author's signature. However, they differ in terms of number of variables in the signature to be embedded, signature variable encoding, signature embedding process, and target phase of HLS for watermarking constraints insertion. This part of the chapter discusses triple-phase watermarking [15], binary signature-based watermarking [16], and in-synthesis watermarking [17] approaches. Let us discuss each of these approaches one by one in brief.

(a) Triple-Phase Watermarking [15]

Sengupta et al. [15] proposed triple-phase watermarking approach to secure DSP hardware accelerator against piracy and false claim of IP ownership threats. This approach leverages the following three phases of HLS to embed a robust watermark: (1) scheduling phase, (2) hardware allocation phase, and (3) register allocation phase. This watermarking is based on the author's robust signature comprising seven variables. Signature combination of desired strength is converted to watermarking constraints and embedded during the aforementioned three different phases of HLS. A complex signature is constituted using following seven variables: "γ," "α," "β," "i," "I," "T," and "!." Now, let us discuss how the three phases of HLS are exploited to embed signature digits pertaining to a specific variable type. The scheduling phase of HLS is leveraged to apply the first phase of watermarking. The "γ" digits in the signature are embedded in this phase. Furthermore, the hardware allocation phase of HLS is leveraged to apply the second phase of watermarking. The "α" and "β" digits in the signature are embedded in this phase. Subsequently, the register allocation phase of HLS is exploited to apply the third phase of watermarking. The "i," "I," "T," and "!" digits in the signature are embedded in this phase. The rules of embedding signature digits pertaining to "γ," "α," "β," "i," "I," "T," and "!"

Signature variables	Embedding Rules
γ	On each occurrence of 'γ' digit in the signature, an operation (opn) of non-critical path with highest mobility is moved into immediate next CS
α	On each occurrence of 'α' digit in the signature, hardware allocation of vendor1 and vendor2 is performed in odd CS to odd and even opns respectively
β	On each occurrence of 'β' digits in the signature, hardware allocation of vendor1 and vendor2 is performed in even CS to even and odd opns respectively
i	On each occurrence of 'i' digits in the signature, an edge is added between node-pair (V_{ij}) of two prime nodes in the CIG
I	On each occurrence of 'I' digits in the signature, an edge is added between node-pair (V_{ij}) of two even nodes in the CIG
T	On each occurrence of 'T' digits in the signature, an edge is added between node-pair (V_{ij}) of odd and even nodes in the CIG
!	On each occurrence of '!' digits in the signature, an edge is added between node-pair (V_{ij}) of node number 0 and any integer nodes in the CIG

Fig. 7.8 Embedding rules for all signature variables in a triple-phase watermarking approach [15]

variables are given in Fig. 7.8. The overview of triple-phase watermarking technique is shown in Fig. 7.9. The security achieved using this watermarking approach has been evaluated in terms of probabilities of coincidence (P_c) and tamper tolerance (T_t). The P_c metric indicates the probability of coincidentally obtaining the same watermark solution in an unwatermarked version. The toughness of finding the embedded signature using the brute-force analysis has been measured as tamper tolerance of the watermark. The time and cost consumed by an attacker to find the exact signature through brute-force search become higher with an increase in the number of combinations of potential signature digits. The watermark embedded using this technique is highly tamper tolerant and capable of providing extremely low P_c because of embedding manifold signature variables at three different phases of HLS.

(b) Binary Signature-Based Watermarking

A dynamic watermarking based on binary signature has been proposed by Koushanfar et al. [16] to protect IP cores against piracy. The watermarking constraints are embedded in a pre-synthesis phase, i.e., register allocation phase of HLS. In order to embed watermark, additional design and timing constraints are imposed during HLS. The additional constraints to be added are obtained by encoding the author's binary signature (i.e., a combination of bits, 0s and 1s). Each bit of the signature is translated into watermarking constraints based on encoding rule and embedded into

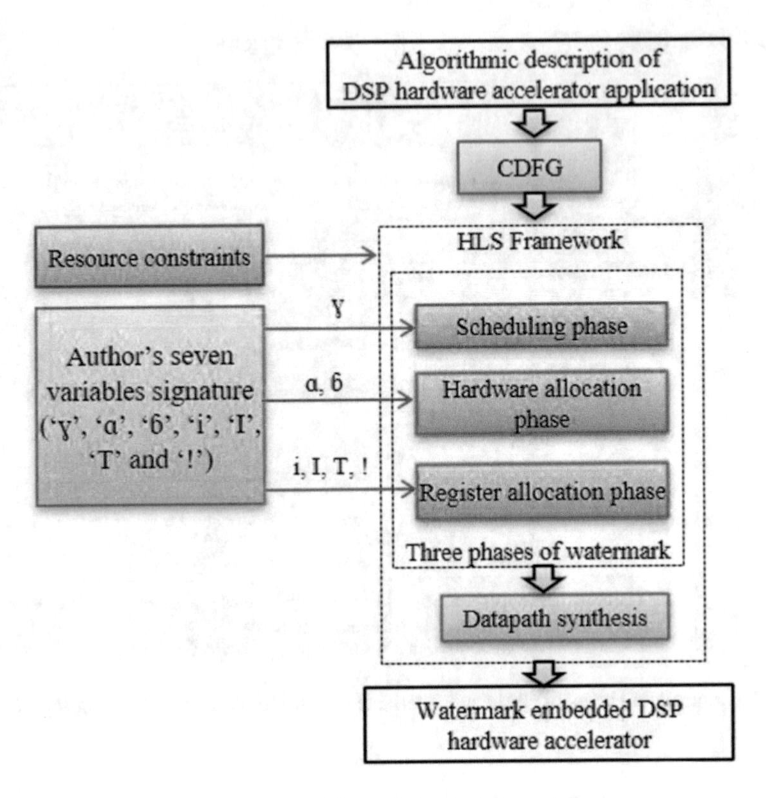

Fig. 7.9 Triple-phase watermarking-based IP protection technique [15]

register allocation phase using CIG framework. The process of embedding binary signature-based watermark is shown in Fig. 7.10. This watermarking approach targets computational intensive circuits such as DSP cores, which are designed using HLS process because of high algorithmic complexity and large size. Let us discuss the overview of watermark embedding process during HLS as shown in Fig. 7.10. A CDFG representation (i.e., an intermediate high-level description) of the design is first scheduled in control steps using a specific scheduling algorithm chosen by the designer based on the user-given area, power, and delay constraints. Next, an interval graph (IG) is formed to show the overlapping of lifetime among storage variables of the design. The overlapping of lifetime between two storage variables is denoted by an edge between two nodes in the graph. Furthermore, the graph is colored using a minimum number of colors to show the allocation of storage variables to minimum registers. To embed the author's signature, extra constraints are imposed during the register allocation of storage variables in terms of additional edges in the IG. For embedding each bit of the author's binary signature, the embedding rules shown in Fig. 7.11 are followed. The security achieved using this watermarking approach has been evaluated in terms of probability of coincidence (P_c) and probability of tampering (P_T).

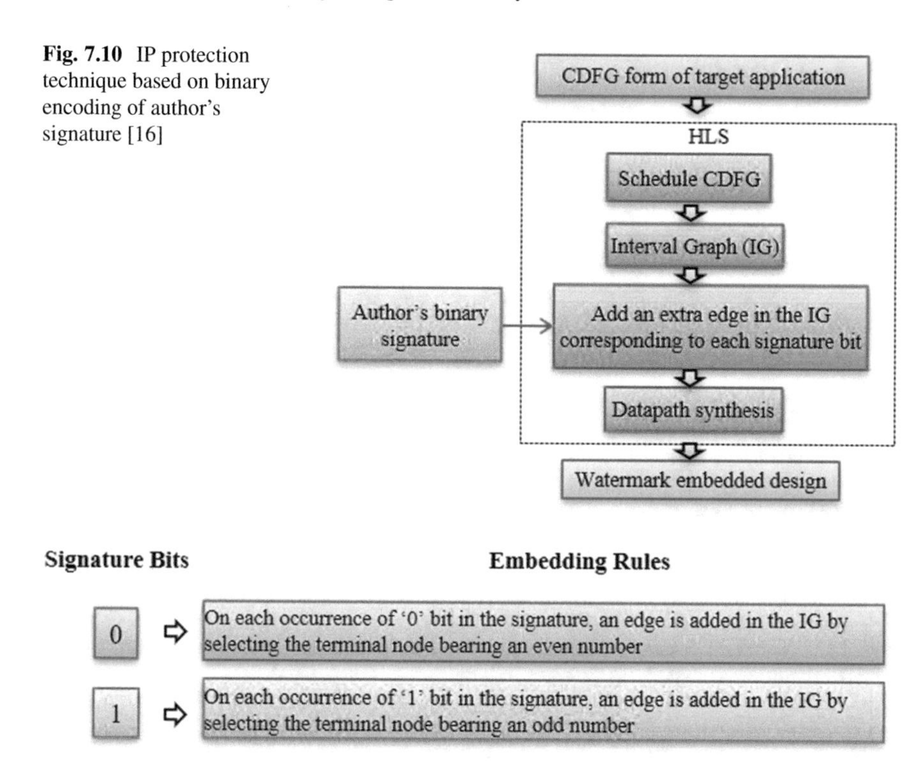

Fig. 7.10 IP protection technique based on binary encoding of author's signature [16]

Fig. 7.11 Embedding rules for signature bits in binary signature-based watermarking

(c) In-Synthesis Watermarking [17]

The watermarking techniques during the in-synthesis phase of HLS process have been proposed by Le Gal and Bossuet [17]. The in-synthesis watermarking approach exploits empty time slots between consecutive high levels of data valid output to implant the owner's signature. Therefore, this technique is aptly suitable for those circuits that produce free output slots. The mathematical relations among input values and intermediate values of circuit are utilized to embed sub-marks. The implanted sub-marks collectively act as full watermark. To identify the embedded watermark, sub-marks are read as output values during the empty time slots of output.

Two methods of implanting in-synthesis watermarking have been proposed by Le Gal and Bossuet [17]. The type of method to be adopted by the designer relies on the desired security level and admissible cost overhead. The methods are (1) random low-cost watermark and (2) cost-less watermark. Both methods of in-synthesis watermarking are shown in Fig. 7.12a, b, respectively. Here, the owner's watermark is automatically inserted to control the design cost overhead.

1. *Random low-cost watermarking:* For watermarking, the intermediate values of datapath are randomly selected and propagated to free output slots by adjusting

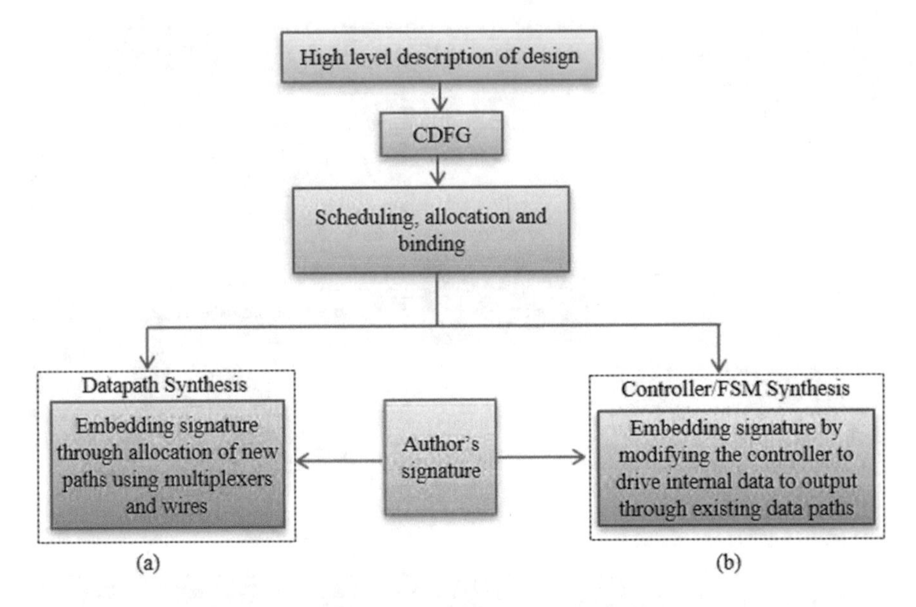

Fig. 7.12 Watermark insertion methods using in-synthesis approach [17]. (**a**) Random low-cost watermark insertion method. (**b**) Cost-less watermark insertion method

the datapath (by adding extra muxes and interconnections). This causes a small increase in the design cost.

2. *Cost-less watermarking:* For watermarking, the controller or Finite State Machine (FSM) is modified to propagate internal values to output slots. This modification does not result in design cost overhead because it does not impact the datapath. Here, the internal values to be propagated to empty output slots are the existing dynamic transient outputs.

7.3 Low-Cost Watermarking Using DSE [14]

During dynamic watermarking-based IP core protection, performing trade-off among design parameters such as area, power, and delay is highly critical. This is because various competitive design solutions are available in the design space, and therefore a low-cost solution must be adopted for embedding watermark. Trade-off among design parameters is performed using the design space exploration (DSE) process [18, 19] to explore an optimal watermarked solution from a set of numerous competitive design alternatives. If a design solution is chosen without performing trade-off, then it may significantly affect execution delay, power, and area of the end watermarked IP design. This is because watermarking employed for various candidate design solutions affects design parameters in a different way. Therefore, important design objectives are considered before generating a final solution for embedding watermark so that a low-cost optimal watermarked design can be yielded. An intelligent DSE process is capable of exploring an optimal watermarked solution, which satisfies the user's constraints of design latency and area, thus ensuring a low-cost solution. In single-phase watermarking approach [14], new watermarked design solutions have iteratively been explored through DSE to generate a low-cost watermarked design. The fitness of the resultant watermarked solution is assessed with respect to the given area and delay constraints. This DSE process is repeated until a watermarked solution of desired quality (low cost) is achieved. The details of the DSE framework for obtaining optimal watermark solution are discussed in the subsequent sub-sections [14].

7.3.1 Motivation for Particle Swarm Optimization (PSO)-Based Design Space Exploration (DSE)

Sengupta and Bhadauria [14] leveraged particle swarm optimization (PSO)-based DSE process, integrated with HLS, to generate an optimal watermark satisfying the user-given area-delay constraints [18]. In contrast to other optimization techniques such as Simulated Annealing (SA), Genetic Algorithm (GA), hybrid GA, and Bacterial Foraging Optimization Algorithm (BFOA), the PSO algorithm is more suitable for DSE. This is because other optimization techniques do not incur sufficient stochasticity into the DSE and are computationally expensive as well. The underlying reasons behind the selection of PSO-based DSE framework, for obtaining low-cost optimal watermark, are as follows [14]:

- In contrast to evolutionary algorithms such as GA, pre-tuned PSO algorithm offers superior results and faster exploration of design space.
- PSO-DSE offers higher adaptability and faster convergence to optimal solution.
- The control parameters of PSO such as inertia weight is capable of achieving a clinical balance of exploration–exploitation in order to yield a High-quality optimal solution, subjected to condition that the inertia weight is linearly decreased within the range of [0.9 0.1] during the DSE.
- Guided or adaptive searching is offered by PSO-based DSE process. This feature facilitates to make a change in direction using velocity vector on detecting an unproductive search path.
- The control parameter of PSO such as acceleration coefficient facilitates the right balance between cognitive and social best of the design solution.

In the following sub-section, we discuss about the swarm particles encoding or initialization and their movement from one design solution (particle position) to another, exploiting the concept of velocity in the PSO-DSE process of a watermarked design solution.

7.3.2 *Exploring Low-Cost Watermarked Solution Using PSO-DSE [14]*

Figure 7.13 shows the PSO-DSE process for obtaining low-cost watermarked design solution. Following are the major steps in the PSO-based DSE: (i) particle encoding, (ii) velocity determination, (iii) velocity clamping, (iv) determination of local and global best positions, (v) cost computation, and (vi) stopping criterion. Let us discuss each step briefly.

(i) *Particle encoding*: The PSO algorithm is mapped to DSE process by encoding the swarm particles. The position of a particle X_i is encoded as below:

$$X_i = (N_{F1}, N_{F2}, \ldots N_{Fj} \ldots N_{FD}) \tag{7.1}$$

where NF_j indicates the number of instances of resource type F_j, and D indicates the dimension of the particle which maps the total type of Functional Unit (FU) resource. For swarm size S, initialization of particles is as follows:

$$X_1 = (min(N_{F1}), min(N_{F2}), \ldots min(NF_j) \ldots min(N_{FD})) \tag{7.2}$$

$$X_2 = (max(N_{F1}), max(N_{F2}), \ldots max(NF_j) \ldots max(N_{FD})) \tag{7.3}$$

$$\begin{aligned} X_3 =&((min(N_{F1}) + max(N_{F1}))/2, (min(N_{F2}) + max(N_{F2}))/2 \\ &\ldots..(min(N_{FD}) + max(N_{FD}))/2) \end{aligned} \tag{7.4}$$

Initialization of remaining particles is as follows:

$$\begin{aligned} X_i =&((min(N_{F1}) + max(N_{F1}))/2{\pm}\theta, (min(N_{F2}) + max(N_{F2}))/2 \\ &\pm\theta \ldots..(min(N_{FD}) + max(N_{FD}))/2{\pm}\theta) \end{aligned} \tag{7.5}$$

where θ is a random integer value between minimum and maximum of a particular resource type.

(ii) *Velocity determination*: The velocity determination step contains the following parameters: (a) inertia weight to consider the effect of previous velocity during evaluation of current iteration velocity and (b) cognitive factor and social factor to balance the exploration process between local and global best particle positions obtained so far using acceleration coefficients variables. The following function is used for updating each dimension d of ith particle's position:

$$F_{di}^{+} = F_{di} + V_{di}^{+} \tag{7.6}$$

where V_{di}^{+} is the new velocity of ith particle, which is determined using following equation:

$$V_{di}^{+} = \beta V_{di} + c_1 d_1 \left[F_{d_{lbi}} - F_{di} \right] + c_2 d_2 \left[F_{d_{gb}} - F_{di} \right] \tag{7.7}$$

where V_{di} is the velocity of ith particle, β is inertia weight, c_1 and c_2 are acceleration coefficients, d_1 and d_2 are random numbers between 0 and 1, $F_{d_{lbi}}$ is the number of FU resource of X_{lbi} (local best position) in dth dimension, and $F_{d_{gb}}$ is the number of FU resource of X_{gb} (global best position) in dth dimension.

(iii) *Velocity clamping*: Clamping of velocity is needed to avoid swarm outburst and control undesired drift.

(iv) *Determination of local and global best positions*: This step determines the best position of each particle based on the fitness/cost function and best among the entire swarm population.

(v) *Cost computation*: This step computes the fitness or the cost of a particle position (design solution) for determining local and global best positions.

(vi) *Stopping criterion*: The stopping criterion enables the termination of the DSE process based on the following conditions: (1) iteration count (k) reaches its maximum number, and (2) global best, among swarm population, is not updated over the last ten (10) iterations.

As shown in Fig. 7.13, the PSO-DSE process is integrated with the watermark embedding process during HLS. The owner's signature is embedded into the design for each particle position or resource configuration or design solution, which is used to schedule the DFG of intended IP core. After embedding watermark for each design solution, the design cost is evaluated based on latency and area. The latency and area information is extracted from the scheduled DFG and design solution, respectively. The cost of the watermarked solution at present particle position is compared with previous particle position to determine the fitness of the particle. For the first iteration, previous particle position does not exist. Therefore, the cost is not required to be compared to determine the local best and global best particle positions. Initial particle positions itself are considered initial local best. Furthermore, new particle positions are determined in the subsequent iterations. However, cost comparison is performed from the second iteration onward. Local particle position is updated if the cost of present ($\mathrm{C}_{\mathrm{i}}^{\mathrm{k}}$) is obtained to be less than the cost of previous position ($\mathrm{C}_{\mathrm{i}}^{\mathrm{k}-}$) as shown in Fig. 7.13. Otherwise, a new position is explored within the design space. After finding new better positions for entire swarm population, the global best is updated. This DSE process is repeatedly continued until stopping criterion is reached. Upon termination of the DSE process, an optimal design solution is found, which gives a minimum cost of watermarked design. Thus, the owner's watermark is embedded into the design simultaneously minimizing the design cost (i.e., minimizing latency and area).

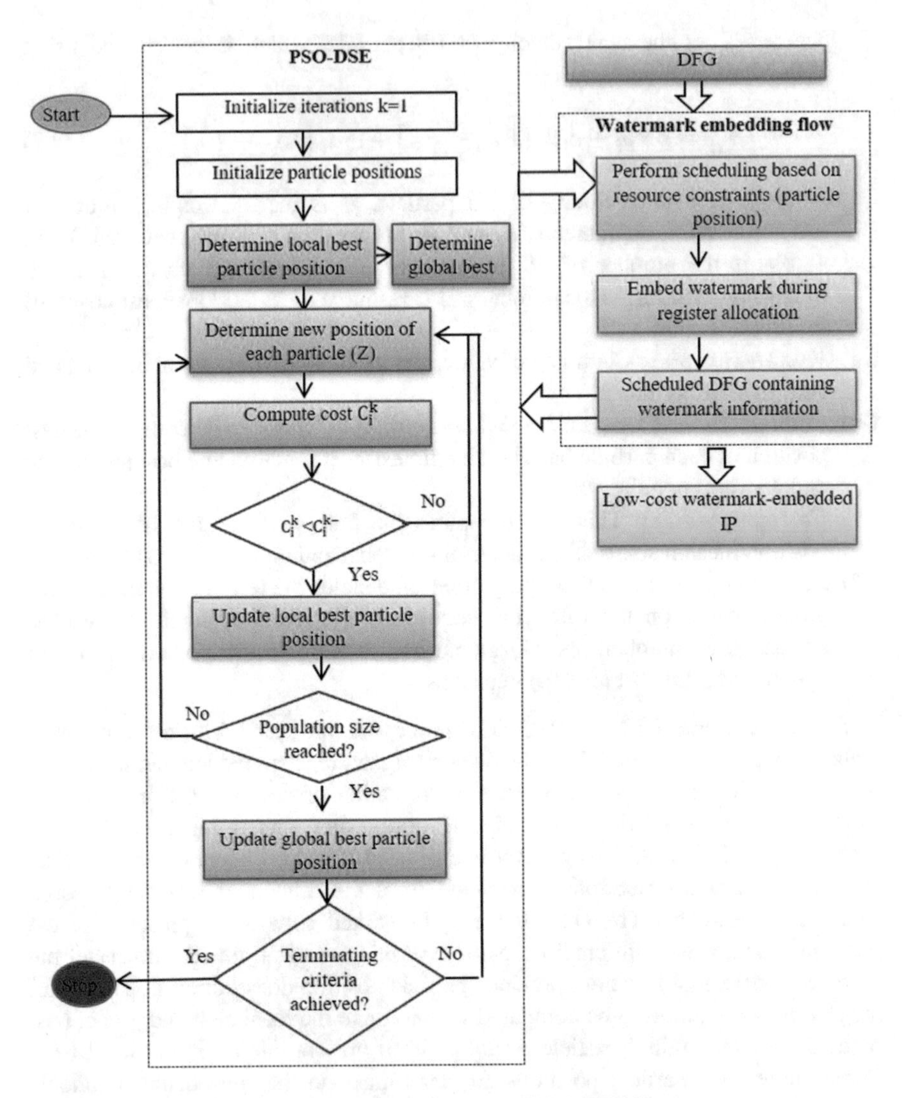

Fig. 7.13 Process of obtaining low-cost solution using PSO-DSE [14]

7.4 Threat Scenarios of False Claim of Ownership and Desirable Properties of Watermarking

The watermarked design of an entity "X" is subjected to the following threat scenarios of false claim of IP ownership from an entity "Y" [15].

(a) *Unauthorized signature insertion:* On the top of watermarked IP that carries signature of entity "X," watermarking constraints can be superimposed by entity "Y," In other words, IP ownership can be claimed fraudulently by entity "Y" after imposing his/her own signature into the watermarked IP of "X." This conflict can be resolved through watermarking. This is because the IP of "X" will contain only his/her own signature, whereas the IP of "Y" would be containing the signature of both "X" and "Y."

(b) *Tampering original signature in the design:* Entity "Y" attempts to generate his/her own IP appearing to be authorized by applying some modifications to the original watermarked IP of entity "X." A robust watermarking thwarts this threat by distributing watermarking constraints throughout the design post-synthesis. This disables the full tampering of all watermarking constraints.

(c) *Unintended signature extraction:* Inverse watermark calculation approach may be adopted by entity "Y" to extract his/her own signature in the watermarked design of "X." Therefore, the existing knowledge of the IP design may randomly be claimed as his/her signature by entity "Y." However, it may be meaningful for a single design and cannot work out for other watermarked designs. Therefore, this way of claiming ownership is weaker in contrast to a claim based on authentic watermark detection.

To protect against the aforementioned threat scenarios, the following properties of watermark are desirable [14, 16]:

1. *IP ownership proof:* The embedded watermark should be able to be scientifically proved by a genuine owner to resolve IP ownership conflict.
2. *Functional correctness:* Original functionality of the IP core should remain intact after embedding watermark.
3. *Low cost:* Embedding watermark should incur minimal change in the design latency, area, and power. Ideally, watermarking scheme should incur zero design cost overhead.
4. *Resilience:* The embedded watermark constraints should not be easily removable. Even a single watermark constraint removal by an adversary should cause considerable quality and performance degradation of the watermarked design. Thereby, toughness of removal of even a single constraint is the degree of resilience.
5. *Fault tolerance:* The fault tolerance is the ability of watermark that its existence does not get distorted on partial removal of signature constraints embedded into the design. Therefore, a fault-tolerant watermark is capable to be detected during forensic detection even if it was partially tampered.
6. *Difficult to locate:* The watermarking constraints should remain unlocatable or perceptually invisible to an attacker in order to make the watermark removal task highly intricate. Only the genuine owner should be able to locate it using specific knowledge of signature variable encoding, signature size, and location of insertion.
7. *Adaptable to CAD tools:* The watermarking algorithm should seamlessly be integrated with the design flow so that any CAD tool can easily adapt it.

7.5 Case Study on DSP Applications

This section discusses the case study of single-phase watermarking [14] and triple-phase watermarking [15] on various DSP applications such as Auto-Regression Filter (ARF), Discrete Wavelet Transform (DWT), Inverse Discrete Cosine Transform (IDCT), Moving Picture Experts Group (MPEG), etc. The case study is discussed based on robustness of watermarking using both the aforementioned approaches.

7.5.1 Case Study of Single-Phase Watermarking [14]

The robustness of single-phase watermarking approach was measured using probability of coincidence (P_c^s). The following metric is used to evaluate the P_c^s [14]:

$$P_c^s = \left(1 - \frac{1}{r}\right)^w \tag{7.8}$$

where w indicates the signature size or the number of constraints embedded, r indicates the number of registers or colors used in the CIG. As evident from Eq. (7.8), P_c^s decreases with the increase in the number of watermarking constraints added. The probability of coincidence signifies the probability of coincidentally detecting the same watermarking constraints in a baseline (non-watermarked) version. The probability of coincidence should be as low as possible. Therefore, a designer targets to achieve lower P_c^s concurrently ensuring minimal design overhead. To achieve lower probability of coincidence, more constraints are added, which results in stronger digital evidence and stronger proof of ownership. Since embedding more constraints lowers the P_c^s value, a lower probability of coincidence indicates the higher robustness of watermarking.

Figure 7.14 shows the variation in probability of coincidence for an increasing number of watermarking constraints. As shown in the figure, the probability of coincidence value becomes lower with the increasing size of constraints. In Fig. 7.14, for computing P using Eq. (7.8), the number of colors/registers (r) in the CIG (before embedding watermark) of DWT, ARF, IDCT, MPEG, IIR, and FIR is 5, 8, 8, 14, 5, and 8, respectively. Furthermore, the number of registers required after single-phase watermarking [14] has been compared with the binary signature-based watermarking [16] for constraint size, W = 60. In addition, the number of registers in baseline design (pre-watermarking) has been compared with the single-phase and binary signature-based watermarking in Fig. 7.15. As shown in the figure, the number of registers required post single-phase watermarking is either lesser or equal to that of binary signature-based watermarking. This indicates that single-phase watermarking offers security at lesser storage hardware (registers) overhead than binary signature-based watermarking, for the same number of watermarking

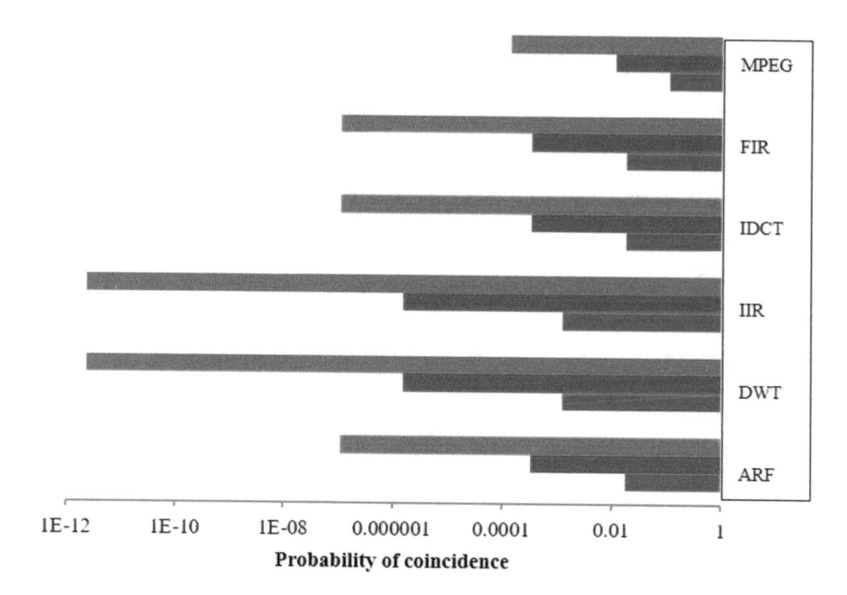

Fig. 7.14 Variation in probability of coincidence for varying constraints size [14]

constraints. Furthermore, register overhead of single-phase watermarking was compared to baseline. As reported in [14], the register overhead compared to baseline is nominal for small size watermark (such as constraint size = 15). However, as constraint size increases, and the register overhead gradually increases (for example, at constraint size = 60, the overhead is higher than at W = 15). The register overhead of single-phase watermarking compared to baseline is shown in Fig. 7.15 for a constraint size of 60.

7.5.2 *Case Study of Triple-Phase Watermarking [15]*

The robustness of triple-phase watermarking approach was measured using the following probability of coincidence P_c^t [15]:

$$P_c^t = \left(1 - \frac{1}{r * \prod_{j=1}^{U} N(F_j)}\right)^w \tag{7.9}$$

where w indicates the number of digits in the signature, r indicates the number of registers or colors used in the CIG, $N(F_j)$ indicates the number of FU resources of type j, and U indicates the number of various FU resource types used in the design. The comparison of probability of coincidence using triple-phase

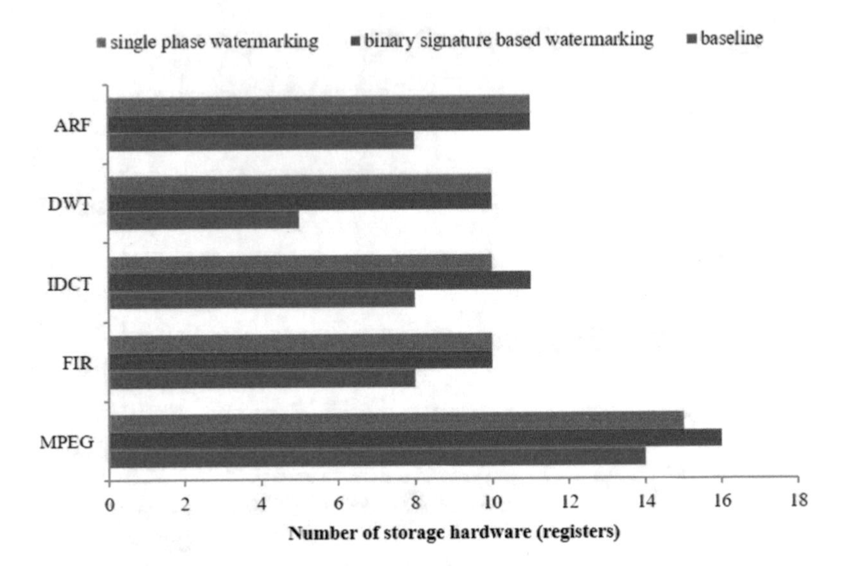

Fig. 7.15 Comparison of single-phase watermarking with respect to baseline and binary signature-based watermarking in terms of the number of register [14]

watermarking (for the number of signature digits W = 80) with respect to single-phase watermarking [14] and binary signature-based watermarking [16] is shown in Fig. 7.16. As shown in the figure, triple-phase watermarking offers significantly lower probability of coincidence compared to single-phase watermarking and binary signature-based watermarking at W = 80. This indicates higher robustness of triple-phase watermarking approach. Furthermore, tamper tolerance ability of triple-phase watermarking is evaluated using the following formula [15]:

$$T_t = E^W \tag{7.10}$$

where E denotes the number of distinct variables in the signature and W indicates the number of digits in the signature. Here, the tamper tolerance ability indicates the maximum number of possible combinations that the owner's signature can have. More the number of variables and digits in the signature, higher is the tamper tolerance ability as evident from Eq. (7.10). This is because the numerous possible signature combinations render it extremely difficult for an attacker to find the exact correct signature combination by applying brute-force attack. Figure 7.17 compares the tamper tolerance ability of triple-phase watermarking [15] with respect to single-phase watermarking [14] and binary signature-based watermarking [16]. As shown in the figure, triple-phase watermarking shows greater tamper tolerance ability than single-phase watermarking and binary signature-based watermarking for the same number of signature digits. This is because, triple-phase watermarking uses seven encoded variables in the signature, which is larger than the number of variables used

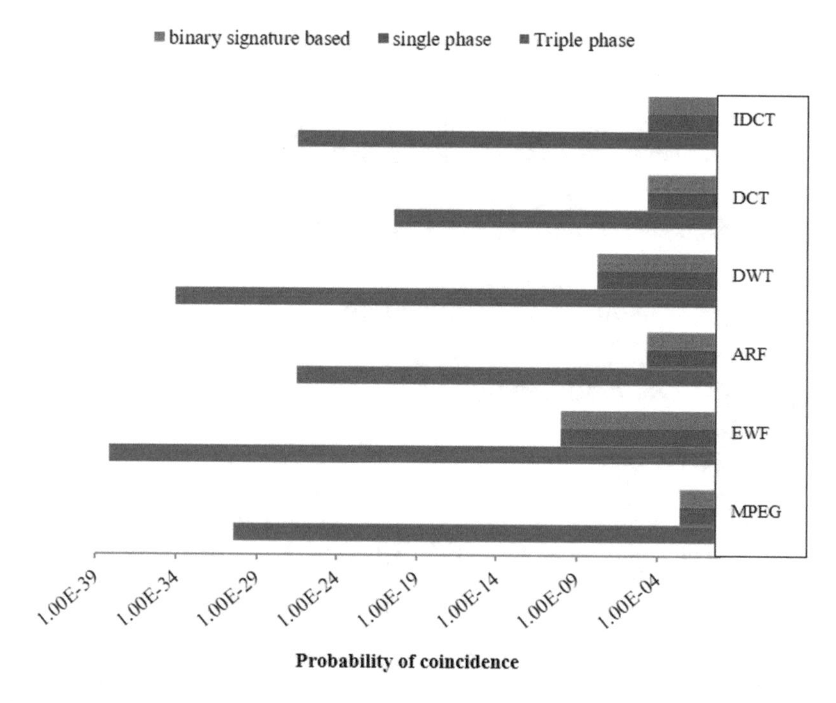

Fig. 7.16 Comparison of triple-phase watermarking with respect to single-phase and binary signature-based watermarking in terms of probability of coincidence [15]

in single-phase watermarking (four encoded variables) and binary signature-based watermarking (two encoded variables).

7.6 Conclusion

Hardware watermarking during HLS is very important to secure hardware of highly complex data-intensive cores such as DSP IPs from piracy and false claim of ownership threats. Moreover, minimizing the impact of watermarking on design overhead is obligatory to ensure a low-cost solution. This chapter highlighted the role of HLS in IP core protection using hardware watermarking against piracy and false claim of ownership. Furthermore, IP core protection using single-phase watermarking has been comprehensively discussed along with brief discussion on contemporary watermarking approaches such as triple-phase watermarking, binary signature-based watermarking, and in-synthesis watermarking. Furthermore, this chapter discussed the importance of low-cost watermarked design, followed by discussion on exploration of low-cost watermarked design using PSO-DSE approach. In summary, this chapter specifically highlighted the following topics:

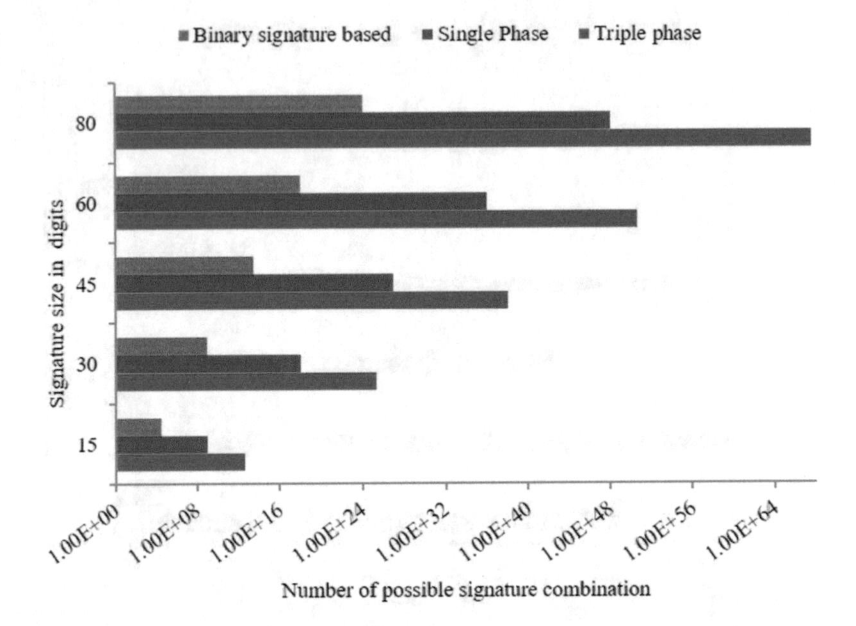

Fig. 7.17 Comparison of triple-phase watermarking with respect to single-phase and binary signature-based watermarking in terms of tamper tolerance (number of possible signature combination) [15]

- Importance of HLS in watermarking.
- Exploitation of register allocation phase of HLS in watermarking.
- Concept of colored interval graph in watermarking.
- Detection of piracy and false claim of IP ownership using watermarking.
- A comprehensive study on single-phase watermarking during HLS.
- Overview of other watermarking approaches such as triple-phase watermarking, binary signature-based watermarking, and in-synthesis watermarking.
- Low-cost watermarking using PSO-DSE.
- Threat scenario and desirable properties of watermarking.
- Robustness assessment of single-phase and triple-phase watermarking approaches through different case studies (DSP applications).

References

1. Schneiderman, R.: DSPs evolving in consumer electronics applications [special reports]. IEEE Signal Process. Mag. **27**(3), 6–10 (2010). https://doi.org/10.1109/MSP.2010.936031
2. Mahdiany, H.R., Hormati, A., Fakhraie, S.M.: A hardware accelerator for DSP system design: university of tehran DSP hardware emulator (UTDHE). In: ICM 2001 Proceedings. The 13th International Conference on Microelectronics, pp. 141–144 (2001). https://doi.org/10.1109/ICM.2001.997507

3. McFarland, M.C., Parker, A.C., Camposano, R.: Tutorial on high-level synthesis. In: Proceedings of the 25th Design Automation Conference, pp. 330–336. ACM, New York (1988)
4. Sengupta, A., Kumar, E.R., Chandra, N.P.: Embedding digital signature using encrypted-hashing for protection of DSP cores in CE. IEEE Trans. Consum. Electron. **65**(3), 398–407 (2019). https://doi.org/10.1109/TCE.2019.2924049
5. Pilato, C., Garg, S., Wu, K., Karri, R., Regazzoni, F.: Securing hardware accelerators: a new challenge for high-level synthesis. IEEE Embed. Syst. Lett. **10**(3), 77–80 (2018). https://doi.org/10.1109/LES.2017.2774800
6. Sengupta, A., Rathor, M.: Enhanced security of DSP circuits using multi-key based structural obfuscation and physical-level watermarking for consumer electronics systems. IEEE Trans. Consum. Electron. **66**(2), 163–172 (2020). https://doi.org/10.1109/TCE.2020.2972808
7. Sengupta, A., Mohanty, S.: IP core and integrated circuit protection using robust watermarking. In: IP Core Protection and Hardware-Assisted Security for Consumer Electronics, pp. 123–170 (2019)
8. Koushanfar, F., Fazzari, S., McCants, C., Bryson, W., Song, P., Sale, M., Potkonjak, M.: Can EDA combat the rise of electronic counterfeiting? In: DAC Design Automation Conference 2012, pp. 133–138 (2012). https://doi.org/10.1145/2228360.2228386
9. Castillo, E., Meyer-Baese, U., Garcia, A., Parrilla, L., Lloris, A.: IPP@HDL: efficient intellectual property protection scheme for IP cores. IEEE Trans. Very Large Scale Integration Syst. **15**(5), 578–591 (2007). https://doi.org/10.1109/TVLSI.2007.896914
10. Plaza, S.M., Markov, I.L.: Solving the third-shift problem in IC piracy with test-aware logic locking. IEEE Trans. Comput.-Aid. Des. Integr. Circuits Syst. **34**(6), 961–971 (2015). https://doi.org/10.1109/TCAD.2015.2404876
11. Cui, A., Chang, C.H., Tahar, S.: IP watermarking using incremental technology mapping at logic synthesis level. IEEE Trans. Comput.-Aid. Des. Integr. Circuits Syst. **27**(9), 1565–1570 (2008). https://doi.org/10.1109/TCAD.2008.927732
12. Colombier, B., Bossuet, L.: Survey of hardware protection of design data for integrated circuits and intellectual properties. IET Comput. Digit. Tech. **8**(6), 274–287 (2014). https://doi.org/10.1049/iet-cdt.2014.0028
13. Ziener, D., Teich, J.: Power signature watermarking of IP cores for FPGAs. J. Signal Process. Syst. **51**(1), 123–136 (2008)
14. Sengupta, A., Bhadauria, S.: Exploring low cost optimal watermark for reusable IP cores during high level synthesis. IEEE Access **4**, 2198–2215 (2016). https://doi.org/10.1109/ACCESS.2016.2552058
15. Sengupta, A., Roy, D., Mohanty, S.P.: Triple-phase watermarking for reusable IP core protection during architecture synthesis. IEEE Trans. Comput.-Aid. Des. Integr. Circuits Syst. **37**(4), 742–755 (2018). https://doi.org/10.1109/TCAD.2017.2729341
16. Hong, I., Potkonjak, M.: Behavioral synthesis techniques for intellectual property protection. In: Proceedings 1999 Design Automation Conference (Cat. No. 99CH36361), pp. 849–854 (1999). https://doi.org/10.1109/DAC.1999.782161
17. L. Gal, B., Bossuet, L.: Automatic low-cost IP watermarking technique based on output mark insertions. Des. Autom. Embed. Syst. **16**(2), 71–92 (2012). https://doi.org/10.1007/s10617-012-9085-y
18. Mishra, V.K., Sengupta, A.: MO-PSE: adaptive multi-objective particle swarm optimization based design space exploration in architectural synthesis for application specific processor design. Adv. Eng. Softw. **67**, 111–124 (2014). https://doi.org/10.1016/j.advengsoft.2013.09.001. http://www.sciencedirect.com/science/article/pii/S0965997813001555
19. Krishnan, V., Katkoori, S.: A genetic algorithm for the design space exploration of datapaths during high-level synthesis. IEEE Trans. Evol. Comput. **10**(3), 213–229 (2006). https://doi.org/10.1109/TEVC.2005.860764

Chapter 8
Encoding of Finite-State Controllers for Graded Security and Power

Richa Agrawal and Ranga Vemuri

8.1 Introduction

Digital system designs usually consist of a datapath and a finite-state controller. For many cryptography and digital signal processing algorithms, the control flow tends to be fairly simple. Such designs are vulnerable to a potent form of side-channel attacks based on power measurements.

Furthermore, finite-state controllers are central to the design of numerous small-scale electronic appliances used in home automation, environment/infrastructure monitoring, health care, and emerging safety-critical systems such as drones and self-driven cars. Some IoT (Internet-of-Things) devices contain small datapaths with the computation/control performed mostly by the control part. These small controller-based devices are extremely vulnerable to side-channel attacks. It is estimated that there will be 50 billion small-scale IoT devices by 2020. These devices typically have limited power and need to be energy efficient [1], which makes sophisticated cryptographic algorithms and hardware protection schemes such as [2, 3] that incur high cost in terms of area and power unsuitable for these systems. Low-cost, low-power defense methods are highly desirable.

Small-scale designs are vulnerable to both data leaks and reverse engineering attacks as long as the attacker is able to isolate the control registers from datapath registers and perform the required power analysis on them. Both invasive and non-invasive reverse engineering attacks have been studied during the past several years [4, 5]. Invasive methods are destructive in nature in addition to being expensive and

R. Agrawal
Synopsys Inc., Mountain View, CA, USA
e-mail: richa.agrwl@outlook.com

R. Vemuri (✉)
University of Cincinnati, EECS Department, Cincinnati, OH, USA
e-mail: ranga.vemuri@uc.edu

S. Katkoori, S. A. Islam (eds.), *Behavioral Synthesis for Hardware Security*,
https://doi.org/10.1007/978-3-030-78841-4_8

laborious to perform [6, 7]. Non-invasive attacks to reverse engineer an FSM are based on characterizing the machine behavior using only input–output values of FSMs and are restricted by memory and time usage [8, 9].

Side-channel attacks are non-invasive attacks [10, 11] that exploit the relationship between the operations of the target device and measurable physical variables [12]. They can use physical parameters such as power consumption, electromagnetic radiation, and response time to reverse engineer the contents of critical registers or the structure of the controller. Side-channel measurements can be used independently or to supplement cryptanalysis attacks [13]. Power analysis attacks use only power consumption information, which makes them cost-efficient, easy, and powerful against low-cost small-scale embedded devices [14–18].

State encoding used while implementing an FSM has a direct correlation with the power drawn by its CMOS implementation. This correlation can be exploited by the attacker to determine the contents of the control register and reverse engineer the structure of the FSM. However, encodings that protect against power analysis attacks tend to increase power consumption. This leads to a trade-off between the level of security desired and the amount of power consumed.

We have introduced a graded security metric based on the concept of reachability equivalence among the states and among the state transitions of an FSM. Given a level of graded security desired, we propose a method to generate constraints on state encoding to meet the specified level of security while minimizing power consumption.

The rest of this chapter is organized as follows. We review the relevant background in Sect. 8.2. We formally introduce the concept of security and secure state encoding in Sect. 8.3. We define reachability equivalences and state transition probabilities in Sect. 8.4. Based on these ideas, we develop a method for secure state encoding in Sect. 8.5. In Sect. 8.6, we introduce the concept of graded security and discuss constraint generation methods for security vs power trade-off. We discuss Boolean constraint generation and solving using satisfiability solvers in Sect. 8.7. We present a detailed experimental analysis in Sect. 8.8. Concluding remarks and directions for future work are in Sect. 8.9.

8.2 Background

8.2.1 Reverse Engineering Attacks on FSMs

Reverse engineering attacks are known to be quite effective for small-scale FSMs [19]. Non-destructive methods for characterizing and reverse engineering state machines have been previously proposed in [8, 9]. Non-destructive attacks can be made more effective by analyzing side-channel information along with the functional output values. Fault-detection attacks already use power [20] and EM side-channel leakage information [21] to determine a known, good baseline and then detect alterations by comparing them to the baseline.

The attacker intends to reverse engineer the FSM structure by attacking the state register of the sequential circuit implementation. The attacker tries to construct the state machine by utilizing the information leaked through the power side-channel during the run-time to predict the changes in the state variables of the FSM implementation realized in CMOS technology. This can be done in two ways:

1. The attacker hypothesizes a state machine, making use of a high-level power model, and generates a predicted high-level power trace for a large input sequence. The attacker then measures the actual power trace and computes its correlation with the predicted power trace for the same input sequence. A high correlation between the two validates the hypothesized state machine.
2. The attacker constructs a state machine. This is possible when each state and each transition have unique power profiles. In this case, the attacker can conduct distinguishing experiments in which different input sequences causing different power draws are applied leading to the incremental reconstruction of the state machine.

   The first type of attacks is called Differential Power Attacks (DPA) and the second type is termed Simple Power Attacks (SPA) [18].

   Power side-channel attacks are based on high-level power models for synchronous sequential circuits. A successful attack is possible only if the states and transitions can be distinguished from one another based on the power measurements. Power analysis attack can also be combined with cryptanalysis to reverse engineer the state machine, which has proven to be quite effective [22]. On the other hand, if the circuit consumes *constant* power in all states and *constant* power during all transitions, then there would be no information leakage through the power side-channel.

8.2.2 Existing Defense Methods

High-level protection schemes hide or mask critical information to make side-channel measurements independent of input data and the device's computational trajectory. In finite state machines, the state registers hide information within their encodings [23]. This information can be subject to power analysis attacks since power consumption profiles can be correlated to data changes in registers and be used to reverse engineer the state machine. Current defense methods include using cryptographic subsystems [3] to encrypt the data flow to/from the device. But for small-scale devices, their large area and power overheads can render them unfeasible for practical use.

Another idea is to design circuits that draw constant power during all states and transitions. This has been implemented by several cell-level defenses against DPA and SPA attacks. Such hardware protection schemes where special power-invariant cells are used work on a low-level implementation. Schemes such as WDDL [24] and MDPL [25] impose significant power and area penalties.

8.3 Secure State Encoding

8.3.1 Finite State Machines

Let us define a finite state machine as $\mathcal{M} = (S, I, T, s_o)$, where S is a finite set of states, I is a finite set of input symbols, $T : S\,X\,I \rightarrow S$ is a state transition function, and $s_o \in S$ is the initial state. Let $s_1, s_2 \ldots s_M$ denote the states, where the total number of states $M = |S|$.

When an FSM is implemented as a synchronous sequential circuit, the states are encoded as a set of Boolean state variables that are stored in flip-flops, say, R delay flip-flops. Let $Q =< q_o, q_1, \ldots q_R >$ denote the Boolean valued vector of these R state variables. Let state encoding function that maps each state to a Boolean vector be denoted by $E : S \rightarrow (b_1, b_2, \ldots b_R)$, where $b_i \in \{0, 1\}$. The mapping must assign a unique vector to every state, that is, $\forall_{s_i, s_j \in S}, E(s_i) = E(s_j) \implies s_i = s_j$. A sequential circuit is an implementation of an FSM if and only if the FSM is in state $s \in S$, then sequential circuit is in state $Q = E(s)$.

Let HW(B) denote the number of ones in a Boolean vector B or the Hamming weight of B. Let HD(B_1, B_2) denote the Hamming distance between two Boolean vectors B_1 and B_2 of the same size, which is equal to the number of positions in which the two vectors differ.

R is referred to as the size or length of the encoding. State encodings in which $R = \lceil log_2(M) \rceil$ are called *minimal length encodings* or simply *minimal encodings*, and this value is denoted as R_{min} for that FSM. Encodings in which $R = M$ and, for all states s, HW(E(s)) = 1, are called one-hot encodings. We will refer to this value of R as R_{max} for that FSM. Given an FSM, encodings range within size $R \in [R_{min}, R_{max}]$.

8.3.2 Hamming Models

Power analysis attacks exploit the information leaked through the power side-channel to reveal the data stored in internal registers. Strong correlation between the power profiles and data stored within the registers shows that variations in power consumption can be used to predict the changes in the state variables which can be used to construct the FSM.

A design implemented using CMOS technology is susceptible to Hamming model-based attacks. The Hamming weight model assumes that the power consumed when a circuit is in state s is correlated to $HW(E(s))$. This is called the Hamming weight of a state under encoding E. An attacker's intuition for this model is based on the observation that in a stable state, the power consumption in a CMOS circuit is dependent on the modes of operation of each transistor, which in turn depends on E(s) for each states.

Table 8.1 Switching distance model for one bit

Transition	SD
$0 \rightarrow 0$	0
$0 \rightarrow 1$	1
$1 \rightarrow 0$	1-δ
$1 \rightarrow 1$	0

The Hamming distance model assumes that the dynamic power consumption in a CMOS circuit depends on the state transitions that are characterized by the switching activity in the state registers. Hence, the power consumed when a circuit transitions from state s_1 to state s_2 is correlated to the Hamming distance of the transition from s_1 to s_2 under encoding E or $HD(E(s_1), E(s_2))$.

The switching distance model (or a modified HD model) assumes that a CMOS gate consumes slightly more power during rise than during fall. The attack model can take this difference into account. For example, the authors of [26] introduced a parameter δ to capture the difference. δ is defined as the normalized difference of the transition leakages: $\delta = (P_{0\rightarrow 1} - P_{1\rightarrow 0})/P_{0\rightarrow 1}$. Nominally, $\delta = 0.17$. In this model, power consumed when a circuit transitions from state s_1 to state s_2 is correlated to $SD(E(s_1), E(s_2))$, where the SD values shown in Table 8.1 for one bit are summed up across all the bits in the encoding. This is called the *switching distance* of the transition from s_1 to s_2 under encoding E.

8.3.3 *Constraints on State Encodings*

We can represent an FSM by a State Transition Graph (STG), in which each state is represented by a unique vertex and all transitions from one state to another are represented by unique edges. For convenience, overloading the notation used for the FSM, we refer to the STG as (S, T, s_o), where S denotes the set of state vertices, $T \subseteq (SXS)$ is the set of directed edges denoting the state transitions, and $s_o \in S$ is the vertex representing the start state. When there is no confusion, we use the terms FSM and its STG interchangeably.

Consider an encoding that ensures that all states have the same Hamming weight and all transitions have the same Hamming distance and switching distance. For example, the one-hot encoding ensures that all states have a Hamming weight of 1, and all transitions have a Hamming distance of 2 and a switching distance of $(2 - \delta)$. Under such an encoding, the FSM implementation is unlikely to leak any information via power traces under the Hamming model attacks. However, this constraint leads to a large encoding size (R_{max}) requirement. Another approach was discussed in Chap. 14 where state encoding ensures Hamming weight and Hamming distance of constants c1 and c2 for all states and transitions [27, 28].

8.4 Quantification of FSM Behavior

We need to ensure that the state encoding is done so as to avoid information leakage through the power side-channel. This chapter proposes a secure encoding method for FSMs with smaller encoding size and lower power and area requirements. But before that, we need to understand and quantify FSM behavior, which exposes the information stored within.

8.4.1 Reachability

Given an FSM with start state s_0, we define reachability of any state s as follows:

A state s is *L-reachable* if there is path of length L from s_o to s in the STG, where L is a non-negative integer. In other words, if there exists an input sequence of length L to take the machine to state s, or s can be reached in L clock cycles from the start state s_0, then a state s is L-reachable.

We use the predicate $R_L(s)$ to denote the L-reachability of state s. It is of course possible that there may be multiple paths to reach a state and, hence, multiple values of L for which a state s is L-reachable.

Let $S_L \subseteq S$ be a subset of states that are L-reachable. Formally, $S_L = \{s \in S | R_L(s)\}$. Let T_L be the set of all transitions in the STG originating from the states in S_L. Formally, $T_L = \{(s_1, s_2) \in T | s_1 \in S_L\}$.

In order to avoid information leakage through the power side-channel and thwart DPA or SPA attacks, all states that are reachable in the same number of cycles and all transitions that can be traversed in same number of clock cycles should have the same power footprint. This leads to the following constraints on the state encoding. For any reachability L,

(1) all states in set S_L should have the same Hamming weight and
(2) all transitions in set T_L should have the same Hamming distance.

Let us consider an example shown in Fig. 8.1:

- For L = 0,3,6..., $S_L = \{S_1\}$.
- For L = 1,4,7,...,$S_L = \{S_2, S_3, S_4\}$.
- For L = 2,5,8..., $S_L = \{S_5, S_6\}$.

An encoding that satisfies the following constraints will thwart an attacker's effort to exploit the power channel using the HW and HD models:

1. HW(S_2) = HW(S_3) = HW(S_4).
2. HW(S_5) = HW(S_6).

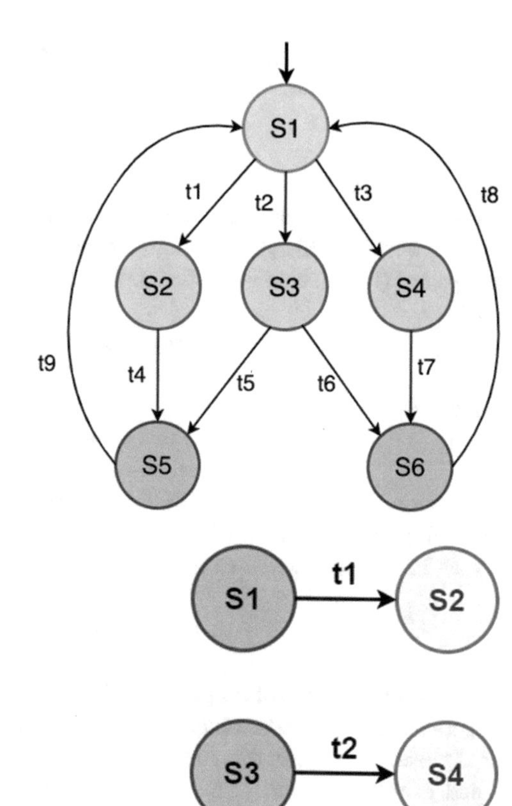

Fig. 8.1 Example: 6-state FSM

Fig. 8.2 State transition pairs of a given FSM

3. $\mathrm{HD}(t_1) = \mathrm{HD}(t_2) = \mathrm{HD}(t_3)$.
4. $\mathrm{HD}(t_4) = \mathrm{HD}(t_5) = \mathrm{HD}(t_6) = \mathrm{HD}(t_7)$.
5. $\mathrm{HD}(t_8) = \mathrm{HD}(t_9)$,

It is not necessary that all states should have the same HW and all transitions should have the same HD, which would simply increase the encoding length. Using the precise requirements based on S_L and T_L sets reduces the encoding length while retaining security.

In order to understand security against switching distance attacks, let us consider two separate state transitions of a given FSM, as shown in Fig. 8.2. Transitions $\{S_1 \rightarrow S_2\}$ and $\{S_3 \rightarrow S_4\}$ are such that states S_1 and S_3 belong to same L-reachable set S_{L1}, and states S_2 and S_4 belong to set S_{L2}. Let their state encodings be such that

$$\mathrm{HW}(S_1) = \mathrm{HW}(S_3) = w_1$$
$$\mathrm{HW}(S_2) = \mathrm{HW}(S_4) = w_2$$
$$\mathrm{HD}(t_1) = \mathrm{HD}(t_2) = d$$
$$\mathrm{SD}(t_1) = z_1,\ \mathrm{SD}(t_2) = z_2$$

Let the number of low to high bit changes (0 to 1) during transition t_1 be denoted by r_1 and the number of high to low bit changes (1 to 0) be denoted by f_1.

$$d = r_1 + f_1 \tag{8.1}$$

By simple math, it can be calculated that during transition t_1, the number of ones in state encoding E(S_2) is equal to the sum of the number of ones in E(S_1) and the number of (0 to 1) changes subtracted by the number of (1 to 0) changes. Therefore,

$$w_2 = w_1 + (r_1 - f_1) \tag{8.2}$$

which implies

$$r_1 = (d + w_2 - w_1)/2$$
$$f_1 = (d - w_2 + w_1)/2$$

Since z_1 is a function of r_1 and f_1, which in turn are functions of d, w_1, and w_2, switching distance of transitions t_1 and t_2 is equal. Hence, transitions in the same T_L set have equal switching distance.

$$z_1 = r_1 + \delta * f_1 \implies z_2 = z_1 \tag{8.3}$$

In general, when for all path lengths i, sets of states $|S_i|$ are a small fraction of $|S|$, then small and secure encodings can be found. This means state encoding length R is closer to R_{min}. In this case, the S_i sets induce a partition on S such that for any i and j, either $S_i = S_j$ or $S_i \cup S_j = \emptyset$. The example in Fig. 8.1 illustrates this. On the other hand, if for some i, $S_i = S$, then all states must have the same HW and all transitions the same HD. In this case, secure encodings tend to be long, that is, R closer to R_{max}. We will discuss these situations next.

8.4.1.1 Conflicts in Reachability

There are several situations that can result in large set size of states for some reachability values, i.e., large S_i sizes for some i values.

1. *States with multiple L-reachability values*: Consider the FSM shown in Fig. 8.3. Given $S1$ as the start state, we try to meet requirement (2) above. Transitions t_1 and t_2 should have the same HD and transitions t_3 and t_4 should have the same HD. For i=3, transitions t_1, t_2, and t_4 should have the same HD. Hence, for $i \geq 3$, $S_i = S$. This forces all states to have the same HW and all transitions the same HD for secure encoding.
2. *Transitions in both directions*: Let us consider the FSM shown in Fig. 8.4. According to requirement (2) above, transitions $\{t_1, t_3, t_5\}$ should have the same HD and transitions $\{t_2, t_4\}$ should have the same HD. However, it is impossible to encode different HD values to t_1 and t_2 since both transitions are between the same pair of states. Similar is the case with transitions t_4 and t_5. This forces all

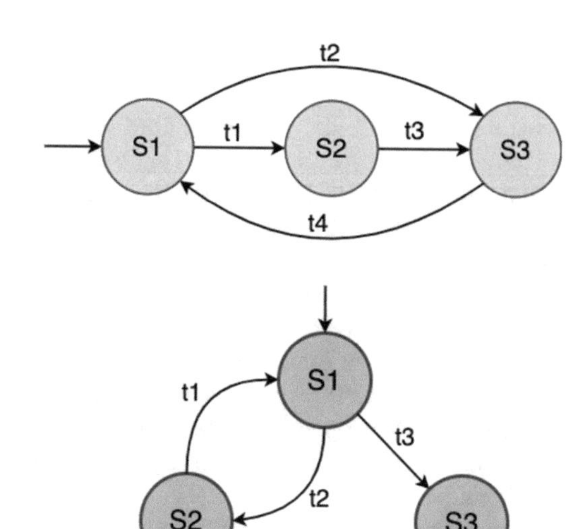

Fig. 8.3 States with multiple L-reachability value

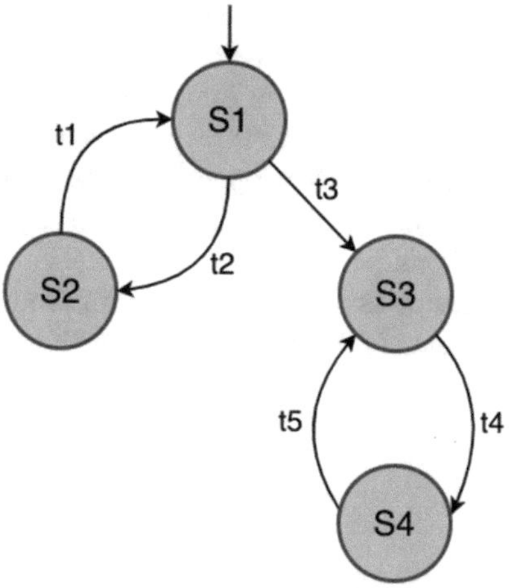

Fig. 8.4 Pair of states with transitions in both directions

transitions in the FSM to have the same Hamming distance, again leading to a longer encoding length.

3. *Self-loop state transition*: Next, consider a self-loop state transition or a transition of the form $T(s, i) = s$. For any encoding E, Hamming distance of the transition $HD(E(s), E(s)) = 0$. This leads to a vulnerability since HD of 0 cannot be achieved for all other transitions that originate out of s to other states. Hence, the requirement (2) above is unenforceable when a state in STG has self-loops as well as other edges. To circumvent this situation, when security is more important than power and area, we replace each state with self-loop with the *functionally equivalent* pair of states [27, 28] as discussed in Chap. 14. This allows flexibility in encoding the resulting pair of states subject to the above constraints.

8.4.2 State Transition Probability

Practical FSMs often consist of cycle and back-to-back transitions discussed in the previous section. These features increase the length of secure encoding, leading to power and area penalties. Hence, we aim to minimize the information leakage through Hamming models by using probabilistic methods to divide states into sets.

Assuming that all primary inputs I are independent of one another and equally likely to occur in every clock cycle, we group together states and transitions which have a high probability of being in the same L-reachability set S_L and T_L. State encoding constraints are generated such that each state (or transition) in a group has the same HW (or HD) as the other states (or transitions) in that group. Since in the absence of information about the FSM structure, an attacker is likely to apply random input data, this heuristic attempts to avoid information leakage in the resulting power trace.

In order to calculate the state transition probability (STP) values, FSMs must be deterministic in nature. Non-deterministic FSMs can have multiple paths or multiple next states for a given state and input pair. They might also have undefined transitions for few inputs to a state, also known as an incomplete FSM. Deterministic FSMs, on the other hand, have exactly one transition defined for every possible state and input pair.

Let us assume that the given FSM has a set of M states and $I_{1,2}$ denotes a subset of input (I) that initiates the transition from state s_1 to state s_2:

$$S = \{s_1, s_2, s_3, \ldots, s_M\}$$

$$I_{1,2} = \{i \in I | (s_1 X i) \rightarrow s_2\}$$

For every state s_i ($i \in [1, M]$), all input sets $I_{i,j}$ ($j \in [1, M]$) must be disjoint in nature. We define predicate $DistinctPath$ for two transitions (s_1, s_2) and (s_1, s_3) originating from the same state s_1:

$$DistinctPath((s_1, s_2), (s_1, s_3)) = I_{1,2} \bigcap I_{1,3} \equiv \emptyset \tag{8.4}$$

Similarly, for every state s_i ($\forall i \in M$), a transition must be defined for every input. Let us define predicate $CompleteFSM$:

$$CompleteFSM(s_1) = \bigcup_{i \in I} I_{1,i} \equiv I \tag{8.5}$$

Both predicates must be satisfied by the given FSM to be determined as a deterministic FSM. The predicates can be checked by a satisfiability solver [29].

For a given FSM, if multiple paths exist for the same state–input pair, it cannot be modified into a deterministic one without changing the functionality of FSM. On the other hand, an incomplete FSM can be transformed into a deterministic FSM. If any transition (or input for a given state) is found missing, the said transition is included as a self-loop transition for that given state of FSM. This completes the FSM and transforms a non-deterministic FSM to a deterministic one without altering its functionality.

STP values of deterministic FSMs can be obtained experimentally or theoretically. We have adapted a theoretical approach to calculate STP, previously proposed in [30]. At any given time step, the FSM must be in one of its states. Therefore, the probability that the FSM is in state s_n can be determined by the following system of equations:

$$P(s_n) = \sum_{m=1}^{M} P(s_m) * P(I_{mn}), \quad n \in \{1, M\} \tag{8.6}$$

This is a set of M linear equations with M unknown state probabilities $P(s_n)$. Out of these M equations corresponding to the M states, any one equation can be derived from the remaining M-1 equations and, hence, is redundant. The final equation required to solve the unknown state probabilities is formulated from the fact that the FSM always has to exist in one of its internal states. Hence,

$$\sum_{m=1}^{M} P(s_m) = 1 \tag{8.7}$$

After solving for state probabilities, state transition probability (STP) for every transition (s_m, s_n) can then be calculated using

$$P((s_m, s_n)) = P(s_m) * P(I_{mn}) \tag{8.8}$$

Example Let us consider the FSM shown in Fig. 8.1 again. Let us assume that the FSM is deterministic in nature, and it is input probabilities are as follows:
$P(I_{1,2}) = P(I_{1,4}) = 1/4$
$P(I_{1,3}) = P(I_{3,5}) = P(I_{3,6}) = 1/2$
$P(I_{2,5}) = P(I_{4,6}) = P(I_{5,1}) = P(I_{6,1}) = 1$
The set of 6 linear equations given by Eqs. 8.6 and 8.7 can be solved to
$P(S_1) = 1/3$, $P(S_2) = 1/12$, $P(S_3) = 1/6$,
$P(S_4) = 1/12$, $P(S_5) = 1/6$, $P(S_6) = 1/6$
Using Eq. 8.8, it leads to state transition probabilities of the FSM:
$P(S_1 \rightarrow S_2) = 1/12$, $P(S_1 \rightarrow S_3) = 1/6$, $P(S_1 \rightarrow S_4) = 1/12$
$P(S_2 \rightarrow S_5) = 1/12$
$P(S_3 \rightarrow S_5) = 1/12$, $P(S_3 \rightarrow S_6) = 1/12$
$P(S_4 \rightarrow S_6) = 1/12$
$P(S_5 \rightarrow S_1) = 1/6$, $P(S_6 \rightarrow S_1) = 1/6$

8.5 Secure Encoding Method

In this section, we describe a heuristic based on state transition probabilities (STP) to partition the states and transitions so as to reduce the encoding length while minimizing the side-channel information leakage. This results in short and secure encodings that are power and area efficient.

8.5.1 Heuristic Algorithm for Secure FSM Encoding

Division of the states of an FSM into the S_L sets depends on the reachability of states in various clock cycles. The aim of this heuristic algorithm is to find the "most probable" clock cycles for reaching states. It can be determined by using the most probable path to reach the given state (s_i) from the start state (s_0). To find the most probable path, we need to follow the path of maximum State Transition Probability.

Consider an STG (FSM) with STPs as *weights* on the edges (transitions). The problem statement then simplifies to finding a path between the start state (s_0) and the given state (s_i) with maximum weights. To solve this problem, a greedy iterative algorithm is used. It is similar to a weighted breadth-first search or a simpler version of Dijkstra's algorithm [31]. It should be noted that unlike Dijkstra's algorithm, weights of the edges are not added to the weights for the next states, since they have already been taken into account during the calculation of the STPs.

The algorithm divides the FSM into a maximum number of sets based on its internal structure. It generates two mappings:

1. SN, which assigns a set number to each state, and
2. ST, which assigns a set number to each transition.

These sets are probabilistic approximations of S_L and T_L, respectively. The algorithm uses a priority-queue or max-heap of state transitions weighted on their STPs. Hence the transition with maximum STP is always on the top of the heap. It should be noted that states with self-loops should either be transformed into a pair of states or those transitions should be ignored before this algorithm is invoked.

8.5.2 Effectiveness of Probabilistic Set Division

After dividing the states and transitions of a given FSM into groups, state encodings are generated following the S_L and T_L requirements discussed previously. The aim of the algorithm is to curb the attacker from uniquely determining the state sequences traversed by their HD or HW footprints or equivalently by the power footprints. To illustrate the effectiveness of the algorithm, consider two MCNC benchmark FSMs, as shown in Fig. 8.5, namely, (i) *DK15*—consisting of 4 states and 12 transitions and (ii) *S8*—consisting of 5 states and 13 transitions.

Let both FSMs be restructured by transforming states with self-loops into a pair of states, as discussed previously. Let us divide both the FSMs into 3 SN sets using Algorithm 8.1. State encodings given in Table 8.2 satisfy the S_L and T_L requirements.

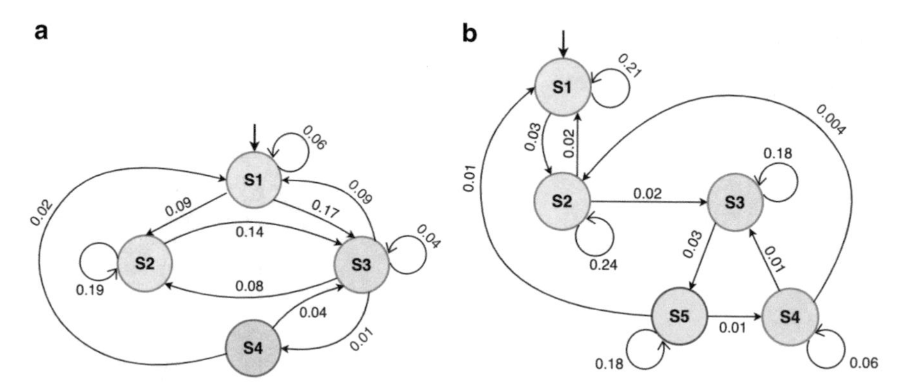

Fig. 8.5 Original DK15 (**a**) and S8 (**b**) MCNC Benchmarks

---

Algorithm 8.1 Partition states and transitions for secure encoding

Input: $STG\ (S, T, s_o)$ and STP values
Output: Mapping $SN : S \rightarrow N$ of states to set numbers
Mapping $ST : T \rightarrow N$ of transitions to set
Procedure:
1: Initialize heap: Max-heap of transitions weighted on STP
2: $SN(s_o) = 0$
3: **for** transition (s_o, s) originating from s_o where $(s_o \neq s)$ **do**
4: push (s_o, s) onto the *heap* with priority $P((s_o, s))$
5: **end for**
6: **while** heap is not empty **do**
7: pop transition (u, v) from the heap
8: **if** $SN(v)$ is unassigned **then**
9: $SN(v) = SN(u) + 1$
10: **for** transition (v,w) from v where $(v \neq w)$ **do**
11: push (v,w) on the heap with priority $P((v,w))$
12: **end for**
13: **end if**
14: **end while**
// ASSIGN SET NUMBERS TO TRANSITIONS
15: **for** transition $t = (u,v)$ in the STG where $(u \neq v)$ **do**
16: $ST(t) = SN(u)$
17: **end for**

If both FSMs are designed using these encodings, then their state registers can assume only two HW values—2 or 4. Similarly, their state transitions can also take only 2 HD values: either 2 or 4. Different HW patterns generated in the state register of restructured DK15 implementation with respect to clock are {2, 2, 4}, {2, 4, 4}, and {2, 2, 4, 4} in varying combinations. This gives the illusion of a generic 4-state FSM as shown in Fig. 8.6. It can assume HWs of (2, 2, 4, 4) and HDs of (2, 4). The HW and HD patterns generated by this generic FSM can be quite similar to the patterns generated by the restructured DK15 benchmark, thereby masking the internal behavior of the target FSM.

Table 8.2 Secure state assignments for restructured DK15 and S8 FSMs

State	Set no.	State encoding
DK15		
S1A	1	01111
S1B	2	00110
S2A	1	11110
S2B	2	10100
S3A	2	01100
S3B	1	11101
S4	3	10001
S8		
S1A	1	01001
S1B	2	11101
S2A	2	11011
S2B	1	11000
S3A	3	01010
S3B	2	10111
S4A	2	01111
S4B	3	10100
S5A	1	00101
S5B	3	10010

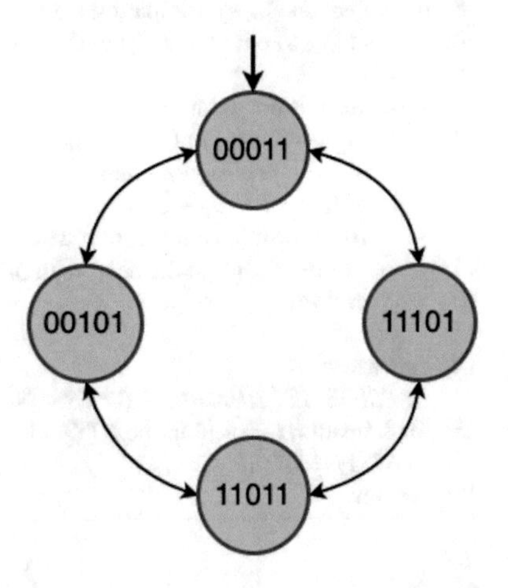

Fig. 8.6 Generic 4-state FSM with state assignment

The HW patterns generated by restructured FSM-S8 implementation are also {2, 2, 4}, {2, 4, 4}, and {2, 2, 4, 4} in varying combinations and can also be represented by the same generic 4-state FSM design. This conceals the internal FSM structure through Hamming models and hence prevents attacker to reverse engineer the FSM.

It should be noted that an FSM with no branches or loops is an interesting special case in terms of reachability. In such an FSM, there is only one state that can be reached in any given clock cycle and expels the ambiguity between states being reached in the same clock cycle. The absence of branches from a current state of

FSM confirms the single next state as the only possible outcome. On the other hand, an FSM with multiple branches and loops contains multiple states that can be reached in the same clock cycle. These are the kinds of FSMs that our solution works best for.

8.6 Encoding for Graded Security

Apart from reducing the encoding size of FSMs, this methodology can also be used to tune the security vs. cost scale. The algorithm provides the designer with an ability to choose the amount of security with respect to cost, which can be in terms of area and/or power. We discuss three techniques to achieve this graded security.

8.6.1 Security vs. Area Trade-Off

Maximum security is ensured when the attack model is unable to distinguish between any two states or any two transitions. This is achieved by encoding all states to have the same HW and all transitions to have the same HD values. That is, all the states are placed in the same S_L set and all transitions are placed in the same T_L set. As the number of set divisions increases, the amount of information leaked about the FSM's structure also increases in general. However, while increasing security, reducing the number of set divisions increases the encoding length. The designer can trade area for security.

Algorithm 8.1 results in a solution with maximum possible set divisions of the FSM to ensure some level of security while containing area and power. Given an STG, let L_{max} denote the number of SN sets produced by Algorithm 8.1. Various values of l, $l \leq L_{max}$, allow the user to trade security for power and area. $l = 1$ yields maximum security by placing all states and transitions in one set. But it also results in maximum power and area.

The algorithm can generate different set divisions (say, for l sets) with a simple modification to the set numbers assigned to states and transitions in Lines 9 and 16, respectively:

```
9:    SN(v)  =  (SN(u)  + 1) mod l
16:   ST(t)  =  SN(u) mod l
```

Forming the states and transitions using the modified algorithm is also likely to generate a cyclic assignment of HW and HD values assumed by the FSM with respect to clock cycles. However, although a perfect cyclic pattern is aimed at, it is not achieved for every FSM due to conflict resolution using STPs.

8.6.2 Security vs. Power Trade-Off

Power consumption in CMOS designs are related to switching activity of gates. A transition with zero Hamming distance implies that the state register did not change its values and none of the CMOS gates switched during that transition. Self-loop transitions in FSMs therefore lead to reduced power consumption. However, near-zero power consumption indicates to the attacker that the machine remained in the current state. In order to save power/area, if we ignore self-loops, then there is a trade-off with security [27]. The states and transitions are divided into the desired number of S_L and T_L sets, as discussed before using Algorithm 8.1.

Secure designs with intact self-loop transitions do not lead to any information leakage against HW power attacks. Against HD and SD power attacks, the proposed compromise might reveal some reachability information in the form of a self-loop transition. However, it does not compromise information regarding other transitions or reachability of rest of the states in any manner. The critical information is secure, and as the results show the consequent impact on security due to the proposed design trade-off is minimal.

8.6.3 Power Reduction with No Security Trade-Off

Dynamic power consumption of a CMOS circuit is proportional to its average switching activity. Switching activity of an FSM state register can be estimated using its state transition probabilities (STPs).

Let us assume that a state encoding E assigns bit vectors of length R to every state. Let $b_n = E(s_n)$ be encoding of state s_n, where b_n^r is the value of rth bit, $r \in [1, R]$. The switching activity (SA_r) of each flip-flop can be calculated using the STP and the toggle density of the flip-flops:

$$SA_r = \sum_{m=1}^{M} \sum_{n=1}^{M} P((s_m, s_n)) * (b_m^r \bigoplus b_n^r) \tag{8.9}$$

Let C be the average capacitance per register bit, f_{CLK} be the frequency of FSM operation, and V_{DD} be the supply voltage. Dynamic power consumption of the FSM can be calculated as

$$P = \frac{1}{2} * V_{DD}^2 * f_{CLK} * C * \sum_{r=1}^{R} SA_r \tag{8.10}$$

The switching activity (SA) of the state register can be estimated as

$$SA = \sum_{r=1}^{R} SA_r = \sum_{m=1}^{M} \sum_{n=1}^{m} (w_{mn}) * HD(E(s_m), E(s_n)) \tag{8.11}$$

where

$$w_{mn} = P((s_m, s_n)) + P((s_n, s_m))$$

$$HD(E(s_m), E(s_n)) = \sum_{r=1}^{R} (b_m^r \bigoplus b_n^r) = HD_{mn}$$

Given an FSM, we are interested in finding state encoding of size $R \in [R_{min}, R_{max}]$ such that its switching activity (SA) is minimal, while maintaining the level of security afforded by the choice of l (number of sets).

In order to obtain a low-power design in addition to security, additional constraints need to be placed on state encodings to minimize the average switching activity of the FSM state register. This can be achieved by assigning lower Hamming distance values to transitions with higher STPs. All transitions in a given set must have the same HD value. If we assign the lowest possible HD value to the set with highest combined STP, then the average SA of the FSM will reduce.

Let the FSM transitions be divided into J number of T_L sets, $T_1, T_2, \ldots T_J$. Since all transitions in every T_L set must have the same HD value, switching activity estimation can be formulated as

$$SA = \sum_{j=1}^{J} (HD_j * \sum_{t \in T_j} P(t)) \tag{8.12}$$

where HD_j is the HD value associated by encodings with T_j and $P(t)$ are the STPs of corresponding transitions in set T_j.

Let us calculate the bounded range of switching activity $[SA_{min}, SA_{max}]$. The lower bound of SA can be achieved with a minimum possible value of HD, which is '0' for a self-loop transition. Though this FSM is of no practical use, theoretically the lower bound of SA is

$$SA_{min} = \sum_{j=1}^{J} (0) * \sum_{t \in T_j} P(t) = 0 \tag{8.13}$$

SA_{max}, the upper bound of SA, on the other hand corresponds to every flip-flop switching during every clock cycle. In other words, the HD of every valid transition is equal to the number of encoding bits R:

$$SA_{max} = \sum_{j=1}^{J} R * \sum_{t \in T_j} P(t) = R * \sum_{m=1}^{M} \sum_{n=1}^{M} P((s_m, s_n)) = R \quad (8.14)$$

These parameters can be used to generate valid encodings that are both secure and have low power consumption.

8.7 State Encoding Generation

8.7.1 Security and Power Constraints

The problem of generating the desired state encodings can be stated in terms of a set of constraints. Algorithm 8.1 divides all the states and transitions of the FSM into an appropriate number of S_L and T_L sets. This is done to achieve a given level of security—both with and without restructuring self-loop transitions. The following set of constraints is applied to find a valid set of state encodings within the range $[R_{min}, R_{max}]$:

1. *Distinct States*: Every state in the FSM must have a unique state encoding.
2. *S_L Constraint*: States with the same S_L set number must have the same Hamming weight.
3. *T_L Constraint*: Transitions with the same T_L set number must have the same Hamming distance.
4. *Switching Activity Constraint*: In order to generate secure and low-power state encodings, this additional constraint to minimize average switching activity is applied. Switching activity (SA) of the state register in the FSM should be as low as possible.

8.7.2 Boolean Approach to Encoding

Given the security and power constraints, the problem of generating the state assignment can be transformed into a Boolean satisfiability (SAT) problem. Let us assume that the given FSM has a set of M states:

$$S = \{s_1, s_2, s_3, \ldots, s_M\}$$

Given the binary encodings $b_m = E(s_m)$ of state s_m of bit-length R, in order to generate secure encodings, all the above defined constraints must be satisfied.

Let us define a predicate $DistinctStates$ for two states s_1 and s_2 being distinct—by defining that Hamming distance between them ($HD(s_1, s_2)$) must be a positive integer:

$$DistinctStates(s_1, s_2) = \sum_{r=1}^{R}(b_1^r \bigoplus b_2^r) \geq 1 \tag{8.15}$$

States s_1 and s_2 of bit-length R being distinct can be defined by predicate $DistinctStates$ in Eq. 8.15.

Let the division of "M" states into "J" S_L sets and transitions into "K" T_L sets be obtained using Algorithm 8.1. Let us define a predicate $EqualHW$ for two states s_1 and s_2 having the same Hamming weight:

$$EqualHW(s_1, s_2) = \sum_{r=1}^{R}(b_1^r) \equiv \sum_{r=1}^{R}(b_2^r) \tag{8.16}$$

For any S_j, all $s_m \in S_j$ ($\forall j \in [1, J]$) should have equal Hamming weight, i.e., every pair in S_j set must satisfy predicate $EqualHW$.

Similarly, let us define predicate $EqualHD$ for two transitions ($s_1 \rightarrow s_2$) and ($s_3 \rightarrow s_4$) with equal Hamming distance:

$$EqualHD((s_1, s_2), (s_3, s_4)) = \sum_{r=1}^{R}(b_1^r \bigoplus b_2^r) \equiv \sum_{r=1}^{R}(b_3^r \bigoplus b_4^r) \tag{8.17}$$

All transitions in T_k set must have equal Hamming distance, i.e., every pair of transitions in T_k ($\forall k \in [1, K]$) must satisfy $EqualHD$.

Next, let us define predicate *CorrectSA* to obtain minimum switching activity in the FSM transitions. SA is defined using Eq. 8.12, and our aim is to minimize it:

$$CorrectSA(SA) = \sum_{k=1}^{l} HD(T_k) \times W_k \equiv SA \tag{8.18}$$

where W_k is the sum of w_{mn}'s of all transitions in T_k.

For a valid solution, we need to find state encodings for a given R-bit encoding and a given SA value, such that it satisfies all the above defined predicates.

8.7.3 SMT Solver for Boolean Constraint Solving

Satisfiability Modulo Theories (SMT) solvers [29] are built upon technologies in two areas: Boolean satisfiability solving and first-order theorem proving. They can be used to solve various problems involving Boolean, integer, and real-valued variables. Z3 is a high performance Satisfiability Modulo Theories solver, which can be used to check the satisfiability of logical formulae over one or more theories. It allows the user to solve first-order formulae by defining them as a set of constraints and has the ability to solve models comprising bit-vector variables and a set of constraints that can manipulate and calculate across those variables. This allows it to solve FSM encoding problems, which makes it beneficial for our purposes.

Boolean constraints can be applied to the SMT solver to find a valid set of state encodings within the range $[R_{min}, R_{max}]$. (1) Restructured secure FSMs transform their self-loop transitions and apply only three constraints: *Distinct States, Equal HW, and Equal HD*, (2) secure encodings for the original FSMs also apply the above three constraints without restructuring their self-loop transitions, and (3) the fourth constraint *"Switching Activity constraint"* is applied to generate secure and low-power encodings with minimized switching activity. For each encoding size R, the SMT solver finds an encoding (if it exists) to yield the minimum possible switching activity.

8.8 Experimental Analysis

All experiments were performed using over 100 BenGen [32] and MCNC [33] benchmarks. All FSMs were tested for their deterministic nature. Few MCNC benchmark FSMs were found to be incomplete and were modified to deterministic FSMs without altering their functionality.

Each FSM was encoded with different encodings, minimal binary, and secure encodings with different levels of security, using restructured method and the two low-power methodologies discussed. These encodings were then converted to Verilog, and the gate-level synthesis in 90-nm CMOS technology was performed using the Synopsys DC Compiler. These gate-level netlists were then converted to Spice using a Verilog-to-Spice converter to perform power simulation. A set of 1000 random input vectors were generated using a stimuli generator, which were then provided to every implementation of every FSM. HSPICE simulations were performed using Nanosim to obtain power traces. Different state encodings of the same FSM result in different implementations and hence have different area and power results.

Best-case attack model data is generated using the same input stimuli, for all three Hamming models. Statistical analysis is then performed to compare different FSM implementations against this attack model data. Hamming weight, Hamming distance, and switching distance data were calculated for every implementation of the FSM for the 1000-vector input stimuli. Mutual Information (MI) between these best-case attack model data and the power traces was calculated using a Perl script. Figure 8.7 presents the entire experimental flow for generating the area, power, and security results. Figure 8.8 shows sizes of the benchmark FSMs in terms of the number of states and transitions. The number of input bits ranged from 1 to 16. We report results for (1) the restructured FSMs with transformed self-loops, (2) the original FSMs without restructuring self-loops, and (3) low-power design for the original FSMs [34].

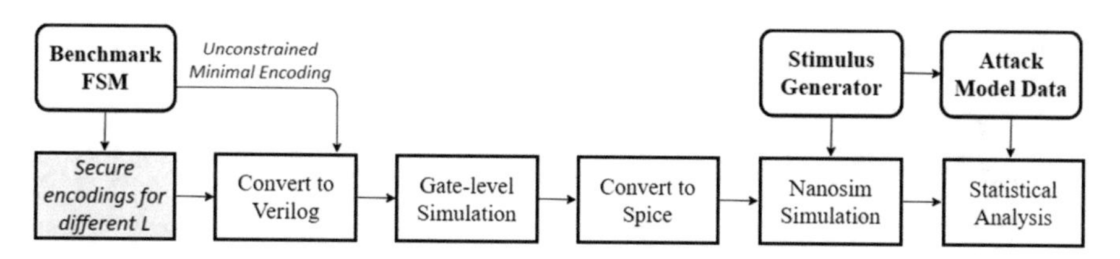

Fig. 8.7 Flowchart used in experiments

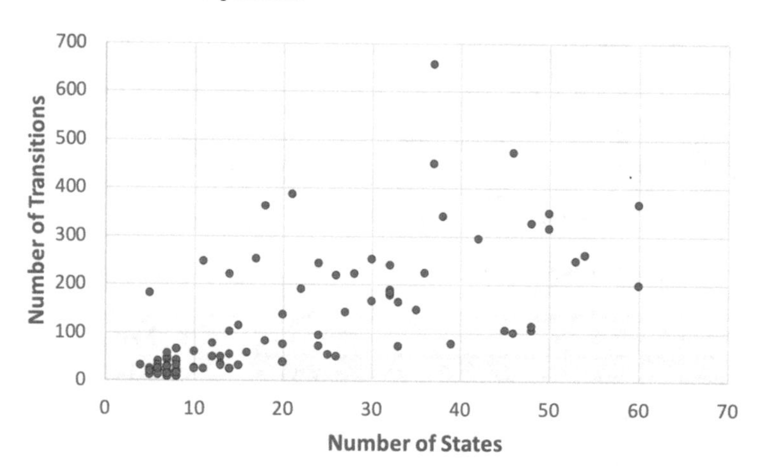

Fig. 8.8 Size of benchmark FSMs

8.8.1 *Security Analysis*

Algorithm 8.1 provides a way to increase security by adjusting the parameter L. Decreasing L increases the encoding length and improves security. We use Mutual Information (MI) (between the power traces and the HW, HD, or SD model) as a distinguisher to measure security ($MI \geq 0$) against all three forms of attacks [35].

Figure 8.9 shows reduction in information leakage (in terms of MI) against Hamming weight attack model as security increases. Since encoding length requirement increases for higher security, the plot demonstrates the basic trade-off between security and area. Similarly, Figs. 8.10 and 8.11 show the improvement in security against Hamming distance and switching distance attacks. In FSMs without restructuring (i.e., original FSMs), perfect security ("zero" MI) can only be achieved against HW attacks due to the presence of self-loops in FSMs. This is because HD and SD attacks can reveal self-loops within the FSMs, no matter how the FSM is implemented. On the other hand, perfect security can be achieved against all three attack models by the restructured FSMs, by achieving "zero" MI for implementations with lower L values. The low-power design for original FSMs shows no further reduction in security and results in similar results as original FSMs.

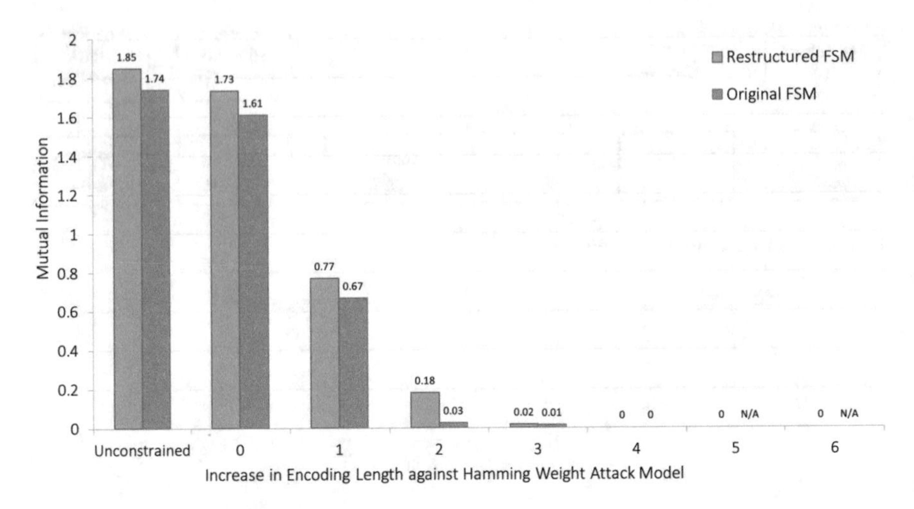

Fig. 8.9 Security against hamming weight attack model

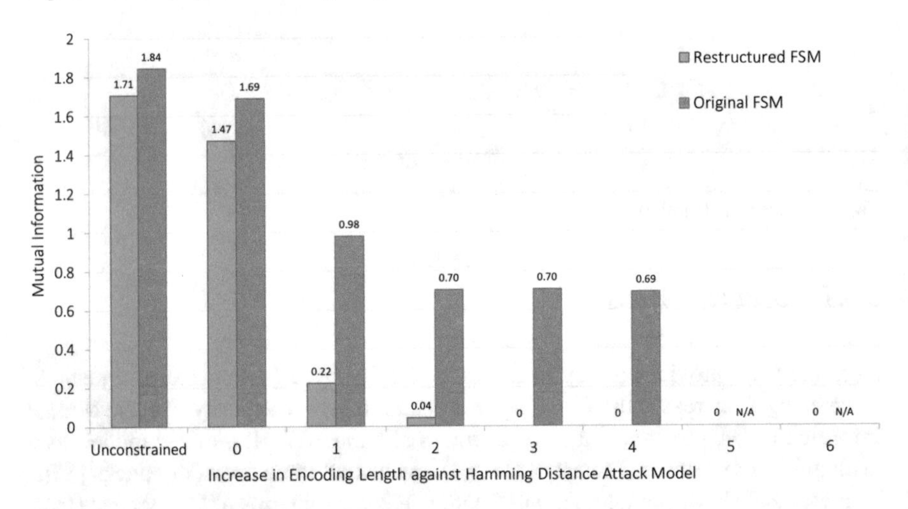

Fig. 8.10 Security against hamming distance attack model

Figure 8.12 shows the increase in the encoding bit-length (with respect to unconstrained minimal length encoding) for maximum possible security, where all states and transitions have the same Hamming weight, Hamming distance, and switching distance for restructured, original, and original low-power FSMs. Difference in encoding length (R) requirement can be seen when loops are restructured. On average, restructuring the FSMs increases the number of states by 60% and transitions by 158%. This results in an average increase of encoding length by 40–70% depending on the level of security (L) chosen, whereas the original FSMs only increase encoding length by 15–40%.

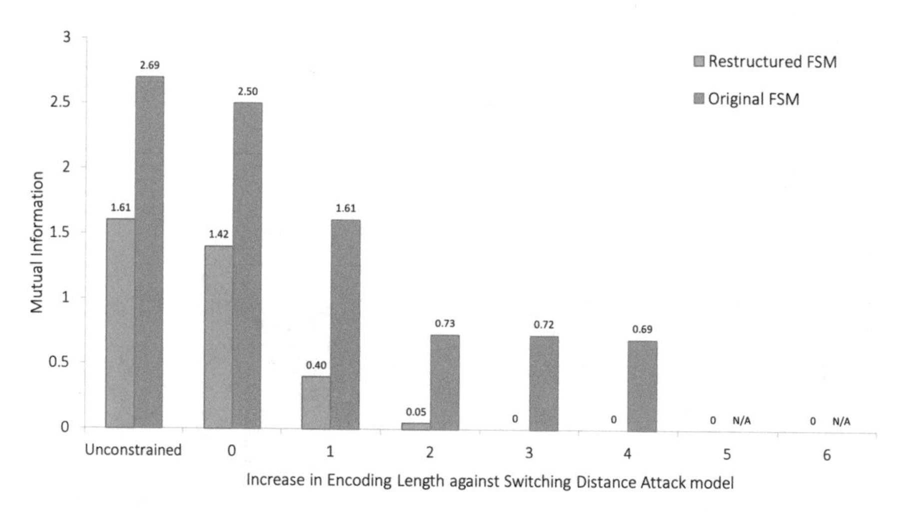

Fig. 8.11 Security against switching distance attack model

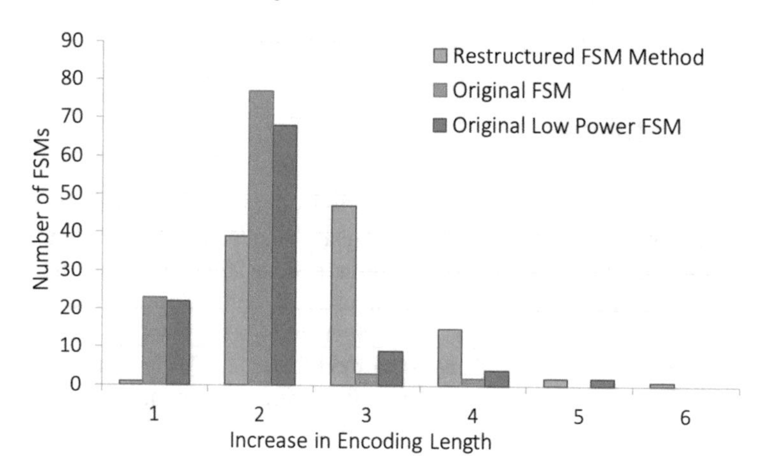

Fig. 8.12 Encoding length for maximum security

Encodings for different bit-lengths are obtained with different L values, ranging between 1 and L_{max} for every FSM (L_{max} being characteristic to every FSM). Figure 8.13 shows the increase in the encoding length from their respective minimal encoding lengths, for different L values. For example, for *dk17*, as L value is reduced from $L_{max} = 6$ to 1, encoding length increased from 1 to 4 bits. It should be noted that, while each L value results in a different state encoding, the length of said encoding may or may not vary with every L. But the trend to trade encoding length for security using L as the control "knob" can be seen for each benchmark.

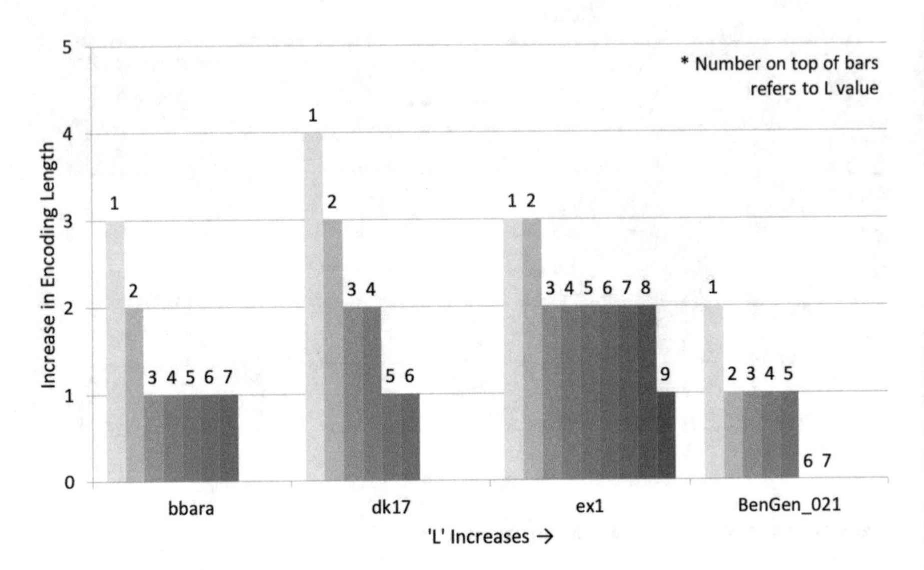

Fig. 8.13 Encoding length for graded security for 4 MCNC benchmarks

8.8.2 *Area Analysis*

Areas of the synthesized FSMs for different encodings depend on their encoding lengths. Increase in security leads to an increase in area, which exhibits the trade-off capability of the proposed method. Figure 8.14 shows the normalized area increase for maximum security for both original and restructured FSMs with respect to the unconstrained minimal length encodings for the original FSMs. On average, for maximum security in restructured FSMs (i.e., $MI(power, HW) = 0$, $MI(power, HD) = 0$, and $MI(power, SD) = 0$), the area increases by a factor of 2.04. But it can be reduced to as low as factor of 1, depending on the desired security level controlled by L.

In the original FSMs, for maximum security (i.e., $MI(power, HW) = 0$), the area increases only by a factor of 1.37. Though, additional low-power switching activity constraint introduces a slight increase in the encoding bit-requirement. 30% of benchmark FSMs observed no increment, while the rest observed a 5% increase on average. The average layout area requirement increased by 4% due to these constraints.

This increase in area is quite low in comparison to existing low-level hardware techniques. For example, MDPL technique [25] increases area requirement by a factor of 4.5 and reduces the speed by half. Whereas WDDL technique [24] increases area by a factor between 3.2 and 3.6, also reducing the speed by half. The proposed method in comparison shows significant area improvement without any penalty in speed.

Figure 8.15 shows the increase in area requirement as the security increase for few benchmarks, illustrating the trade-off capability of the proposed method.

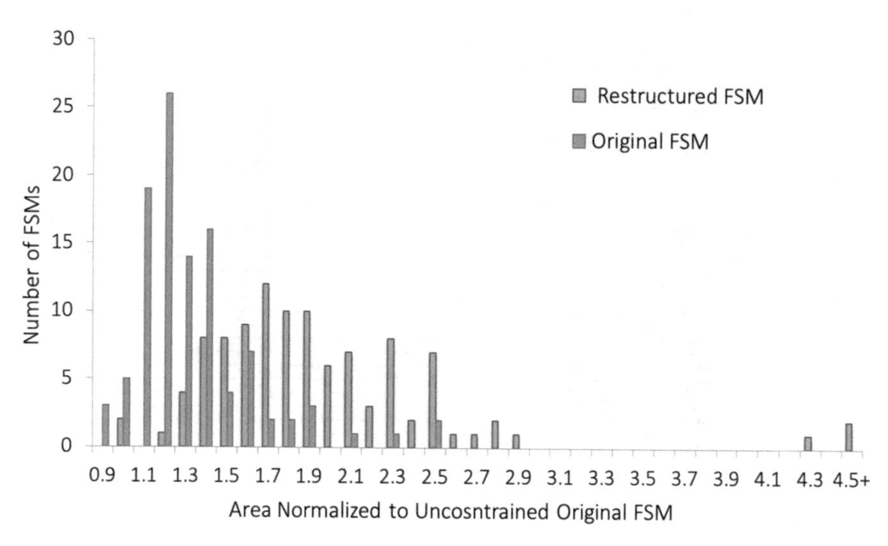

Fig. 8.14 Normalized area for maximum security

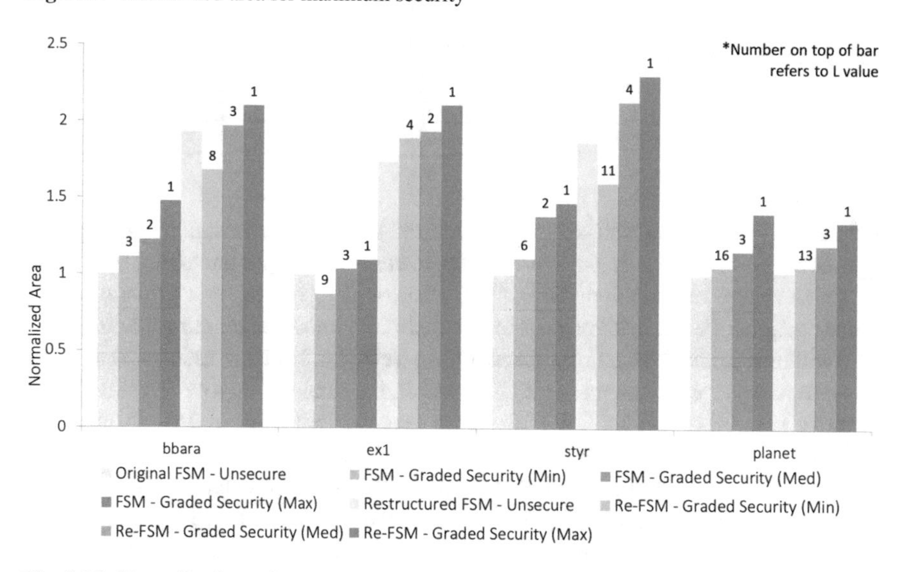

Fig. 8.15 Normalized area for graded security

8.8.3 Power Analysis

The average power consumption of FSMs is determined using NanoSim simulations in our experiments. Low-power techniques discussed in Sects. 8.6.2 and 8.6.3 implemented along with security measures observed a reduction in average power consumption. Figure 8.16 shows the average normalized power consumption with respect to unconstrained minimal length encodings. Due to increase in encoding

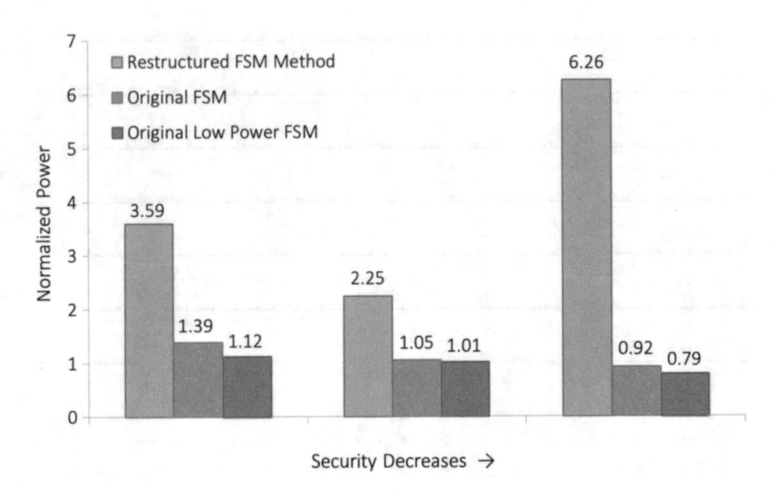

Fig. 8.16 Normalized power for graded security

Table 8.3 Power comparison of different encoding strategies as L Increases

L-value	Original low power vs. original	Original low power vs. minimal
one	−16.29%	+3.71%
L_{median}	−15.50%	+7.84%
L_{max}	−18.51%	−7.97%

length and synthesized area, secure restructured FSMs consume much higher power than secure original FSMs. On average, the power consumption increased by a factor of 3.4 in secure restructured FSMs [36], whereas it increased by a factor of 1.6 in secure original FSMs, with respect to unconstrained minimal length encodings.

Switching activity constraint discussed in Sect. 8.6.3 further reduced the power consumption for every benchmark FSM within the range of 0–40%. For maximum security (i.e., $MI(power, HW) = 0$), an average reduction of 15% is observed in benchmark FSMs. Table 8.3 shows additional power reduction due to switching activity constraints applied to secure original FSMs, as L increases. In over 30% benchmark FSMs, this technique observes power reduction with respect to unsecure unconstrained minimal FSMs. It should be noted that these low-power techniques do not result in any security trade-off and have no substantial increase in synthesized area.

Figure 8.17 compares power profiles for various security levels in a single *STYR* MCNC benchmark FSM (consisting of 30 states and 93 transitions). As can be seen, restructured secure designs have much higher power consumption than that of the original secure designs. Power consumption further decreases when additional low power constraints are applied, even though encoding length increases quite a bit for maximum security (i.e., $MI(power, HW) = 0$).

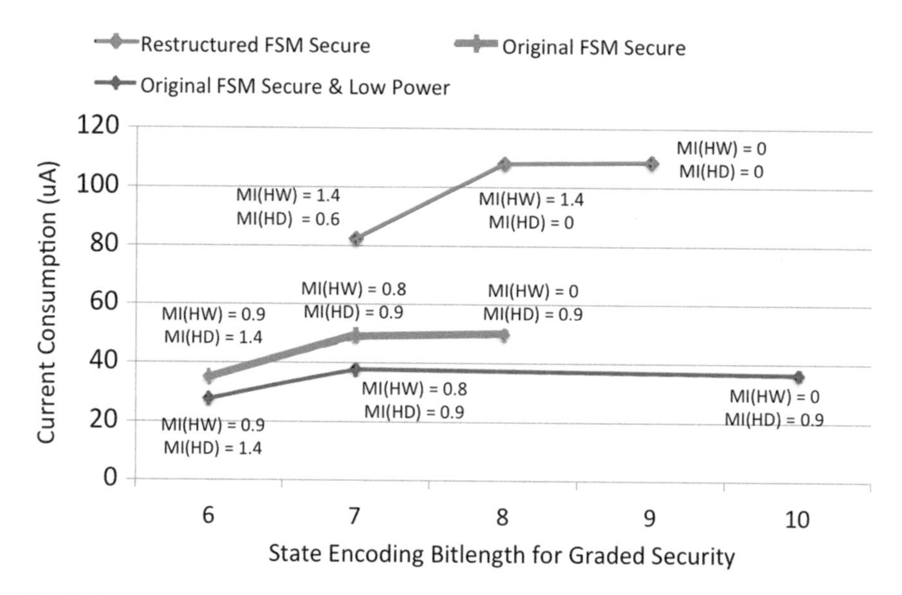

Fig. 8.17 Power vs encoding length of *STYR* FSM

8.9 Conclusion

This chapter discussed design methodology with user-controlled security metric to defend against FSM reverse engineering using power measurements. The methodology uses constrained state assignment to generate secure encodings while ensuring low power consumption. Experimental results show that, for restructured FSMs, security against HW models ranged between 34–100%, 62–100% against HD model, and 56–100% against SD model. Encoding length of typical benchmarks increased in the range of 40–73% depending on the security level. By restructuring the FSMs, MI of zero could be obtained against all three SD, HD, and HW attack models.

Using constrained state encodings, two types of low-power improvements, with and without security trade-off, could be achieved. For the original FSMs in our experimental set, the security ranged between 18 and 100% against HW model, between 30 and 58% against the HD model, and between 27 and 60% against the SD model. Increase in encoding length for a typical benchmark was in the range of 15–55% depending on security level and power optimization. Power consumption was reduced by more than 50% for all FSMs with respect to restructured FSMs, and an additional low-power constraint introduces further reduction in power ranging 4–20% depending on the security level.

The proposed method provides protection only against Hamming model-based power attacks. These attacks, when combined with cryptanalysis attacks, can prove to be very powerful. Cryptanalysis attacks, which we have neglected in this work, use input–output values of the FSMs to determine the functionality of the FSM.

Exploring encodings to prevent combined power and cryptanalysis attacks is a possible direction for future research.

Acknowledgments The authors would like to thank Prof. Mike Borowczak, University of Wyoming for his prior work in this research area and generous input for this work. We would also like to thank Mr. Pravesh Agrawal, Microsoft, for the many helpful discussions and constructive feedback.

References

1. Dofe, J., Frey, J., Yu, Q.: Hardware security assurance in emerging IoT applications. In: Proceedings of 2016 IEEE International Symposium on Circuits and Systems (ISCAS), pp. 2050–2053. IEEE, Piscataway (2016)
2. Aljazeera, K.R., Nandakumar, R., Ershad, S.B.: Design and characterization of L-block crypto-core. In: 2016 International Conference on Proceedings of Signal Processing, Communication, Power and Embedded System (SCOPES), pp. 166–172. IEEE, Piscataway (2016)
3. Bahnasawi, M.A., Ibrahim, K., Mohamed, A., Mohamed, M.K., Moustafa, A., Abdelmonem, K., Ismail, Y., Mostafa, H.: ASIC-oriented comparative review of hardware security algorithms for internet of things applications. In: Proceedings of 2016 28th International Conference on Microelectronics (ICM), pp. 285–288. IEEE, Piscataway (2016)
4. Masalskis, G., et al.: Reverse engineering of CMOS integrated circuits. Elektronika ir elektrotechnika **88**(8), 25–28 (2008)
5. Ferrigno, J., Hlaváč, M.: When AES blinks: introducing optical side channel. IET Inf. Secur. **2**(3), 94–98 (2008)
6. Chikofsky, E.J., Cross, J.H.: Reverse engineering and design recovery: a taxonomy. IEEE Softw. **7**(1), 13–17 (1990)
7. Torrance, R., James, D.: The state-of-the-art in semiconductor reverse engineering. In: Proceedings of 2011 48th ACM/EDAC/IEEE Design Automation Conference (DAC), pp. 333–338. IEEE, Piscataway (2011)
8. Brutscheck, M., Schmidt, B., Franke, M., Schwarzbacher, A.T., Becker, S.: Identification of deterministic sequential finite state machines in unknown CMOS ICs. In: IET Irish Signals and Systems Conference (2009)
9. Uting, S., Brutscheck, M., Schwartzbacher, A., Becker, S.: FPGA based optimisation and implementation of nondestructive identification procedures. In: Proceedings of International Solid State Circuits Conference (2011)
10. Zhou, Y.B., Feng, D.G.: Side-channel attacks: ten years after its publication and the impacts on cryptographic module security testing. IACR Cryptology ePrint Archive, 2005 p. 388 (2005)
11. Tria, A., Choukri, H.: Invasive Attacks. In: van Tilborg, H.C., Jajodia, S. (eds.) Encyclopedia of Cryptography and Security, pp. 623–629. Springer, Boston (2011)
12. Verbauwhede, I.: Secure Integrated Circuits and Systems. Springer, Berlin (2010)
13. Renauld, M., Standaert, F.X.: Algebraic side-channel attacks. Inscrypt **6151**, 393–410 (2009)
14. Ambrose, J.A., Ragel, R.G., Jayasinghe, D., Li, T., Parameswaran, S.: Side channel attacks in embedded systems: a tale of hostilities and deterrence. In: Proceedings of Sixteenth International Symposium on Quality Electronic Design, pp. 452–459. IEEE, Piscataway (2015)
15. Akkar, M.L., Bevan, R., Dischamp, P., Moyart, D.: Power analysis, what is now possible.... In: Proceedings of International Conference on the Theory and Application of Cryptology and Information Security, pp. 489–502. Springer, Berlin (2000)
16. Gebotys, C.H., Gebotys, R.J.: Secure elliptic curve implementations: an analysis of resistance to power-attacks in a DSP processor. In: Proceedings of International Workshop on Cryptographic Hardware and Embedded Systems, pp. 114–128. Springer, Berlin (2002)

17. Mangard, S., Oswald, E., Popp, T.: Power Analysis Attacks: Revealing the Secrets of Smart Cards. Springer Science & Business Media, New York (2008)
18. Kocher, P., Jaffe, J., Jun, B.: Differential power analysis. In: Proceedings of Advances in Cryptology CRYPTO'99, pp. 789–789. Springer, Berlin (1999)
19. Smith, J., Oler, K., Miller, C., Manz, D.: Reverse engineering integrated circuits using finite state machine analysis. In: Proceedings of 50th Hawaii International Conference on System Sciences, pp. 2906–2914 (2017)
20. Potkonjak, M., Nahapetian, A., Nelson, M., Massey, T.: Hardware trojan horse detection using gate-level characterization. In: 46th ACM Proceedings of Design Automation Conference, 2009, pp. 688–693. IEEE, Piscataway (2009)
21. Gandolfi, K., Mourtel, C., Olivier, F.: Electromagnetic analysis: concrete results. In: Proceedings of International Workshop on Cryptographic Hardware and Embedded Systems, pp. 251–261. Springer, Berlin (2001)
22. Vamja, H., Agrawal, R., Vemuri, R.: Non-invasive reverse engineering of finite state machines using power analysis and boolean satisfiability. In: Proceedings of 2019 IEEE 62nd International Midwest Symposium on Circuits and Systems (MWSCAS), pp. 452–455. IEEE, Piscataway (2019)
23. Yuan, L., Qu, G.: Information hiding in finite state machine. In: Proceedings of International Workshop on Information Hiding, pp. 340–354. Springer, Berlin (2004)
24. Tiri, K., Verbauwhede, I.: A logic level design methodology for a secure DPA resistant ASIC or FPGA implementation. In: Proceedings of the Conference on Design, Automation and Test in Europe-Volume 1, p. 10246. IEEE Computer Society, Washington (2004)
25. Popp, T., Mangard, S.: Masked dual-rail pre-charge logic: DPA-resistance without routing constraints. In: Proceedings of International Workshop on Cryptographic Hardware and Embedded Systems, pp. 172–186. Springer, Berlin (2005)
26. Peeters, E., Standaert, F.X., Quisquater, J.J.: Power and electromagnetic analysis: improved model, consequences and comparisons. Integration **40**(1), 52–60 (2007)
27. Borowczak, M., Vemuri, R.: S* FSM: a paradigm shift for attack resistant FSM designs and encodings. In: Proceedings of ASE/IEEE International Conference on BioMedical Computing, pp. 96–100. IEEE, Piscataway (2012)
28. Borowczak, M., Vemuri, R.: Enabling side channel secure FSMs in the presence of low power requirements. In: Proceedings of 2014 IEEE Computer Society Annual Symposium on VLSI, pp. 232–235. IEEE, Piscataway (2014)
29. De Moura, L., Bjørner, N.: Z3: an efficient SMT solver. In: Proceedings of International Conference on Tools and Algorithms for the Construction and Analysis of Systems, pp. 337–340. Springer, Berlin (2008)
30. Tsui, C.Y., Monteiro, J., Pedram, M., Devadas, S., Despain, A.M., Lin, B.: Power estimation methods for sequential logic circuits. IEEE Trans. Very Large Scale Integr. Syst. **3**(3), 404–416 (1995)
31. Dijkstra, E.W.: A note on two problems in connexion with graphs. Numer. Math. **1**(1), 269–271 (1959)
32. Jozwiak, L., Gawlowski, D., Slusarczyk, A.: An effective solution of benchmarking problem: FSM benchmark generator and its application to analysis of state assignment methods. In: Proceedings of Euromicro Symposium on Digital System Design, pp. 160–167. IEEE, Piscataway (2004)
33. Yang, S.: Logic synthesis and optimization benchmarks user guide: version 3.0. Microelectronics Center of North Carolina (MCNC) (1991)
34. Agrawal, R., Vemuri, R., Borowczak, M.: A state machine encoding methodology against power analysis attacks. J. Electron. Test. **35**(5), 621–639 (2019)
35. Gierlichs, B., Batina, L., Tuyls, P., Preneel, B.: Mutual information analysis. In: Cryptographic Hardware and Embedded Systems–Cryptographic Hardware and Embedded Systems, pp. 426–442 (2008)
36. Agrawal, R., Vemuri, R.: On state encoding against power analysis attacks for finite state controllers. In: 2018 IEEE International Symposium on Hardware Oriented Security and Trust (HOST), pp. 181–186. IEEE, Piscataway (2018)

Chapter 9
State Encoding Based Watermarking of Sequential Circuits Using Hybridized Darwinian Genetic Algorithm

Matthew Lewandowski and Srinivas Katkoori

9.1 Introduction

As circuit technology dives deeper into nanometer (nm) technology nodes, the associated costs of fabrication become astronomical. To conceptualize the amount of money required consider Taiwan Semiconductor (TSMC) as they recently announced the planning of a new facility for only the 5 nm fabrication process that has an expected eight year completion date (2029) and cost of $12 Billion [1]. The result of such costs has led even the most well recognized companies, such as Intel and AMD, to employ a "semi-fabless" or "fabless" business model; one wherein some or all of their designs are fabricated through vendors like TSMC instead of building their own facilities.

While this fabless offshore manufacturing has proved fruitful for China, as their semiconductor industry grew from $178M to $5.4B [2], it has otherwise proved detrimental to not only the trust of foundry [3, 4] but also circuits being exported from China. Specifically, the U.S. Armed Services Committee estimated in 2011 that counterfeit Integrated Circuits (ICs) from China were causing a revenue loss of $7.5B per year [5]. In addition to this, defense contractor Raytheon Technologies reported that counterfeit ICs had made their way into the Electromagnetic Interference (EMI) filters for Forward Looking Infrared (FLIR) utilized by the Hellfire missile [5]. To emphasize the significance and impact that a single counterfeit component in the supply chain could have been, the 2020 U.S. Department of Defense (DoD) budget requested just for Hellfire missiles alone [6] was roughly 13% of the $5716.3M budget requested solely for munitions.

M. Lewandowski (✉) · S. Katkoori
University of South Florida, Tampa, FL, USA
e-mail: mlewando@usf.edu; katkoori@usf.edu

S. Katkoori, S. A. Islam (eds.), *Behavioral Synthesis for Hardware Security*,
https://doi.org/10.1007/978-3-030-78841-4_9

However, another important take away from this is that it is a common practice for companies to utilize third-party vendors/Intellectual Property (IP) cores and/or "off-the-shelf" components when developing contracted systems; as it would not be feasible/practical to design and manufacture every aspect of the system in-house. Thus, it is important for methodologies to exist that allow for the incorporation of trust and security in the event of litigation, or counterfeiting, for both hardware as a whole and IP core designers. Conversely, counterfeiting ICs is not simply making a "knock-off" replacement like many would consider, rather, it involves many different [7]: (1) recycled, (2) remarked, (3) overproduced, (4) defective, (5) cloned, (6) forged documentation, and (7) tampered.

First, a *recycled* IC simply denotes one that was physically de-soldered from used/dead equipment and marketed as new. Second, *remarked* is when the surface of an IC, that displays manufacturer and original part or lot number, is in itself resurfaced to remove the original markings in order to put new product information and misrepresent the actual IC. Third, *overproduced* ICs are those which were produced after meeting the contractual amount of ICs for a purchasing party, i.e., if we were to contract some fabrication facility to produce 20,000 ICs and they ultimately fabricate 30,000 and sell the extra 10,000 these are considered as counterfeit. *Defective* ICs are those which have failed post-fabrication quality assurance testing and are intentionally sold as "functional" units to consumers or vendors. *Cloned* ICs are those wherein the functionality of some original IC is duplicated, this could be via means of reverse engineering or theft of Hardware Description Language (HDL) level code used for synthesizing functional layouts. *Forged Documentation*, intuitively, is simply falsifying potential certifications or compliance forms for an IC, for example, certain equipment may need to be certified in radiation hardening for use in either space or medical applications, if documents are falsified then catastrophic failures may occur. Lastly, *tampered* simply denotes an IC that has additions or modifications to the logic and circuits within the chip, for example, the addition of an antenna into some IC design will normally go undetected but could have been inserted in-order to leak important information otherwise intended as secret, perhaps, security codes.

To combat several specific instances pertaining to IP theft/counterfeiting (recycling, cloning, overproduction, remarked), we present a watermarking method employed prior to the synthesis process of sequential circuits that takes advantage of the state assignment process to integrate a reproducible watermark signature into the state codes assigned; such that, the post-synthesis physical design will allow for recovery/extraction of the watermark signature that was embedded into the state codes during synthesis of the sequential circuit. Thus, this method is intended to enable IP designers or manufacturers to potentially embed watermarks specific to the revisions or models produced, while also enabling ownership verification during litigation in the event suspicion of overproduction or cloning arises.

In the remainder of this chapter, we will provide the background on related methods from literature, detail the method we present at length, examine the testing setup and results when compared against the state-of-the-art, and finally perform an in-depth analysis on these results while accounting for synthesis overhead and anomalies and identifying their root causes.

9.2 Background

When considering the methods for watermarking Finite State Machines (FSMs), or sequential circuits, there are currently four genres of works: (1) *State* [8–11], (2) *Input/Output* (I/O) [12], (3) *Edge* [13–17], and (4) *Encoding* [18–21]. For clarity, we note that works [18–21] directly relate and led-up-to the work centric to state encoding based watermarking presented in this chapter.

9.2.1 State Based Watermarking

This is the simplest form of watermarking [8–11] and involves the addition of an entirely extra FSM that is gated by a secret key, such that, after this private sequence is input then an FSM of otherwise entirely unexpected behavior is exhibited (to an end user). Thus, only an IP core designer would know the secret key required to put the system in a state of "verification" based behavior and allow the ability to prove ownership in litigation, as these secret key values would not be inputs expected for given states during normal operation. To illustrate this we provide a simple example in Fig. 9.1 and note that the I/O of edges or states is arbitrary and omitted. However, the important information that should be gathered from Fig. 9.1 is that the consumer intended behavior is denoted by *S0-S4* while the secret key sequence known to the IP core designer and the hidden functionality is denoted by states *K0-K1* and *W0-W4*, respectively. For further clarity, the only edges expected during normal operation for $S4$ in Fig. 9.1 are: $S4 \rightarrow S3$, $S4 \rightarrow S1$, and $S2 \rightarrow S4$, with edges between $\{K0, K1\}$ denoting hidden functionality.

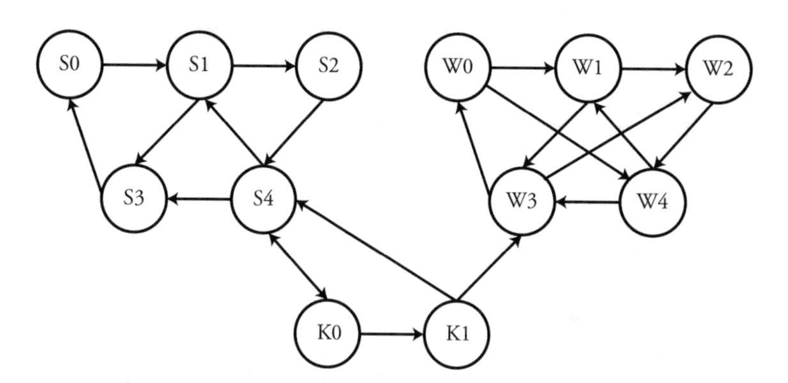

Fig. 9.1 State based watermarking example (S = State, K = Key, W = Watermark)

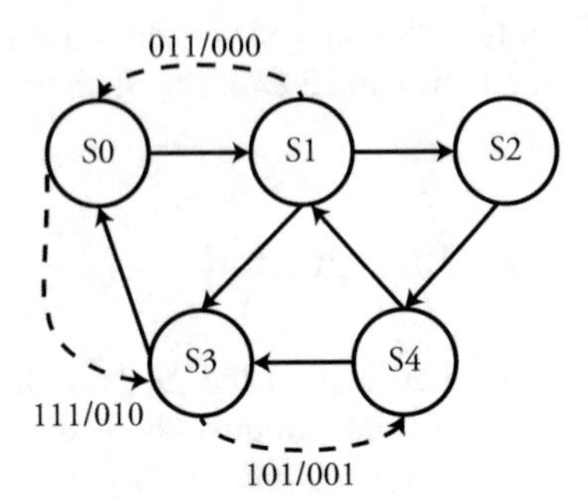

Fig. 9.2 I/O based watermarking example (Dashed = Watermark, Solid = Original FSM)

9.2.2 I/O Based Watermarking

This method [12] takes advantage of the unspecified behavior of Incompletely Specified FSMs (ICFSMs) by adding a series of edges that will reproduce the watermark via I/O, in that, the subset of input and corresponding output bits reconstructs the watermark sequence. To illustrate this method we provide an example using the watermark sequence of "011_000_111_010_101_001" in Fig. 9.2. For clarity, if an FSM is not an ICFSM but a Completely Specified FSM (CSFSM), or an FSM which all possible input/output/state combinations have been explicitly declared, the addition of input bits is required to expand the FSM size and create otherwise unspecified behavior that can be used for inserting the watermark.

9.2.3 Edge Based Watermarking

While this type of watermarking [13–17] is similar to I/O based watermarking, the significant difference that differentiates the two methods is the use of coinciding edges, allowing for a potentially seamless integration of the watermark. However, just as seen in the I/O method, when the I/O of an FSM is completely specified and does not, or cannot, fully accommodate the intended watermark then this method will similarly increase the number of input bits to achieve integration. Considering the same watermark sequence used in Fig. 9.2, the potential difference of this method and target methodology is shown by Fig. 9.3.

9.2.4 State Encoding Based Watermarking

The last method of watermarking [18–21] is state encoding based, which utilizes the state code values and edge/bit addition for watermark accommodation. Unlike the integration schemes seen in *State*, *I/O*, and *Edge*, this method integrates

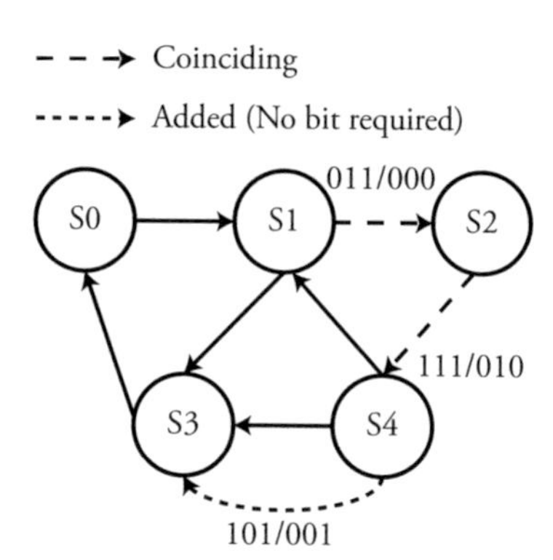

Fig. 9.3 Edge based watermarking example

the watermark solely at the state register. While the addition of edges and bits may be necessary for inducing subgraph isomorphism between a watermark and FSM, they do not directly alter or impact the FSM output under any operating condition; in that, similar to all previously discussed methodologies the edges added are never intended to be traversed unless the watermark is being reproduced; the exception to this is coinciding edges used for watermark integration. To better illustrate how this method works, we consider a sequence of: "0000_0011_1100_1010_1111_1001_0011" (0_3_C_A_F_9_3). From this watermark sequence, we first construct a graph (Fig. 9.4a) that will be mapped to states in the original FSM (Fig. 9.4b). For simplicity, if we assume that the mapping solution found is just a one-to-one mapping, $\{W0 \ldots W5\} \rightarrow \{S0 \ldots S5\}$. The resultant watermarked FSM can be illustrated as shown in Fig. 9.4c. For clarity, the process is normally performed via subgraph matching and isomorphic induction for producing mapping solutions. While the I/O of the FSM edges in Fig. 9.4c are arbitrary, the manner in which the edge from $9 \rightarrow 3$ will be added is not, and this method operates by specifically giving the first available unspecified input combination with an unaltered output value; in the event an FSM is a CSFSM and an edge must be added, then the number of input bits will be increased by one and the first available unspecified input combination will be used.

9.3 Methodology

Having briefly covered the existing watermarking methods, here we present the exact method employed in this work for state encoding based watermarking and provide an in-depth detailing of all components for the framework. We begin by providing an overview of the method in Fig. 9.5 that shows the components and processes to be detailed throughout this section.

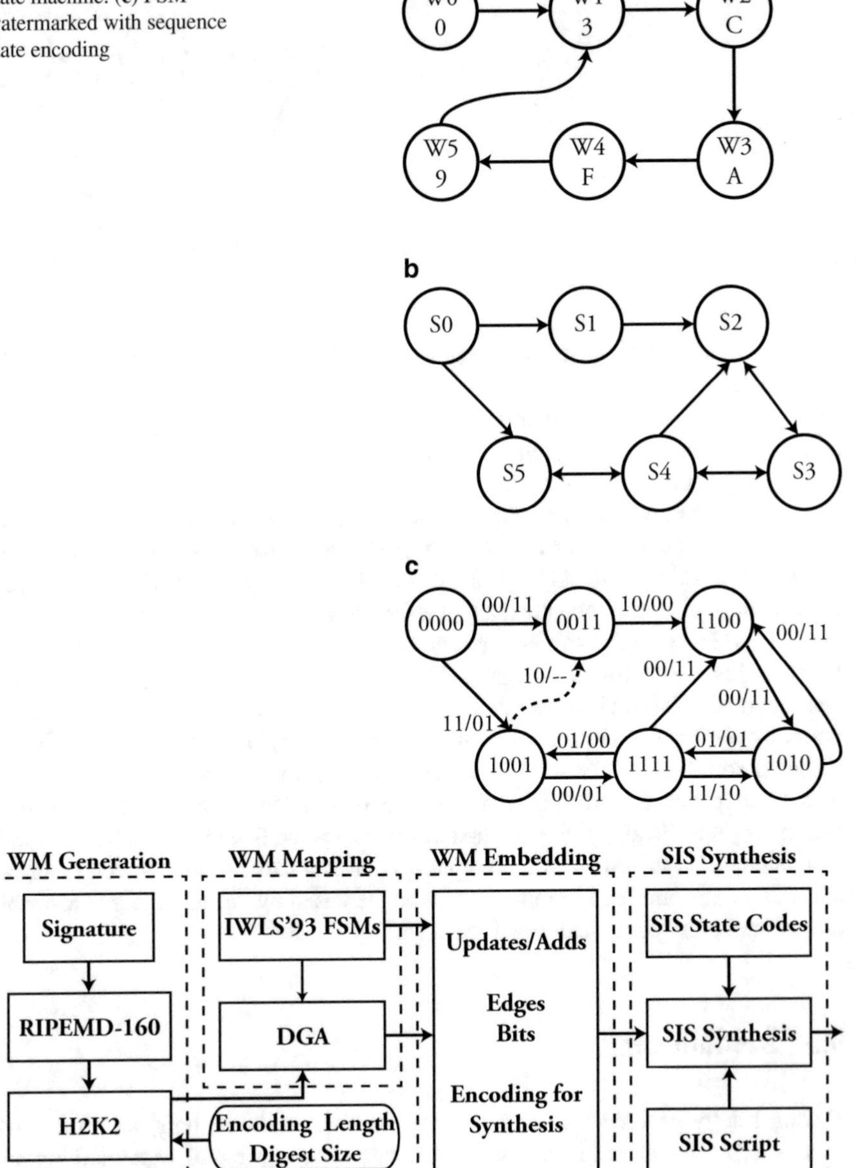

Fig. 9.4 (**a**) Watermark sequence graph. (**b**) Finite state machine. (**c**) FSM watermarked with sequence state encoding

Fig. 9.5 Methodology overview

9.3.1 Signatures

The beginning of every watermarking method starts with the signature, like that of an artist's painting, a way to show ownership and origin. Conversely, when considering computing and the vast amount of data one could choose from, a signature cannot simply be processed or utilized considering its raw binary form. This is because the density of bit subsets contained within any given digital media will result in a completely connected signature graph. Thus, to reduce the watermark graph complexity we need an additional method that still allows for the signature to be uniquely represented while reducing the amount of information representing it; wherein the solution lies with hash functions.

9.3.1.1 Signature Hashing

While hashing the signature will reduce the amount of information representing it, one major consideration arises when employing such a technique: "which hash to use." There are many options for hashing that are widely available and standard in use via OpenSSL: SHA (and variants), MDC2, RIPEMD-160, MD4, MD5, Blake2b, and Canonical [22]. To determine the answer to this question several considerations were made for selecting the best and most appropriate hash algorithm to employ: (1) security, (2) length, and (3) standardization. While security and length are directly related, *standardization* was considered for the ability of making the methodology widely available and easy to utilize. Additionally, standardization of the hash function was important for removing the potential for debating which hash may have been employed on a watermarked FSM using this technique. Conversely, when considering *length* we knew that options such as SHA-385 and SHA-512, while secure, would produce an overabundance of bits and in the worst case produce the same completely connected signature graphs we intended to avoid. Thus, we needed a hash algorithm that would produce a minimal length string while still providing provably high *security*. To evaluate and quantify the secureness of a given hash we considered its resilience to both *collision* and *preimage* attacks in addition to surveying literature for both the popularity of attacks against the algorithm and number of unique attack methods. For clarity, a *collision* attack is defined as attempting to find two unique signatures that produce the same hash string, and a *preimage* attack is reproducing the exact file that matches the hash string.

Ultimately we found the solution was to employ the use of RIPEMD-160, as the hash algorithm is a standard hashing function in the tools that would produce a minimal length signature hash (160-bit/20-Byte) while maintaining provably high [23, 24], as the time complexity for performing standard *collision* and *preimage* attacks on RIPEMD-160 is known to require 2^{80} and 2^{160}, respectively. Additionally, the additional survey of literature produced little popularity of unique attacks at the [25, 26], further aiding in the decision to employ the use of RIPEMD-

160. We also note that since the time this implementation choice was made only several new attack works have appeared in [27, 28].

Signatures

Now that we have selected a signature and processed it into a RIPEMD-160 hash, we can continue forward into the last step of the watermark generation phase (from Fig. 9.5). The "H2K2" (Hash-2-K2) process denotes a simple custom tool for processing the RIPEMD-160 hash and generating a watermark signature graph.

9.3.1.2 Hash-2-K2

This tool processes the signature hash based on multiple variables that ultimately further alter, and/or reduce, the overall complexity of the constructed graph. These input parameters, *Digest* and *Encoding Length*, dictate the number of total bits from the hash to process (64/128/160) and the number of bits that represent a symbolic state in the graph, respectively. The resulting graph is produced in a standardized State Transition Graph (STG) format named Kiss2 (K2) [29].
To illustrate this process, consider a signature that produces the RIPEMD-160 hash:

91efebee4bd48a62f338f244560b668771fa338e

For simplicity, a digest size of 32-bits and an encoding length of 4-bits are initially used as the input parameters to the H2K2 tool. The result of using these parameters will effectively have the tool consider the first eight values from the hash (91ef_ebee) as symbolic states in the constructed graph, which is shown by Fig. 9.6a. To demonstrate the impact encoding length has on complexity, Fig. 9.6b shows the same digest but now specifies the encoding length parameter as 8-bits. While it can be seen by Fig. 9.6b that graph complexity has been significantly reduced, we note that the underlying trade-off is that we are now enforcing a larger state encoding value, such that, if we watermark using this graph on some arbitrary 4-state FSM we increase the state register size by 400% since the original FSM would only require a 2-bit state register minimally.

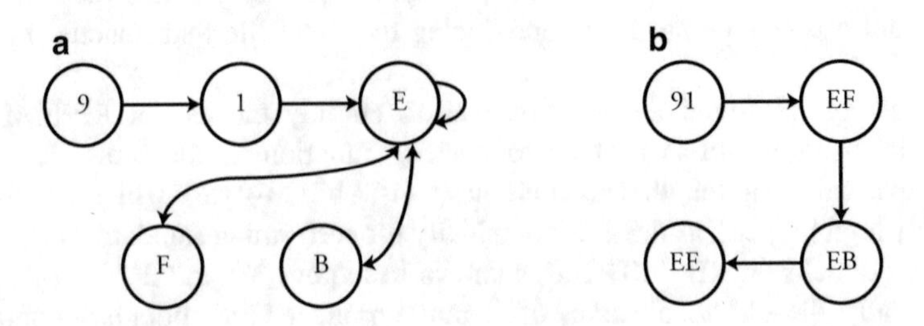

Fig. 9.6 (**a**) Digest = 32, Encoding Length = 4. (**b**) Digest = 32, Encoding Length = 8

Finally, once a desired signature has been hashed and its graph is generated, the watermark generation phase is concluded and the technique progresses forward into the largest, most complex, phase of watermark mapping.

9.3.2 *Watermark Mapping*

When considering methods for properly mapping the watermark graph to the FSM several less than desirable approaches [18–20] led up to the method we present in this section. However, before we detail the method it is important to first understand the difficulties involved in this process, specifically that mapping the watermark states to those in an FSM directly equates to the Subgraph Isomorphism (SGI) problem. Additionally, the decision making process involved in determining where to add edges can potentially relate to the Hamiltonian Completion problem for given cyclic signatures; wherein both of these problems have known computational complexities of Non-deterministic Polynomial (NP)-Complete [30, 31]. As previous attempts of greedy heuristic and simulated annealing methods proved to produce less than ideal results, a new approach was needed. Thus, by building off the standard framework of Genetic Algorithms (GAs) we constructed an enhanced algorithm that would come to be called the Darwinian GA (DGA); while seemingly redundant to those versed in Evolutionary Algorithms (EAs), as they are already inspired by Darwin's evolutionary principles, the DGA we present takes the inspiration significantly further and evolves the naïve traditional GA framework. To fully illustrate these changes and conceptual enhancements provided by the DGA over traditional GAs, we will first provide a brief background on GAs.

9.3.2.1 Genetic Algorithms

As mentioned, genetic algorithms are those which implement a standardized framework that allows for the simulation or mimicry of evolution and natural selection in its simplest form. To aid in the explanation process, we can begin simply viewing the flow of operations, shown in Fig. 9.7, that are performed by a standard GA and detail each item.

Parameters in Fig. 9.7, there is a standard set of inputs to every GA, to concisely list and define these parameters we provide them in Table 9.1. Following, the *Initialization* process simply generates a given number of initial solutions that ultimately seed the process performed in the main loop portion of Fig. 9.7, where the total number of solutions generated is specified by the population size parameter. Following initialization each solution is evaluated in the *Fitness* step, where some function unique to the problem is used that allows for a quantifiable measure of solution quality. For example, if we want to randomly add two numbers and determine which solution from the population was closest to the value 525 without exceeding the value, then a fitness function of $F(x) = (A+B)-525$ would allow us

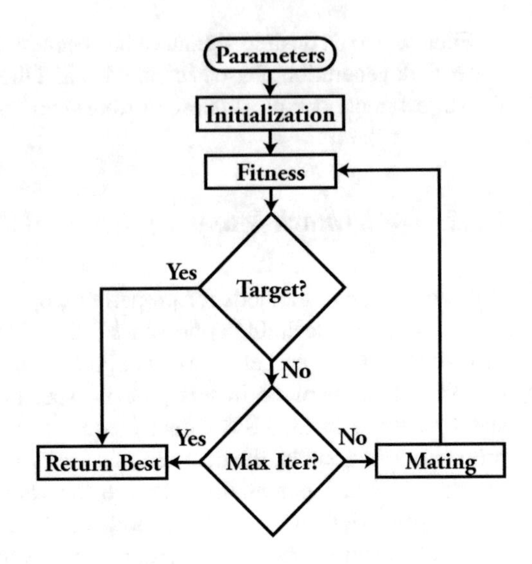

Fig. 9.7 Traditional GA flow

Table 9.1 Traditional GA parameters and descriptions

Parameter	Description
Population size	Total number of individual solutions
Target	The fitness value desired for completion
Fitness	Function quantifying the quality of a solution
Generations	The total number of times the loop in Fig. 9.7 is performed
Elitism	Top (E%) percent of generational survivors
Mutation rate	The chance at which a generated solution is altered randomly

to quantify the distance from our target (525) and express the quality of solutions; we would intuitively know that any negative fitness value is unacceptable and we can easily rank all valid non-negative solutions in ascending order. Once having calculated the fitness for each member of the population, we can easily examine if the desired *Target?* (525) was obtained and either continue on in the flow or simply complete and return the solution found. In the event we did not reach the target then we continue to check if we have executed the maximum number of iterations (*Max Iter?*), or generations, specified; in the event we have the current best is returned, if not then the mating process beings.

The mating process in traditional GAs, shown by Fig. 9.8, while seemingly complex in comparison to the overall flow is rather simplistic. The most complex portions of the *Mating* process are parent selection and child creation. However, starting with *elitism* this sub-process determines a fixed number of solutions from the current population that are guaranteed to survive and continue into the next generation, wherein the remaining population space will be composed of newly generated solutions. The process of generating a single new solution as a new

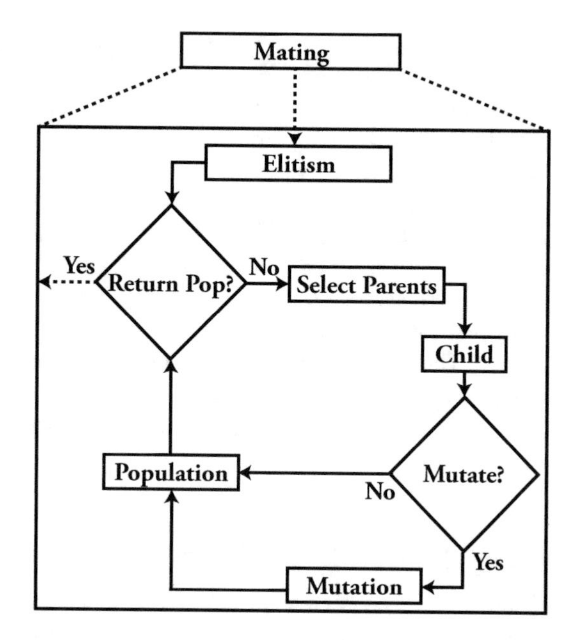

Fig. 9.8 Traditional GA mating process

member of the next generation population begins by employing some parental selection method [32] (roulette, stochastic universal, linear rank, etc.). While we do not detail the exacts of each method here, we do note that conceptually the driving factor behind parent selection is to reduce the likelihood of solution convergence to a local optimum rather than the global. Similarly, child creation is a simplified process that utilizes *crossover operators* to produce a single solution unified by fixed subsets from the parents. If we consider our 525 example this could be rather simple; however, if we expand a solution from $(A+B)$ to $(A+B+C+D+E+F)$ then depending on the operator [33] implemented a child could have any number of parental combinations for the values of $A - F$. The last operation, *mutation*, is the simplest in GAs, considering again the 525 example, we randomly select a value from $A - B$ or $A - F$ and regenerate its value.

9.3.2.2 Darwinian Genetic Algorithm

As previously mentioned it was our belief that the traditional GA was over simplified and did not properly reflect those things observable in nature, that it was a naïve approach to evolution and natural selection. We first begin by introducing the new concepts applied and supporting reasoning. For clarity, we denote which concepts are either changes to the traditional GA model or new parameters that influence model flow and operation by "item: [model|parameter]" for the time being, and these will become more clear when the DGA overview is illustrated.

> Non-Island Multi-Populations: [Model]

While the traditional GA model can be parallelized using an "Island" model [34], one that processes the population in parallel sub-blocks, we consider a different "Island" model per se; such that, we conceptualize our "Island" based on concepts of *allied species* from Darwin's "On the Origin of Species" [35] and as a physical island. Consider the scenario wherein a given population of bird species solely exists on some physical island, wherein on occasion a migratory allied, same genus, bird species migrates to this island for some period of time. We now have two naturally and genetically variant populations with the possibility of mating (variation due to nature). Thus, rather than one population processed in sub-blocks, we consider two unique and independent populations that have the possibility to mate; wherein one population is offset and regenerated in a migratory sense.

> Population Offset: [Parameter]

This is simply the concept that a migratory species would not immediately be present at the said island, such that, after some number of generations (representation of passing time) a migratory allied species would arrive at the island. Thus, we allow for the island native species to evolve independent over time before being introduced to a migratory allied species.

> Population Crossover (X-Mating): [Model and Parameter]

As mentioned, we consider an allied *same genus* migratory species that has the potential to mate with the native island population. Thus, each independent species (native/migratory) population evolves with the chance to mate between populations. For clarity, we note that while this is a change to the GA model, it is also constrained by the associated *population crossover rate* parameter of each species (native/migratory), or the probability that child generated is the result of one parent from each unique population.

> Microscopic Elitism: [Model]

This is applied during the mating process for child creation and is in-part used as the crossover operator rather than one of the existing methods [33]. In nature we know that the fitness of single genes can cause life altering syndromes and conditions [36, 37]; however, naturally we do not evaluate the individual member of a population based on these single genetic dispositions. Thus, we employ *microscopic elitism*, and evaluate the fitness of single genes in order to utilize a

Punnett square methodology of genetic crossover; one that more accurately models the process observed in nature. Additionally, when considering child creation this method operates by first selecting the genes deemed fit, as they are the dominant genes, in the event a single gene is unique to each parent with a same fitness or dominance then the gene passed to the child is randomly selected. Following, the un-selected genes deemed unfit for each parent are considered in child generation, e.g., if gene(4) of each parent is unfit and has not been covered then gene(4) of the child is randomly selected from one parent. Lastly, the remaining uncovered genes are randomly assigned.

> Life Span: [Parameter]

While the model chosen for implementing the life span of a population in itself was naïve, we opt to allow for the total re-initialization of a population after some number of generations have passed, regardless the generation in which the solution was conceived; in that, then we know it is rather inaccurate of nature for single members of a population to live for potentially thousands of years [38, 39] but do note that some species like the *immortal jellyfish* can regress in their life cycle based on environmental sustainability [40] and thus have greater longevity than most species.

> Multi-Objective Fitness: [Model and Parameter]

This simply incorporates the concept of giving multiple points of evaluation for members of a population, for example, if we consider our own subjective selection criteria for our personal likes and dislikes then we may say something smells bad or has a poor texture. Similarly, we provide another quantifiable function for evaluating the quality of individuals. Additionally, just as seen in the traditional GA model, each objective has its own unique fitness function and corresponding target value for which the individual is measured or ranked.

> Floating Elitism: [Model and Parameter]

This is the concept that elitism is not a fixed point, we observe this in nature regularly from the behavior of animals and their struggle for dominance and greater hierarchical status [41, 42]. We know that a lower tier member may challenge higher ranking members in order to obtain greater status. We can see that the elitism of members "floats" based on the fitness of the entire population and is not a fixed point system. Thus, the process of mating is based off of a *floating elitism* value rather than a fixed point value. Additionally, as we utilize two unique and independent populations each ultimately has their own respective floating elitism value.

> Population Mutation: [Model]

While elitism observably floats, we also know from the 1960s "Mouse Utopia" experiment performed by Calhoun [43] that even under environmentally ideal settings there existed a static subset of elite members dubbed "the beautiful ones" that resided in the upper levels of the experimental enclosure, free from the societal tribulations of those at ground level. Simply put, the experiment more or less showed that when all members of an environment can be seen as "elite" and have all of their resource/mating needs met then the population must adapt or will simply die off. Thus, in addition to the normal (floating elitism) and cross mating processes we opted to employ in a system which models this scenario, i.e., we consider the subset of members inside the fixed elitism range free of societal tribulations and those outside this range to those under constant adaptation/change, or mutation in this case, wherein this process is repeated until the floating elitism dictates that the population is no longer entirely composed of elite individuals.

> Death Rate: [Parameter]

This incorporated concept is derived from that of neural networks and is similar to learning rate, in that, here it is a fixed additive value to floating elitism but is used to aid in determining the number of unfit members that will be re-populated in the current generation. Similar to floating elitism, each individual population (native/migratory) has a unique death rate parameter.

> Child Rate: [Parameter]

This is based on the knowledge that not every species produces a single off-spring or even off-spring at all, for example, consider the human race and birth rate information from the Center for Disease Control (CDC) [44] that simply shows how non-static birth rate is. Thus, we opt to upper bound the number of children that parents can produce for each population during mating and randomly select the number of children that two given parents will create. We then employ a natural selection process on the children by selecting the "pick of the litter," or that with the best fitness, to be the child added to the re-population subset.

> **Mutation Gene Size: [Model and Parameter]**

While mutation is generally performed by only altering the value of a single gene, we consider this as a "not always the case" scenario; for example, consider the exclusion zone of Chernobyl, it was found that species inside the zone have elevated rates of genetic mutation [45]. Thus, we can expect that an environment of a population may not always be ideal and geographical locations have the ability to cause both variation due to nature and man; such that, the number of mutations a single member incurs is not static. Rather, we upper bound the number of genes that can be mutated and when mutating an individual randomly selects the number of genes to mutate.

> **Maximum Mutations: [Parameter]**

This is simply an upper bounding on the number of times a single member of the population can undergo mutation during the *population mutation* process; i.e., how rapidly an individual can adapt.

> **Stochastic Darwinian Approach: [Model]**

The most important part of accurately modeling nature is the lack of predictability and human intervention. Thus, when constructing the DGA we opted to remove as much human influence as possible and employ a stochastic model; one wherein everything that possibly can be randomly dictated is done so to induce random variation of members and species.

Now having covered the newly introduced concepts and parameters we can better illustrate how they first impact the overall flow of the DGA in comparison to the tradition GA, we show the DGA overview in Fig. 9.9a. Further, we also expand and illustrate population mutation, cross mating, and gene mutation in Figs. 9.9b, 9.10a, and b, respectively.

9.3.3 DGA and Subgraph Isomorphism

Here, we now clarify the last few pieces of information on how the DGA was utilized in this work to perform the subgraph matching and assignment process. First, the fitness function used for evaluating member quality is a simplified adjacency list check process for a given state mapping. We verify which adjacencies are present given the current mapping of the state under evaluation, leading to each state

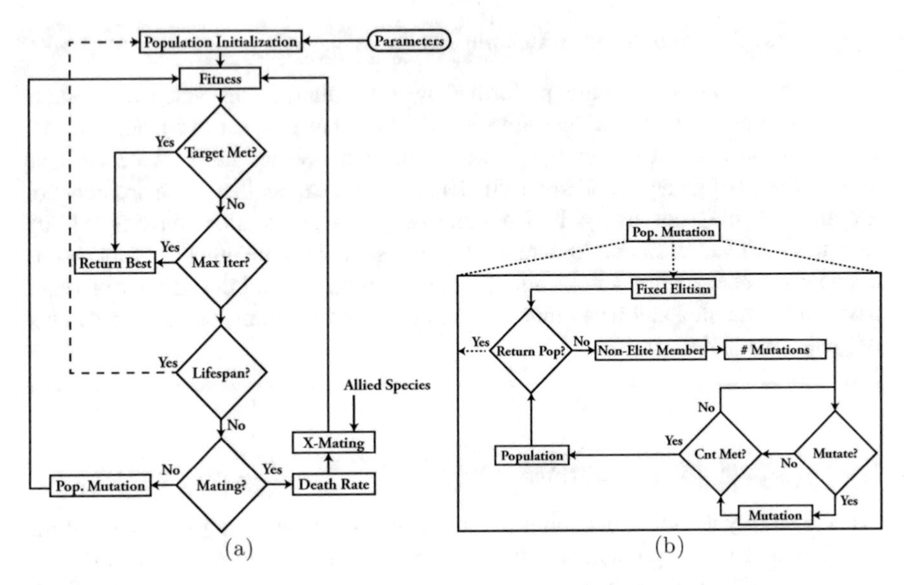

Fig. 9.9 Hardware Trojan triggering models. (**a**) DGA overview and (**b**) DGA population mutation

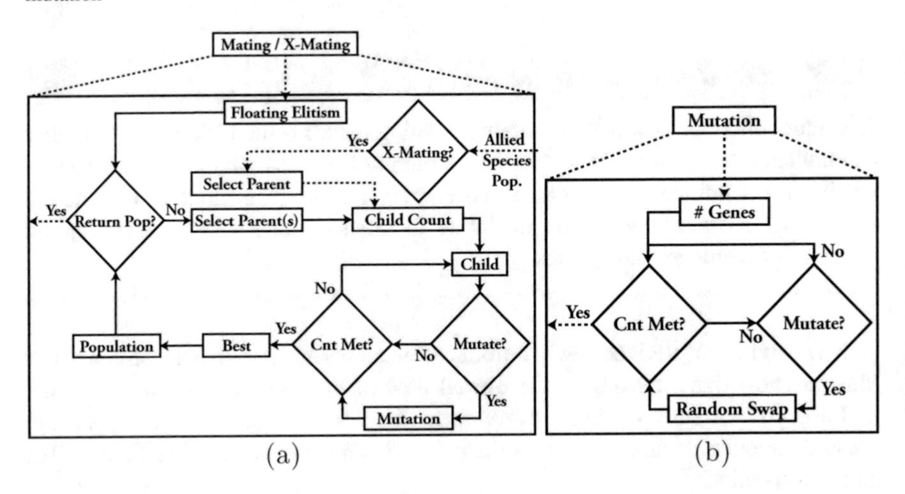

Fig. 9.10 Hardware Trojan triggering models. (**a**) DGA X-Mating (cross mating) and (**b**) DGA gene mutation

having its own microscopic elitism based on the calculated fitness of the mapping that was randomly generated for $W(x) \rightarrow S(y)$. Second, because during edge fitness calculations the number of missing edges and the appropriate locations for edge insertion were stored, we were able to calculate the fitness of bits that were required for integrating the watermark and provided the multi-objective fitness of both ranking solutions by bits and edges to be added when considering targets of zero for both. Additionally, we treated the search process, intuitively, as finding the signature graph within the original FSM, or querying the watermark graph from the target FSM.

9.3.4 Addition of Edges and Modification of Encodings

When the DGA returns its best solution, the framework then utilizes information contained within the data structure to appropriately increase the number of bits, add edges, and update state encodings where applicable. The manner in which non-enforced, or free-encoded, (unwatermark) state encodings are updated is simply in a First Come First Serve (FCFS) manner; for example, if we have a mapping of $\{0, 1, 2, 3\} \rightarrow \{1, 0, 3, 5\}$ for $[S0 - S6]$ then $S2 \rightarrow 100$, $S4 \rightarrow 101$, and $S6 \rightarrow 110$. Similarly, when considering the addition of an edge at a given state we employ a FCFS manner, in that, if the states behavior is explicitly declared for input combinations $[000 - 100, 110 - 111]$ then we add an edge from $S(x) \rightarrow S(y)$ with the input combination of 101 and the output is set to don't care conditions for all bits. Lastly, the addition of bits will simply extend the input information by that necessary and update all input combinations in the Kiss2 STG, the significant difference being that added watermark edges are extended at the Most Significant Bit (MSB) with "1" while existing edges are MSB extended with "0." However, we note that while FCFS is naïve, additional methods of edge insertion and free-partial state encoding assignment could be utilized based on the potential characteristics of a given FSM and synthesis techniques.

9.4 Experimental Setup and Results

A subset of the International Workshop on Logic Synthesis (IWLS)'93 [46] benchmark suite that overlapped with the state-of-the-art [16] was used for testing the presented DGA. While the benchmark suite is designed for evaluating synthesis tools some appropriate actions were required to be taken on the FSMs before they could be used in the watermarking process, as some designs potentially had disconnected states, and to ensure that the FSMs were optimized as close to possible of the values and method in [16]. Thus, we only utilized the algebraic SIS synthesis script in combination with the Mississippi State University (MSU) gate libraries for the optimization process that was loosely detailed in [16] for obtaining the subset of FSMs to be watermarked. The information pertaining to the FSMs is provided in Table 9.2 alongside the DGA parameters used in Table 9.3. For clarity, σ and ρ in Table 9.2 denote states and transitions, respectively. Table 9.3 provides the population size and lifespans in the form of "native/migratory."

The watermark that was used is the hash sequence previously shown in Sect. 9.3.1.2; however, we do note that the watermark was digested at sizes of

Table 9.2 IWLS'93 benchmark Subset [21]

FSM	FSM_σ	FSM_ρ	I/CSFSM
s1488	48	251	CSFSM
s1494	48	250	CSFSM
s208	18	153	CSFSM
s27	6	34	CSFSM
s386	13	64	CSFSM
s510	47	77	CSFSM
s820	25	232	CSFSM
s832	25	245	CSFSM
bbara	10	60	CSFSM
dk15	4	32	CSFSM
ex1	20	138	ICSFSM
ex4	14	21	ICSFSM
planet	48	115	CSFSM
s1	20	107	CSFSM
sand	32	184	ICSFSM
scf	121	286	CSFSM
sse	16	56	ICSFSM
styr	30	166	CSFSM

Table 9.3 DGA parameters used from [21]

Parameter	Value
Runs	[18:144]
Population size(s)	64/64
Population lifespan(s)	64/32
Allied species offset	1
X-Mate rate	50%
Maximum generations	1317602
Generation break	256
Fixed elitism rate	15%
Mutation rate	25%
Mutation gene size	25%
Maximum mutations	2
Death rate	0.5%
Child rate	3
Target edge/bit fitness	(0/0) 100%

64 and 128-bits in order to compare against [16] but the encoding length was swept from 2 to 16 bits, such that, every watermark encoding length possible could be used to evaluate the method across all DGA runs with the best selected for each digest. We provide the results of watermarked FSMs, using the subset from Table 9.2 with the hash from Sect. 9.3.1.2, compared against [16] in Table 9.4. We note that the N, L, Lit, A, and D in Table 9.4 denote the digest size, encoding length, sum-of-product (SOP) literals, area, and delay.

Table 9.4 Comparison of DGA and [16]

FSM	N	L	Lit.	A	D	%Lit	%A	%D
S1488	64	3	952	11,064	22.6	−1.7	1.8	−6.6
S1488	128	7	973	11,040	26	−0.8	−1.1	11.1
S1494	64	2	994	11,280	22.2	5.2	3.9	−8.3
S1494	128	7	950	10,784	23	−9.5	−8.7	−5.7
S208	64	2	149	1936	10.8	−9.1	−2.4	−8.5
S208	128	2	134	1872	9.4	−21.2	−6	−20.3
S27	64	2	80	992	10.6	56.9	15.9	43.2
S27	128	2	52	776	7.2	2	−9.3	−2.7
S386	64	3	212	2560	14	−14.5	−15.6	0
S386	128	3	238	2920	13.8	−7.8	−5.2	−4.2
s510	64	7	377	4536	19.4	−26.4	−23	9
s510	128	2	380	4448	18.6	−30.3	−29.1	4.5
S820	64	2	438	5240	19.6	−5.4	−4.8	0
S820	128	3	453	5320	19.4	−16	−16.1	−3
S832	64	5	430	5256	19.4	−11.5	−7.1	−4
S832	128	5	440	5232	18	−15.1	−13.9	−5.3
bbara	64	3	114	1512	12.8	–	25.2	39.1
bbara	128	3	114	1512	12.8	–	16	25.5
dk15	64	2	122	1408	10	–	−2.8	−3.8
dk15	128	2	119	1432	10	–	−8.7	−10.7
ex1	64	5	353	4432	17.6	–	−14.4	−5.4
ex1	128	3	353	4432	17.6	–	−23.3	−8.3
ex4	64	2	126	1736	10.4	–	-16.2	-7.1
ex4	128	3	159	1944	10.8	–	0	−1.8
planet	64	2	940	11,064	23.4	–	9.1	10.4
planet	128	3	937	10,872	20.8	–	−0.9	2
s1	64	2	538	6064	21.6	–	28.3	28.6
s1	128	2	515	6024	20.4	–	7.7	1
sand	64	2	752	8360	20.8	–	−9	−2.8
sand	128	7	718	8304	20.2	–	−11.6	−18.5
scf	64	2	1190	13,984	25.4	–	−1	-2.3
scf	128	3	1163	13,664	25.8	–	−7.8	−1.5
sse	64	2	192	1368	12.2	–	−20.4	−10.3
sse	128	4	192	2368	12.2	–	−18.2	−14.1
styr	64	2	816	9072	23.4	–	0.4	14.7
styr	128	3	814	9056	21.6	–	4.6	10.2
AVG	64	–	–	–	–	−0.8	−2.0	4.2
AVG	128	–	–	–	–	−12.3	−6.9	−2.2
MIN	64	–	–	–	–	−26.4	−23	−10.3
MIN	128	–	–	–	–	−30.3	−29.1	−20.3
MAX	64	–	–	–	–	56.9	28.3	43.2
MAX	128	–	–	–	–	2.0	16	25.5

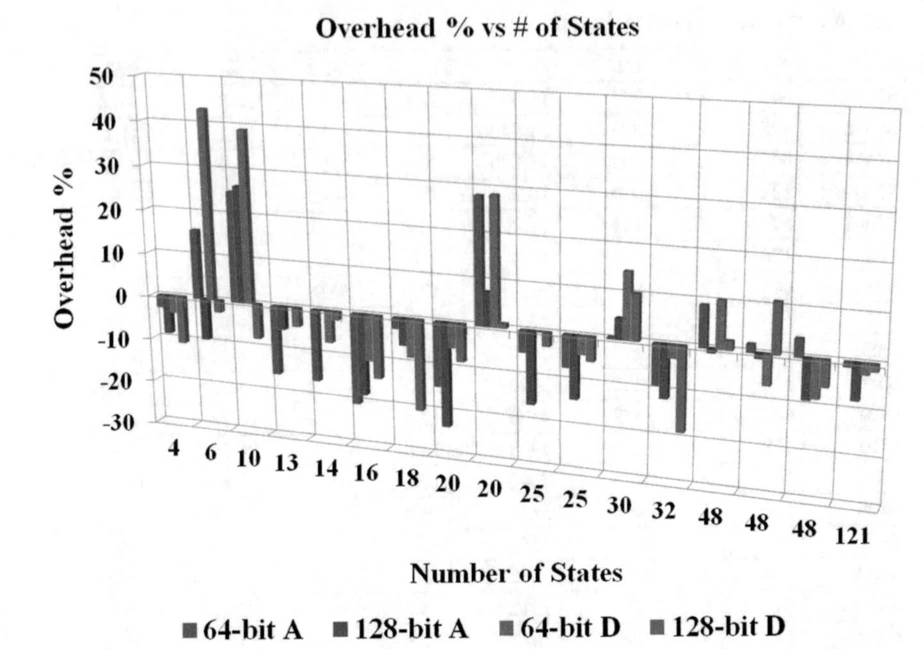

Fig. 9.11 Results from Table 9.4 (A = Area and D = Delay)

From Table 9.4 we can observe that on average this method showed improvements over [16] across nearly all metrics, with the exception of delay when using a 64-bit digest. To better illustrate this, the results from Table 9.4 are provided visually by Fig. 9.11. Additionally, it can be seen that in nearly all cases of excessive (>10%) overhead it can be attributed to the 64-bit digest, with only few instances where the 128-bit watermark digest exceeded tolerable overhead in terms of only either area or delay. Thus, we should examine these instances and determine the root causes of overhead.

9.5 Overhead Analysis

Table 9.5 shows the outliers in Table 9.4 with excessive (>10%) overhead. Additionally, Table 9.6 expands the post-watermarking FSM information with N, L, ρ, and I, denoting again the digest, encoding length, edges, and input bits. For further clarity, when considering $\Delta\rho$ and ΔI in Table 9.6 these are representative of additive edges or bits to the value of ρ or I presented. For example, when considering the 64-bit digest of *bbara* the total number of edges and bits in the post-watermarked FSM are 65 and 5, respectively.

Table 9.5 Overhead outliers

FSM	FSM_σ	FSM_ρ	I/CSFSM
s27	6	34	CSFSM
bbara	10	60	CSFSM
planet	48	115	CSFSM
s1	20	107	CSFSM
styr	30	166	CSFSM

Table 9.6 Overhead outliers I expanded

FSM	N	L	FSM_ρ	$FSM_{\Delta\rho}$	FSM_I	$FSM_{\Delta I}$
s27	64	2	96	0	4	0
bbara	64	3	60	5	4	1
bbara	128	3	60	12	4	1
planet	64	2	115	8	7	1
s1	64	2	107	5	8	1
s1	128	2	107	6	8	1
styr	64	2	166	2	9	10
styr	128	3	166	14	9	10

However, when examining the data in Table 9.6 something becomes immediately apparent; *s27* had excessive (>10%) overhead with zero additional edges and bits. While it is intuitively known that the state assignment process can impact the synthesis outcome of a sequential circuit, it would seem unlikely that this is entirely the root cause. We say this for several reasons: (1) when considering *s27* from Table 9.4 we can see that encoding lengths for both 64 and 128-bit digest sizes are the same, (2) but only the 128-bit digest improves upon the synthesis, (3) employing a watermark with 128-bit digest at the same encoding length will simply result in a more complex watermark signature graph and it is unlikely that a potentially completely connected graph would embed better than one that is not completely connected, and (4) we are able to effectively rule out the FCFS methodology for edge addition as *s27* was the only FSM with perfect isomorphism and nearly all other FSMs showed improvement with edge and bit addition. However, we will examine the mapping solutions produced by the DGA and subsequent state encoding values assigned for more insight.

Upon examination of the specific DGA runs for the best results found for *s27*, we do find something peculiar; such that, even though the mappings were different, as shown by Table 9.7, the mapping employed on the 128-bit digest was never produced during any of the runs under a 64-bit digest. This is peculiar because of graph properties and transitivity of isomorphism. If we consider any two watermarks of equivalent encoding length, then a watermark of smaller digest ($G(a)$) size will exhibit subgraph isomorphism to the graph resulting from a larger digest ($G(b)$). Specifically, if $G(a)$ is isomorphic to $G(b)$ and $G(b)$ is isomorphic to $FSM(x)$ then $G(a)$ is also isomorphic to $FSM(x)$. To illustrate this, Fig. 9.12a and b show the 64 and 128-bit signature watermark graphs using an encoding length of two; it can be seen that as expected the 128-bit digest produces a completely connected

Table 9.7 Mappings for *s27* (N = Digest, L = Encoding Length)

(N = 64, L = 2)		(N = 128 L = 2)	
Query	Target (encoding)	Query	Target (encoding)
1	S2 (001)	1	S3 (001)
2	S0 (010)	2	S1 (010)
3	S3 (011)	3	S0 (011)
4	S1 (100)	4	S2 (100)
N/A	S4 (000)	N/A	S4 (000)
N/A	S5 (101)	N/A	S5 (101)

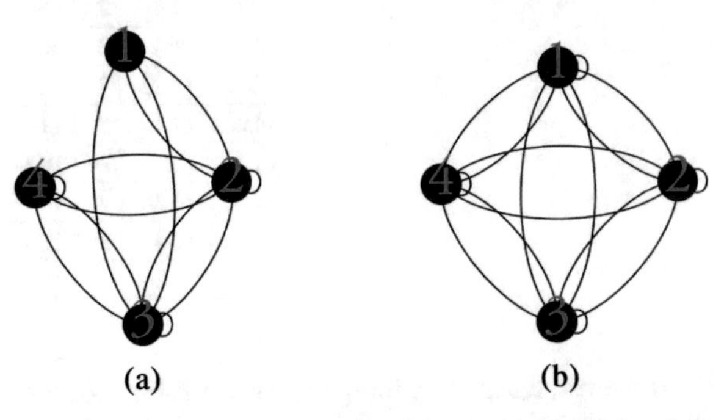

Fig. 9.12 (**a**) Signature graph (Digest = 64, Encoding Length = 2). (**b**) Signature graph (Digest = 128, Encoding Length = 2)

Table 9.8 New s27 encoding result found via DGA

FSM	N	L	Lit.	A	D	%Lit	%A	%D
s27	128	2	28	568	6.8	−45.1	−33.6	−8.1
S0(000), S1(001), S2(011), S3(010), S4(100), S5(101)								

graph, similarly it can be seen that the graph produced using a 64-bit digest is indeed a subgraph of the 128-bit digest (this becomes more intuitive when considering the hash string and substring variants).

Thus, if the overhead is specifically due to the enforced watermark graph encoding values, and not those freely assigned, we should simply be able to modify the state codes for the watermarked *s27* 64-bit digest to those assigned in the 128-bit and observe a significant reduction in the overhead. Conversely, re-running DGA with these watermark graphs produced two entirely new sets of results, again with 128-bit out-performing not only 64-bit but the existing result in Table 9.4 and this data is provided in Table 9.8.

Thus, even though we did not expect the reordering of state encoding values for only a given subset of states for the watermarked FSM to heavily impact the overhead results it has shown to do just that, even when employing a FCFS method for freely encoded states ($S4/S5$ when considering this example). In addition to this we also examined this new result under different combinations of freely encoded states ($S4/S5$) in order to effectively measure the impact of employing a FCFS assignment process; these results are compared against that from [16] for *s27* when using a 128-bit digest in Table 9.9.

As a result of the data shown in Table 9.9, we can see that employing a FCFS assignment of free states did not prove detrimental to the overhead for *s27*;

Table 9.9 Impact of freely encoded states ($S4/S5$) on overhead

FSM	N	L	Lit.	A	D	%Lit	%A	%D
s27	128	2	28	568	6.8	−45.1	−33.6	−8.1
S0(000), S1(001), S2(011), S3(010), S4(100), S5(101)								
s27	128	2	39	632	7	−23.5	−26.2	−5.4
S0(000), S1(001), S2(011), S3(010), S4(100), S5(111)								
s27	128	2	42	752	10	−17.6	−12.1	35.1
S0(000), S1(001), S2(011), S3(010), S4(101), S5(111)								
s27	128	2	71	944	11.2	39.2	10.3	51.4
S0(000), S1(001), S2(011), S3(010), S4(111), S5(100)								
s27	128	2	61	896	9.8	19.6	4.7	32.4
S0(000), S1(001), S2(011), S3(010), S4(101), S5(100)								

however, we do note that a larger FSM with a signature more sporadically integrated may cause overhead to be incurred. This ultimately means that encoding based watermarking becomes a multi-metric problem of orderings for both the enforced state codes from the signature graph and the set of freely encoded states that are not part of the watermark mapping set. Further experiments will need to be conducted in order to examine the impact edges and bits added account for, in a singular sense. Regardless, it can be seen from Table 9.9 that even if we were to simply swap the encoding values of $S4$ and $S5$, then the overhead would go from a significant decrease to increase.

Additionally, because DGA is a non-deterministic algorithm, we find that a significant number of runs must be performed using the DGA in order to ensure that the solution space is entirely explored, as shown from the original results that a 64-bit solution with the same mappings produced for the 128-bit was never found, and that even now a new mapping solution was found that further decreases overhead over the method compared. Similarly, we can expect that the other outliers incur the same encoding problem which caused the overhead generated to be so significant. Lastly, because of this we can safely assume a top-down Monte Carlo analysis approach should be employed when using this method, such that, the ideal setting would be to root as much of the watermark into a design as possible while also gaining the benefits from encoding maps produced; i.e., the state encoding scheme provides both security and synthesis enhancements. In addition to this, starting in a top-down manner allows a single solution space to be thoroughly explored for a given digest using variable encoding lengths with the knowledge that any desirable results produced may be further beneficial at a lower digest due to the transitive nature of the signature graphs when considering the equivalent encoding length.

9.6 Conclusion

From the results, we can conclude that this work out-performs the previous best edge based watermarking method and is a viable framework that can be used for watermarking sequential circuits. Conversely, even though some cases were shown to have significant overhead we were able to determine the causes and devise strategies for overcoming these obstacles; as now the method of encoding based watermarking is not only a multi-metric state assignment problem but also that of SGI and Hamiltonian Completion. In any event, the DGA also proves to be a viable option for taking on these problems but acknowledge that the non-determinism requires a large number of runs in a Monte Carlo fashion to properly and thoroughly explore the solution space of possible state encoding mappings of the watermark to FSM. While more experimental work needs to be performed for properly measuring the impact of single edges/bits added, we can safely say that here we have shown the impact of not using the FCFS manner we opt to employ and how it can seriously hinder design and resulting overhead.

References

1. TSMC: TSMC Announces Intention to Build and Operate an Advanced Semiconductor Fab in the United States (2020). https://pr.tsmc.com/english/news/2033
2. Ernst, D., Naughton, B.: Global Technology Sourcing in China's Integrated Circuit Design Industry: A Conceptual Framework and Preliminary Findings (2012). https://www.eastwestcenter.org/sites/default/files/private/econwp131_0.pdf
3. TSMC: TSMC Will Vigorously Defend its Proprietary Technology in Response to Global-Foundries Complaints. https://pr.tsmc.com/english/news/2006
4. TSMC: TSMC Files Complaints Against GlobalFoundries in U.S., Germany and Singapore for Infringement of 25 Patents to Affirm its Technology Leadership and to Protect its Customers and Consumers Worldwide. https://pr.tsmc.com/english/news/2009
5. The committee's investigation into counterfeit electronic parts in the department of defense supply chain: hearing before the committee on armed services, United States senate, one hundred twelfth congress, first session, November 8, 2011 (2012). https://www.govinfo.gov/content/pkg/CHRG-112shrg72702/pdf/CHRG-112shrg72702.pdf
6. DoD: Program acquisition cost by weapon system: United states department of defense fiscal year 2020 budget request (2019). https://comptroller.defense.gov/Portals/45/Documents/defbudget/fy2020/fy2020_Weapons.pdf
7. Counterfeit Integrated Circuits: Detection and Avoidance. Springer, Cham (2015)
8. Torunoglu, I., Charbon, E.: Watermarking-based copyright protection of sequential functions. In: Proceedings of the IEEE 1999 Custom Integrated Circuits Conference (Cat. No.99CH36327), pp. 35–38 (1999). https://doi.org/10.1109/CICC.1999.777239
9. Oliveira, A.L.: Robust techniques for watermarking sequential circuit designs. In: Proceedings 1999 Design Automation Conference (Cat. No. 99CH36361), pp. 837–842 (1999). https://doi.org/10.1109/DAC.1999.782155
10. Subbaraman, S., Nandgawe, P.S.: Intellectual property protection of sequential circuits using digital watermarking. In: First International Conference on Industrial and Information Systems, pp. 556–560 (2006). https://doi.org/10.1109/ICIIS.2006.365790
11. Nguyen, K.H., Hoang, T.T., Bui, T.T.: An FSM-based IP protection technique using added watermarked states. In: 2013 International Conference on Advanced Technologies for Communications (ATC 2013), pp. 718–723 (2013). https://doi.org/10.1109/ATC.2013.6698210
12. Torunoglu, I., Charbon, E.: Watermarking-based copyright protection of sequential functions. IEEE J. Solid-State Circ. **35**(3), 434–440 (2000). https://doi.org/10.1109/4.826826
13. Abdel-Hamid, A.T., Tahar, S., Aboulhamid, E.M.: A public-key watermarking technique for IP designs. In: Design, Automation and Test in Europe, Vol. 1, pp. 330–335 (2005). https://doi.org/10.1109/DATE.2005.32

14. Abdel-Hamid, A.T., Tahar, S., Aboulhamid, E.M.: Finite state machine IP watermarking: a tutorial. In: First NASA/ESA Conference on Adaptive Hardware and Systems (AHS'06), pp. 457–464 (2006). https://doi.org/10.1109/AHS.2006.40
15. Abdel-Hamid, A.T., Tahar, S.: Fragile IP watermarking techniques. In: 2008 NASA/ESA Conference on Adaptive Hardware and Systems, pp. 513–519 (2008). https://doi.org/10.1109/AHS.2008.73
16. Cui, A., Chang, C.H., Tahar, S., Abdel-Hamid, A.T.: A robust FSM watermarking scheme for IP protection of sequential circuit design. IEEE Trans. Comput.-Aided Des. Integr. Circ. Syst. **30**(5), 678–690 (2011). https://doi.org/10.1109/TCAD.2010.2098131
17. Cui, A., Chang, C.H., Zhang, L.: A hybrid watermarking scheme for sequential functions. In: 2011 IEEE International Symposium of Circuits and Systems (ISCAS), pp. 2333–2336 (2011). https://doi.org/10.1109/ISCAS.2011.5938070
18. Lewandowski, M., Meana, R., Morrison, M., Katkoori, S.: A novel method for watermarking sequential circuits. In: 2012 IEEE International Symposium on Hardware-Oriented Security and Trust, pp. 21–24 (2012). https://doi.org/10.1109/HST.2012.6224313
19. Lewandowski, M.: A Novel Method For Watermarking Sequential Circuits. Graduate Theses and Dissertations (2013). http://scholarcommons.usf.edu/etd/4528
20. Meana, R.W.: Approximate Sub-Graph Isomorphism For Watermarking Finite State Machine Hardware. Graduate Theses and Dissertations (2013). http://scholarcommons.usf.edu/etd/4728
21. Lewandowski, M., Katkoori, S.: A Darwinian genetic algorithm for state encoding based finite state machine watermarking. In: 20th International Symposium on Quality Electronic Design (ISQED), pp. 210–215 (2019). https://doi.org/10.1109/ISQED.2019.8697760
22. Canonical: http://manpages.ubuntu.com/manpages/bionic/man1/dgst.1ssl.html
23. Mendel, F., Pramstaller, N., Rechberger, C., Rijmen, V.: On the collision resistance of RIPEMD-160. In: Katsikas, S.K., López, J., Backes, M., Gritzalis, S., Preneel, B. (eds.) Information Security, pp. 101–116. Springer, Berlin, Heidelberg (2006)
24. Ohtahara, C., Sasaki, Y., Shimoyama, T.: Preimage attacks on step-reduced RIPEMD-128 and RIPEMD-160. In: Lai, X., Yung, M., Lin, D. (eds.) Information Security and Cryptology, pp. 169–186. Springer, Berlin, Heidelberg, (2011)
25. Zou, J., Wu, W., Wu, S., Dong, L.: Improved (pseudo) preimage attack and second preimage attack on round-reduced Grøstl. Cryptology ePrint Archive, Report 2012/686 (2012). https://eprint.iacr.org/2012/686
26. Mendel, F., Nad, T., Scherz, S., Schläffer, M.: Differential attacks on reduced RIPEMD-160. In: Gollmann, D., Freiling, F.C. (eds.) Information Security, pp. 23–38. Springer, Berlin, Heidelberg (2012)
27. Liu, F., Dobraunig, C., Mendel, F., Isobe, T., Wang, G., Cao, Z.: Efficient collision attack frameworks for RIPEMD-160. In: Boldyreva, A., Micciancio, D. (eds.) Advances in Cryptology – CRYPTO 2019, pp. 117–149. Springer International Publishing, Cham (2019)
28. Wang, G., Liu, F., Cui, B., Mendel, F., Dobraunig, C.: Improved (semi-free-start/near-) collision and distinguishing attacks on round-reduced RIPEMD-160. Designs, Codes and Cryptography **88**(5), 887–930 (2020). https://doi.org/10.1007/s10623-020-00718-x
29. Yang, S.: Logic Synthesis and Optimization Benchmarks User Guide Version 3.0 (1991)
30. Karp, R.M.: Reducibility among combinatorial problems. In: Complexity of Computer Computations: Proceedings of a symposium on the Complexity of Computer Computations, pp. 85–103 (1972). https://doi.org/10.1007/978-1-4684-2001-2_9
31. Garey, M.R., Johnson, D.S.: Computers and intractability: A Guide to the Theory of NP-Completeness. W.H. Freeman, New York (1979)
32. Jebari, K.: Parent selection operators for genetic algorithms. Int. J. Eng. Res. Technol. **12**, 1141–1145 (2013)
33. Kora, P., Yadlapalli, P.: Crossover operators in genetic algorithms: a review. Int. J. Comput. Appl. **162**, 34–36 (2017)
34. Corcoran, A.L., Wainwright, R.L.: A parallel Island model genetic algorithm for the multiprocessor scheduling problem. In: Proceedings of the 1994 ACM Symposium on Applied Computing - SAC 94 (1994). https://doi.org/10.1145/326619.326817

35. Darwin, C.: On the origin of species by means of natural selection, or, the preservation of favoured races in the struggle for life (1859)
36. Corballis, M.C.: Left brain, right brain: facts and fantasies. PLoS Biol **12**(1), e1001767 (2014). https://doi.org/10.1371/journal.pbio.1001767. http://www.ncbi.nlm.nih.gov/pmc/articles/PMC3897366/. PBIOLOGY-D-13-03341[PII]
37. Muhle, R., Trentacoste, S.V., Rapin, I.: The genetics of autism. Pediatrics **113**(5), e472–e486 (2004). https://doi.org/10.1542/peds.113.5.e472
38. Mayne, B., Berry, O., Davies, C., Farley, J., Jarman, S.: A genomic predictor of lifespan in vertebrates. Sci. Rep. **9**(1), 17866 (2019). https://doi.org/10.1038/s41598-019-54447-w
39. Petralia, R.S., Mattson, M.P., Yao, P.J.: Aging and longevity in the simplest animals and the quest for immortality. Age. Res. Rev. **16**, 66–82 (2014). https://doi.org/10.1016/j.arr.2014.05.003
40. Ceh, J., Gonzalez, J., Pacheco, A.S., Riascos, J.M.: The elusive life cycle of scyphozoan jellyfish - metagenesis revisited. Sci. Rep. **5**, 12037 EP – (2015). https://doi.org/10.1371/journal.pbio.1001767. https://doi.org/10.1038/srep12037. Article
41. Mech, L.D., Boitani, L.: chap. Wolf Social Ecology, pp. 1–34. University of Chicago Press, Chicago, IL (2003). http://pubs.er.usgs.gov/publication/87253
42. Mech, L.D.: Alpha Status, dominance, and division of labor in Wolf packs. Can. J. Zool. **77**, 1196–1203 (1999). http://www.npwrc.usgs.gov/resource/2000/alstat/alstat.htm
43. Calhoun, J.B.: Death squared: the explosive growth and demise of a mouse population. Proc. R. Soc. Med. **66**(1 Pt 2), 80–88 (1973)
44. Martin, J.A., Hamilton, B.E., Osterman, M.J., Driscoll, A.K., Matthews, T.: Births: final data for 2015. In: National Vital Statistics Report, U.S. Department of Health and Human Services, Centers for Disease Control and Prevention, vol. 66, pp. 1–70 (2017). https://www.cdc.gov/nchs/data/nvsr/nvsr66/nvsr66_01.pdf
45. Mousseau, T.A., Moller, A.P.: Genetic and ecological studies of animals in Chernobyl and Fukushima. J. Hered. **105**(5), 704–709 (2014). https://doi.org/10.1093/jhered/esu040
46. McElvain, K.: IWLS'93 Benchmark Set: Version 4.0. Tech. rep. (1993). https://rb.gy/rhs0p2

Part III
Hardware Trojan: Modeling, Localization, and Defense

Chapter 10
Hardware Trojan Localization: Modeling and Empirical Approach

Sheikh Ariful Islam and Srinivas Katkoori

10.1 Introduction

The proliferation of embedded systems in our daily lives, including national interests, is largely backed by the success of the nanometer scale Integrated Circuit(IC). Semiconductor manufacturing has become a viable business since 1960 and has projected revenue of $542.64 billion by 2022 [1] while adding around 40% extra manufacturing cost for 28 nm technology [2]. As we integrate more transistors within a single die, manufacturing such devices requires an expensive fabrication facility [3, 4]. Due to offshoring the IC design to overseas foundries for manufacturing, the original IC designer does not have centralized control to achieve trust of multiple parties involved from design to manufacturing. This has led to IC subversion and attack surface has increased as we go deeper into electronics supply chain. Furthermore, the same offshore companies manufacture the products that require higher assurance guarantee. Hence, in this loose control of system design, human lives can be endangered by exploiting safety-critical systems and exposing personal identity.

In the long electronics supply chain with untrusted entities, IC has become prone to various security threats, due to the large attack space in extended hardware design and life cycle. Various malicious modifications can be performed on hard-

---

This work was done as part of the first author's dissertation research at USF.

S. A. Islam (✉)
University of Texas Rio Grande Valley, Edinburg, TX, USA
e-mail: sheikhariful.islam@utrgv.edu

S. Katkoori
University of South Florida, Tampa, FL, USA
e-mail: katkoori@usf.edu

S. Katkoori, S. A. Islam (eds.), *Behavioral Synthesis for Hardware Security*,
https://doi.org/10.1007/978-3-030-78841-4_10

ware to achieve an attacker's objective with maximum damage. Similarly, due to sophisticated attack vectors and inherent attacker capabilities, covert manipulations in hardware are easy to perform and the corresponding vulnerabilities are higher than software counterparts. We call such manipulation as Hardware Trojan (HT) that can impact the underlying system in multiple ways (e.g., leaking sensitive information, disabling/malfunction critical parts of system, preventing legitimate access to systems, etc.). Due to multiple (and possibly untrusted) parties (IP vendors, integrator, foundries), it is difficult to define "one solution fits all." Similarly, HTs largely vary by their activation mechanism, physical locations, and payload [5]. Furthermore, the complexity of modern IC has grown bigger which makes it harder to possess HT-free IC.

As attacker wants HT to go unnoticed and undetected during traditional testing and manufacturing stages, she designs HTs that have minimal impact on design parameters, hence making side-channel effect negligible. In the early design phase, we need a tool that can make a good guess of possible locations of HT irrespective of golden designs and HT behavior. Similarly, the tool should complement traditional error-prone testing scheme and reduce the search time in localizing potential HT. Hence, the tool should avoid low-level expensive simulation, be easy-to-use, and scalable. In light of the above problems, we provide a scalable analytical framework that can avoid the exhaustive search of the input space and localize potential malicious HT early-on. We include the following contributions in this chapter:

- We present a modeling approach on how word-level statistics can help to model rare activity nets and their location in an architecture.
- We empirically demonstrate that different input statistics (bit-width, active region) are useful to analyze HT susceptibility.
- We propagate the technology independent statistical information to localize HT in short duration as precisely as possible.

10.2 Background

Stealthy hardware Trojan undermines the integrity of computing systems. Irrespective of the nature of a HT, it has two components: (a) trigger, and (b) payload. The trigger could be combinational or sequential, and it contains the set of input conditions that would characterize the net as "rare [6, 7] and/or less-rare [8]". Upon triggering these nets, payload (hidden behavior) would be activated at a given time. By nature, HT does not make fundamental modifications in function and for malicious activity, trigger conditions can be satisfied following certain event or set of events of attacker choice. For example, in combinational HT in Fig. 10.1a, the edge between (N8 and N11) and (N9 and N12) is found to have rare activity, which has been used to activate HT payload between (N12 and N13). In Fig. 10.1b, a

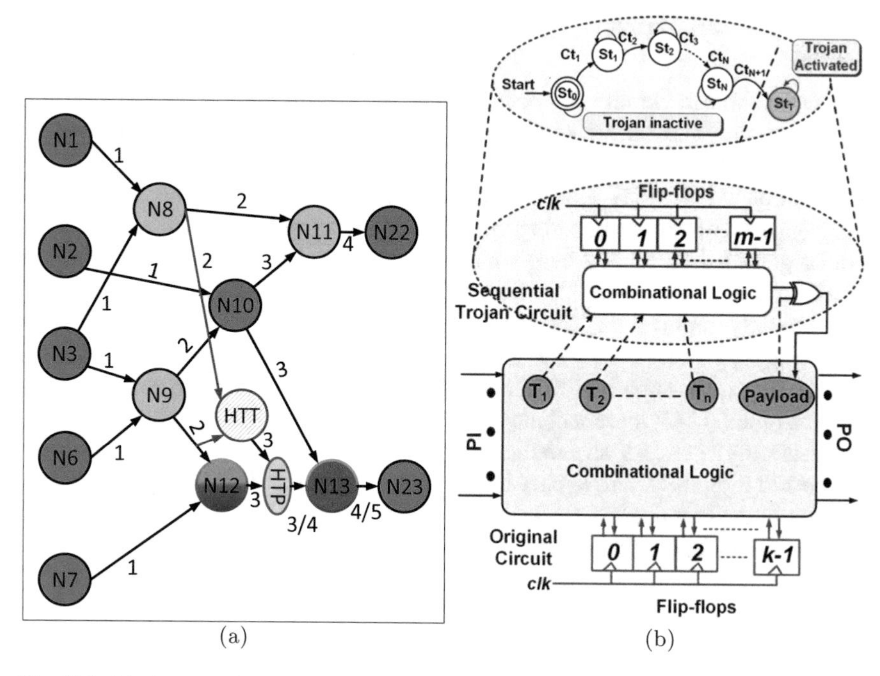

Fig. 10.1 Hardware Trojan triggering models. (**a**) Combinational HT model and (**b**) sequential HT model [9]: Copyright ©2017, Springer

sequence of rare events (e.g., state transitions, k-bit synchronous counter) can be utilized for sequential HT.

An intelligent attacker has sufficient amount of knowledge of traditional testing, verification, and validation procedure. Hence, she designs HT in such a way that will remain stealthy and its behavior will not be exhibited during the normal operation. Further, the size of the HT (<0.1% of the total area [10]) is adjusted accordingly so that any possible change in design parameter (timing, power, area) is insignificant. Similarly, HT can be inserted at any stage of design ecosystem. For example, during high-level design phase, most of the approved employees have access to specifications (micro-architecture, HDL code). Any unhappy employee can make certain modifications in the design which would bypass code review, implementation, and integration phase. Similarly, complex trigger conditions can be introduced at the gate-level for increased stealthiness. A layout designer has complete access to the netlist to compromise the (to be) manufactured product. As an example, she can remove the security verification module or change it in the GDSII file. Hence, the final product would be fabricated without HT detection circuitry.

10.3 Related Work

In general, we can classify the defense and detection techniques for HT into two categories (pre- and post-silicon). In most cases, the previous works focused on HT characterization based on testing. However, there are several works [6, 11–13] proposed on avoiding HT effect during runtime. Similarly, the authors in [14, 15] also suggested reverse engineering mechanism to localize HT. All these approaches require golden designs (HT-free) which is hard given the broad landscape of HT behavior. Hence, in the next section, we briefly summarize methods to identify (and possibly remove) HT at the pre-silicon stages (RT- and gate-level). We provide a brief overview of the existing works that fall under three categories: (a) simulation-free approach, (b) compact test vector generation approach, and (c) industrial tool(s). All these techniques assume certain HT behavior (triggering via rare nets only) whereas an intelligent attacker can combine rare nets with less-rare nets to trigger HT and bypass the existing detection techniques. Therefore, for completeness to capture HT behavior, we need to focus on both rare and/or less-rare nets for HT susceptibility analysis.

10.3.1 Modeling Technique

Although HT triggering mechanisms are non-trivial, we classify all previous techniques for the test pattern generation toward HT susceptibility into two broad categories, namely, statistical- and probabilistic-modeling.

Statistical-modeling techniques use random input vectors to simulate the circuit followed by categorizing rare nodes from non-rare nodes for a given threshold. The authors in [16] presented MERO that performs functional simulation under random test sequence to limit the search space, hence to improve the coverage (Trigger and Trojan). However, the technique does not account for data correlation to closely approximate [17] the bit-level signal transitions. The following work of MERO [18] extended HT detection sensitivity by including genetic algorithm and ignoring correlations that appear at the primary inputs. A practical, compact test vector generation scheme is proposed in [19] that still employs the random vector based circuit simulation. The authors in [20] sped up bit-level simulation by considering statistical parameters, the consideration of particular operand and its size is missing in the simulation environment. An information theoretic approach to generate test vector for the hard-to-reach region is proposed in [21].

Probabilistic-modeling techniques account for user-guided signal probabilities to estimate the switching event of each net in the design. Assuming a golden chip is available, the authors in [22] proposed a technique to generate a unique signature for Circuit Under Test (CUT). Given the geometric distribution of input signal, the authors in [23] inserted dummy scan flip-flops to equate the probabilities of "1" and "0," thus improving HT activation time. Insertion of 2-to-1 MUXes at selective places to avoid the limitation of reconvergent fanout is proposed in [23] which also considers only one form of transition.

10.3.2 Simulation-Free HT Identification

The authors in [24] provided a detailed treatment of statement hardness to score RTL code. Based on the available HT dataset, the authors in [25] proposed subgraph matching technique; however, it cannot capture emerging HT [26]. To verify information flow, the authors in [27] proposed partial Automatic Test Pattern Generation (ATPG) scheme. Similarly, at RT-level, the authors in [28] proposed RTLIFT for security verification while at the gate-level, GLIFT [27, 29] monitors data propagation and ensures if such propagation follows security properties. There are two formal verification techniques, Proof-Carrying Code (PCC) [30] and Bounded Model Checking (BMC) [31], to detect HT. An industrial tool, JasperGold® [32] applies "taint propagation" technique to catch any unintentional design behavior.

10.4 Threat Model

In traditional HT-based attack model, we assume that a rogue agent embeds additional logic and ensures that it would only be activated under rare conditions. In our proposed technique, there are two parties involved as shown in Fig. 10.2. A trustworthy designer develops RTL IP using benign tool. An attacker accepts the design during system integration and delivers that packaged design to the end-user during post-silicon. An attacker can also choose a subset of RTL IPs during integration and makes the detection approach harder. Using the domain knowledge of relevant IPs, an attacker can embed HT in module library instead of original design. As an attacker, she has two objectives in our proposed threat model:

1. Embedded vulnerable logic will have minimal impact on design parameters (power, performance, and area).
2. Location of HT should bypass detection mechanisms and generate higher misclassification rate.

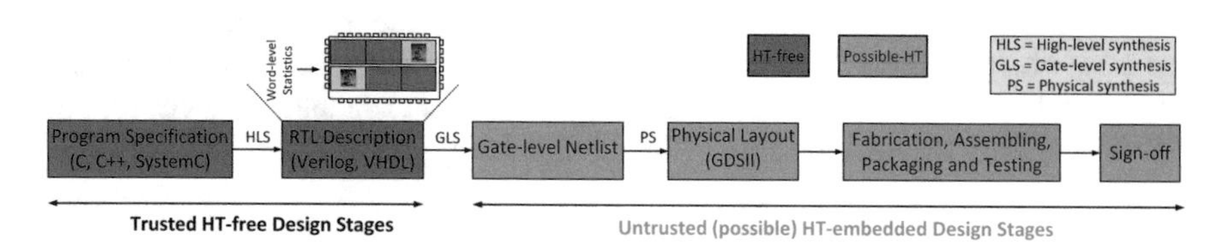

Fig. 10.2 Attack model for HT susceptibility in RTL IP core

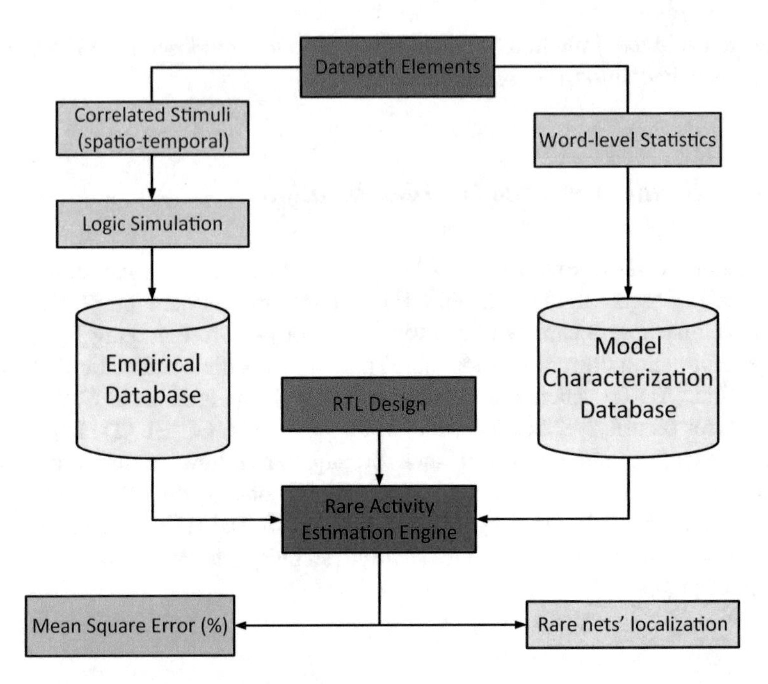

Fig. 10.3 HT susceptibility analysis in RTL IP core

Further, an attacker goal can be classified into two broad categories:

1. *Targeted attack*: Compromise the integrity of device on-field or reduce the lifetime of device under attack.
2. *Non-targeted attack*: No "visible" consequence of the attack (e.g., key leaking via backdoor or covert channel).

10.5 Proposed Approach

In this section, first, we present a modeling approach to capture the rare nets using word-level statistics of the inputs for analyzing HT vulnerability as shown in Fig. 10.3. We describe a technique of how to decompose the design effectively into a subset of basic arithmetic modules to estimate the rare triggering nets. Secondly, we characterize macro-blocks (empirically and analytically) and then use these characteristics to identify "low activity" nets and "local regions" without expensive low-level simulation. Finally, we validate and evaluate the DSP IPs for HT vulnerability over a wide range of input signal statistics.

10.5.1 Modeling Approach

The model parameters we have selected are free from any distribution and signal representation type. For a signal with non-zero mean with Gaussian distribution, word-level statistics can be expressed as follows:

$$(\mu_A, \sigma_A, \rho_A, BP_0, BP_1)_A = f(X_{t-1}^N, X_t^N, BW) \tag{10.1}$$

where, X_t^N is a signal of N-bit at time t in the time interval, $(-\frac{t}{2}, \frac{t}{2}]$. μ_A, σ_A, ρ_A, and (BP_0, BP_1) denote the average, standard deviation, temporal correlation, and breakpoints of X_t^N. Depending on the architecture and distribution type, the accuracy loss can be significant which can be improved by higher characterization time. Regardless of signal distribution, word-level statistics reduce the complexity of expensive low-level simulation with a greater flexibility in the estimation. The probability that X_t^N can assume logic-1 is given as follows [33]:

$$p_i^N = Pr(N_i = 1) = \sum_{\forall x \in \chi_i} \frac{1}{\sigma\sqrt{2\pi}} e^{-(x-\mu)^2/2\sigma^2} \tag{10.2}$$

Hence, the value of p_i^N can be measured once we calculate the parameters of Eq. (10.1). We can express mean, variance, and spatio-temporal auto-correlation of X_t^N as follows:

$$\mu_X = E[X_t^N] \tag{10.3}$$

$$\sigma_X = \sqrt{E[X_t^2] - E^2[X_t]} = \sqrt{E[X_t^2] - \mu^2} = p_i - p_i^2 \tag{10.4}$$

$$\rho = \frac{E(X_t^N, X_{t-1}^N) - \mu_X^2}{\sigma_X^2} = \frac{cov(X_{t-1}^N, X_t^N)}{var(X^N)} \tag{10.5}$$

For transition activity (logic-1 to logic-0 or logic-0 to logic-1), the switching activity can be expressed in terms of correlation (ρ) as follows:

$$\alpha_i = 2\sum_{i=0}^{N-1} p_i(1 - p_i)(1 - \rho_i) = 0.5(1 - \rho_i) \tag{10.6}$$

We can further estimate the switching activity of MSB (α_{msb}) according to [34]:

$$\alpha_{msb} = \frac{1}{\pi}\cos^{-1}(\rho) \tag{10.7}$$

Alternatively, the relation between correlation of MSB (ρ_{msb}) and word-level correlation (ρ) can be expressed as follows:

$$\rho_{msb} = \frac{2}{\pi} \sin^{-1}(\rho) \tag{10.8}$$

Breakpoints Estimation According to Dual Bit Type (DBT) model [35], we can partition a signal into three regions (LSB, linear, and MSB) to estimate transition activity. The characteristics of those regions are as follows:

- LSB ($0 \leq i \leq BP_0$): Maximum switching activity, near zero temporal correlation.
- Linear region ($BP_0 \leq i \leq BP_1$): A linear increase in ρ and lower switching activity.

From [36], we can deduce the following expressions for breakpoints:

$$BP_0 = \lfloor \log_2[2\sigma(1 - \rho_{msb})] \rceil \tag{10.9}$$

$$BP_1 = \lfloor \log_2[6\sigma_x \sqrt{(1 - \rho_{msb})}] \rceil \tag{10.10}$$

Using Eqs. (10.9) and (10.10), we can express the correlation and switching activity of each bit as follows:

$$\rho_i = \begin{cases} 0, & (i < BP_0) \\ \frac{\rho_{BP_1}(i - BP_0 + 1)}{BP_1 - BP_0}, & (BP_0 \leq i \leq BP_1 - 1) \\ \rho_{msb}, & (i \geq BP_1 - 1) \end{cases} \tag{10.11}$$

$$\alpha_i = \begin{cases} 2p_i^1(1 - p_i^1), & (i \leq BP_0) \\ 0.5 + (\alpha_{msb} - 0.5)\frac{i - BP_0}{BP_1 - BP_0}, & (BP_0 < i < BP_1) \\ \alpha_{msb}, & (i \geq BP_1) \end{cases} \tag{10.12}$$

10.5.2 Empirical Estimation

To empirically evaluate module architectures under correlated test sequence, we propose the framework shown in Fig. 10.4 that incorporates probabilistic dependencies of input data. This framework allows HTs to produce maximum switching under such input sequence. On the contrary, random vectors cannot capture correlation. Islam et al. [17] describes the multiplier-accumulator requirements of DSP cores. However, the same framework can also be applied to generate stimuli for other datapath components.

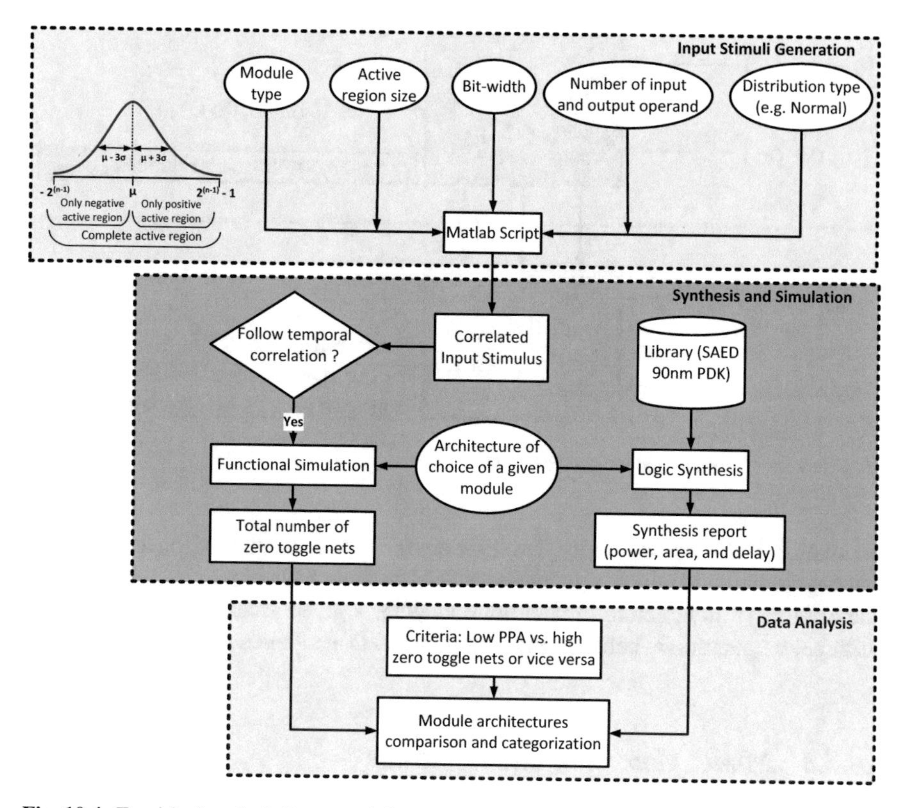

Fig. 10.4 Empirical analysis framework for HT vulnerability

For a signal of normal distribution with statistical parameters (μ and σ), we can analyze bit transition profile as shown in Fig. 10.5. In generating input stimuli, we have considered different active regions depending on the value and sign of the signal. For any active region, the total number of bits present in "sign" can be expressed as follows in terms of input statistics:

$$N_{sign} = \lfloor active\_regions(\%) * bit\_width \rfloor = \log_2 \left(\frac{\mu_X}{3\sigma_X} + 1 \right) \tag{10.13}$$

We also further classify the input value range into three classes (also known as dynamic range) as follows:

- Negative to positive ($-2^{N-1} \leq X_t^N \leq 2^{N-1} - 1$)
- Only negative ($-2^{(N-1)}$ to 0)
- Only positive (0 to $2^{(N-1)} - 1$)

Depending on possible input transitions, these three dynamic ranges include all sign transitions in the sign bit. Thus, it allows a higher accuracy to cover all types of signal transitions for two-input adder and multiplier architectures in the

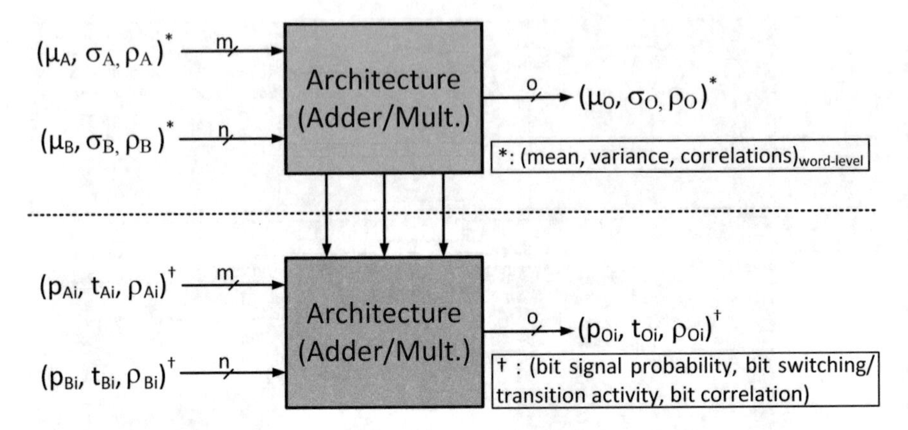

Fig. 10.5 Input stimuli generation technique

estimation of the signal activity. We carry out the functional simulation and consider the toggle activity which also includes intermediate transitions. We characterize the arithmetic architecture to determine whether a given architecture of particular arithmetic operation is better in PPA but poor in HT resilience or vice versa.

10.5.3 Model Validation and Accuracy

During analytical modeling, we first calculate the breakpoints. As discussed previously, the bit(s) beyond BP_1 shows minimal switching and could manifest as HT triggering nets. Alternatively, under rare circumstances, one can focus on the LSB region when we need to observe higher toggle activity. Such a framework is shown in Fig. 10.6. First, we partition the architecture into multiple sub-modules which are later divided into bit-slice units. Once we find out the BP_1 position, all the sub-module(s) from BP_1 to sign-bit position are accounted for and the nets from those modules form rare nets. For different input bit-width and correlations, we observe different BP_1 positions leading to a different count of nets under particular triggering probability.

If there are m modules in the architecture of type i where $1 \leq i \leq m$ and each type has n nets, we can estimate all such rare nets for set of modules, j $(1 \leq j \leq i)$ using

$$T_{rare} = R(m, i, j) = \sum_{j=1} j * m_j \tag{10.14}$$

Given two operands of different bit-widths, we can calculate the error (e and $\overline{e}$) as the difference between simulation (P_{sim}) and estimation (P_{est}) of rare triggering nets:

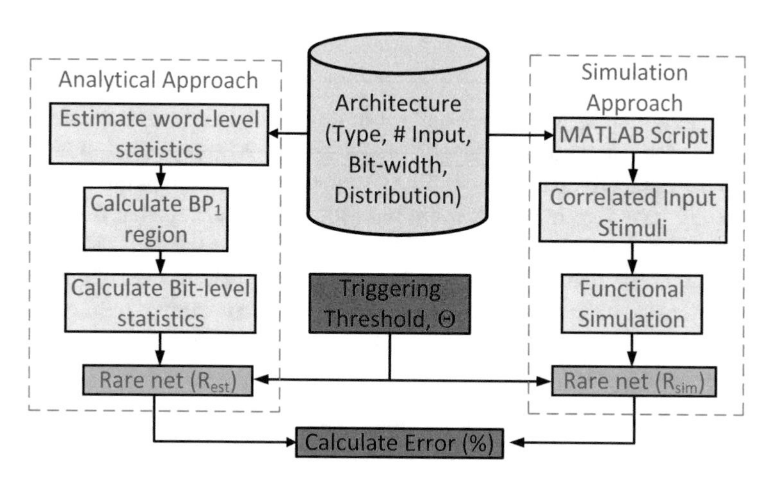

Fig. 10.6 Model validation framework

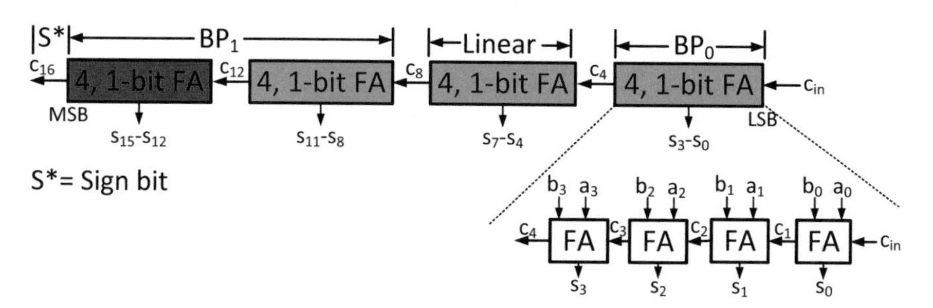

Fig. 10.7 Unit and block-wise decomposition of 16-bit RCA

$$
\begin{aligned}
e &= |\frac{P_{sim} - P_{est}}{P_{sim}}| \\
\bar{e} &= \frac{1}{n}\sum_{i=1}^{n} e_i
\end{aligned}
\tag{10.15}
$$

Motivational Example of Rare Nets Activation We decompose 16-bit Ripple Carry Adder (RCA) into three regions and four blocks as shown in Fig. 10.7. Given input statistics, breakpoints can be estimated as follows:

1. BP_0 (0th to 3rd)
2. linear (4th to 7th)
3. BP_1 (8th to 15th)

As mentioned earlier, we will observe the rarest triggering event in BP_1 region, hence, refer all nets in third and fourth block instances as rare nets.

10.5.4 Propagation of Word-Level Statistics

Given the design statistics at the primary inputs, we can estimate the statistics of each intermediate node and primary outputs by propagating signal statistics. We assume basic datapath operators (addition, subtraction, multiplication, division, and multiplexing) have two inputs of equal bit-width while for register, it is one input. We model the datapath in RTL design as Directed Acyclic Graph (DAG) where the nodes represent the operations and edges denote the relations between operations. We use the cost metric proposed in [17] to choose a competing architecture from all available ones. During the propagation of word-level statistics, we classify the operations into two categories:

- *Fully specified operation*: Input parameters are readily available.
- *Partially specified operation*: We decompose it into a set of cliques and index each of them (cluster) maintaining the dependency. Then, we solve each of the clusters iteratively according to the index. As it is common in DSP core to have feedback paths, we only propagate the information from the current cluster to the next ones if there is dependency.

As an example, consider the scheduled datapath of HAL architecture shown in Fig. 10.8. It has four (4) multiplication and two (2) subtraction operations. We assume each primary input is 8 bits wide. Once simulated with input streams, we calculate the total rare triggering nets and accordingly estimate the model efficiency by measuring error between simulation and analytical approaches. We pinpoint three module instances (SUB1_0, MULT0_0, and Mux5) for possible locations of HT given a triggering threshold less than 10^{-4} (Fig. 10.8b). Further, their locations in the layout are close enough to increase HT impact which confirms that majority of rare transitions happen in BP_1.

10.6 Experimental Evaluation

In this section, we present the empirical and analytical evaluation for six adder and four multiplier architectures. We then report the effectiveness of both approaches on six DSP IP cores. In addition to dominating datapath elements (adder and multiplier), we also include the statistics of remaining datapath components for which we have not found any competing architecture. We provide an automated approach to deal with HTs having any of two triggering types: (a) always-on; and (b) triggered by current inputs. We utilize in-house MATLAB script to generate test vectors for a given active region, dynamic range, and operand size. The module architecture is simulated with the test vectors which provide the nets having toggle activity of attacker's choice from Switching Activity Interchange Format file. The same module IP is also synthesized without any optimization to measure the design metrics. The empirical estimations are compared with nets found by analytical

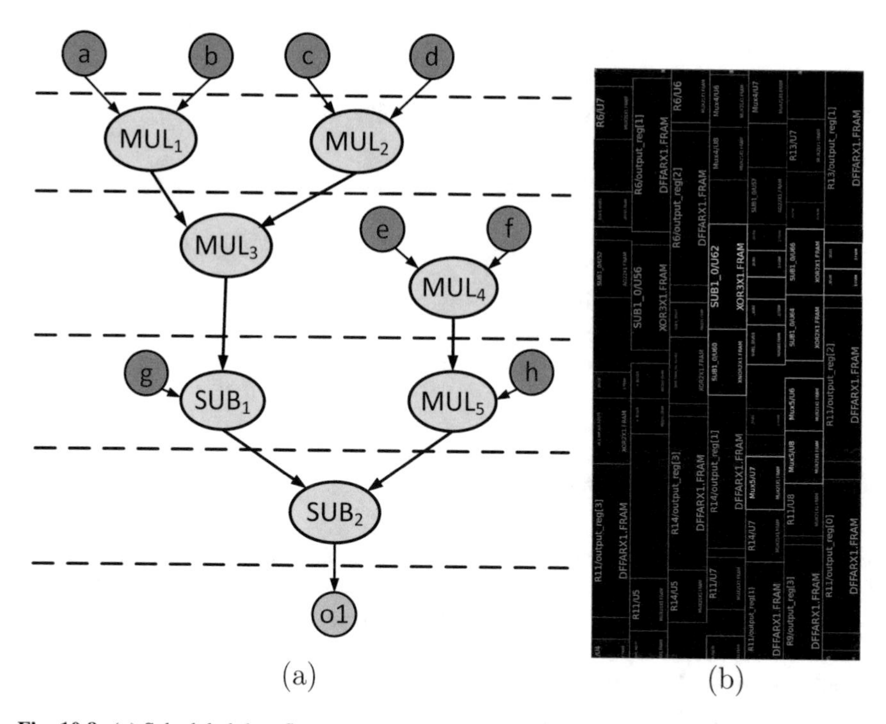

Fig. 10.8 (**a**) Scheduled data-flow graph for HAL architecture and (**b**) leaf cells of floor-planned HAL as location of HT

approach. Accordingly, we estimate the mean square error which can be made more accurate with inexact delay model. To verify the accuracy of the proposed approach, we propagate the statistics of input signals in DSP cores and calculate the error in estimating the rare nets.

IP Modules with Two Input Operands

- `RCA` : Ripple Carry Adder—Input bit-width (8-, 16-, and 32-bit)
- `CLA`: Carry Lookahead Adder—Input bit-width (8-, 16-, and 32-bit)
- `CKA` : Carry Skip Adder—Input bit-width (8-, 16-, and 32-bit)
- `CSA` : Carry Select Adder—Input bit-width (8-, 16-, and 32-bit)
- `KSA` : Kogge–Stone Adder—Input bit-width (8-, 16-, and 32-bit)
- `HA` : Hybrid Adder—Input bit-width (8-, 16-, and 32-bit)
- `Array Multiplier`—Input bit-width (8- and 16-bit)
- `Dadda Multiplier`—Input bit-width (8- and 16-bit)
- `Vedic Multiplier`—Input bit-width (8- and 16-bit)
- `Booth Multiplier`—Input bit-width (8- and 16-bit)

DSP IP Description

- `Diffeq`: Differentiation Equation Solver
- `FFT`: 8-point Fast Fourier Transform algorithm

- `FIR`: 5-point Finite Impulse Response algorithm
- `Lattice`: Single input linear prediction filter
- `Wavelet`: Discrete wavelet transformation algorithm

EDA Tool Description

- `Synopsys VCS-MX`: Functional simulation for 10,000 correlated input test vectors
- `Synopsys Design Compiler`: Logic synthesis for 90nm PDK

Miscellaneous Consideration

- `Input distribution`: Normal
- `Rare net`: Toggle activity between 0 to 10^{-6}

10.6.1 Adder Architecture Analysis

Empirical Results We tabulate the synthesis results of six different adder architectures in Table 10.1. In terms of power and area, RCA performs better than the rest of adders while KSA and CLA outperform in case of delay improvement. Similarly, we report the switching activity of adder architectures for 16-, 32-, and 64-bit in Tables 10.2, 10.3, and 10.4, respectively. We indicate three dynamic ranges (only negative, only positive, and negative to positive) by (−0), (0+), and (+−), respectively. For complete dynamic range, we observe less-rare switching nets compared to other two dynamic ranges in case of 16- and 32-bit. The same observation goes to 64-bit where we see more threshold toggle nets. For RCA and CLA architectures, we observe the same toggle nets; however, CLA is preferred over RCA in terms of carry propagation. Then, we vary the active region from 50% to 98% for each adder architecture under three different bit-widths. We tabulate the results in Table 10.5. Comparing all other architectures, KSA is most susceptible while CLA is least susceptible to HTs even for large operand sizes.

Analytical Estimation For correlation (ρ) value of 0.5, we provide error (%) for 8- and 16-bit adder architectures in Tables 10.6 and 10.7, respectively. We can observe

Table 10.1 Synthesis results of adder architectures of three different bit-widths. The units of power, area, and delay are μW, μm^2, and ns, respectively

Adder	16-bit			32-bit			64-bit		
design	Power	Area	Delay	Power	Area	Delay	Power	Area	Delay
RCA	9.21	794.67	11.17	15.67	1765.33	17.04	22.28	3450.38	31.36
CLA	16.09	1283.41	7.90	28.24	2616.52	9.73	67.09	6289.98	10.22
CKA	13.34	966.96	9	16.444	2181.38	20.34	25.27	4502.57	37.77
CSA	15.60	1621.04	9	34.20	3510.91	13	61.96	7012.29	15.12
KSA	27.97	1829.30	7.46	60.79	4550.14	8	91.12	7535.55	9.23
HA	10.73	1068.01	12.45	25.79	2343.68	10.5	58.90	5771.06	12.5

Table 10.2 Comparison of switching activity (TC_{sel}) of 16-bit adder architectures under three dynamic regions

Arch. name	Dynamic region type	Rare activity nets					
		TC = 0	TC < 10e−6	TC < 10e−5	TC < 10e−4	TC < 10e−3	TC < 10e−2
RCA	(−0)	7	7	7	7	7	7
	(0+)	7	7	7	7	7	7
	(−+)	1	1	1	1	1	1
CLA	(−0)	7	7	7	7	7	7
	(0+)	7	7	7	7	7	7
	(−+)	0	0	0	0	0	0
CKA	(−0)	16	16	16	16	16	16
	(0+)	16	16	16	16	16	16
	(−+)	3	3	3	3	3	3
CSA	(−0)	36	36	36	36	36	36
	(0+)	36	36	36	36	36	36
	(−+)	18	18	18	18	18	18
KSA	(−0)	26	26	26	26	26	26
	(0+)	26	26	26	26	26	26
	(−+)	1	1	1	1	1	2
HA	(-0)	27	27	27	27	35	36
	(0+)	27	27	27	27	35	36
	(−+)	2	2	2	2	11	13

as input bit index (BP_1) is close to sign-bit region, the analytical estimation results are very close to that of functional simulation. When input bit index approaches near to LSB region, there is a limited range of random activity in which case simulation also considers glitch activities. We also present analytical estimation error (%) for 8-, 16-, 32-, and 64-bit adder in Tables 10.6, 10.7, and 10.8, respectively, for correlation (ρ) of 0.99. For CLA, we unroll carry equations to closely approximate the rare nets which increases the delay. This has been reflected on Table 10.9 when the correlation is equal to 0.99. For CKA, we achieve shortest time in carry propagation, however, it requires us to include the nets present in carry skip logic. Hence, for CKA, the average error is around 1.35%.

10.6.2 Multiplier Architecture Analysis

Empirical Results We tabulate the synthesis results of multiplier architectures of 8- and 16-bit in Table 10.10. Dadda multiplier exhibits low PPA characteristics. It is also true for low TC_0 as we sweep the activity region and it is reflected in

Table 10.3 Comparison of switching activity (TC_{sel}) of 32-bit adder architectures under three dynamic regions

Arch. name	Dynamic region type	Rare activity nets					
		TC = 0	TC < 10e−6	TC < 10e−5	TC < 10e−4	TC < 10e−3	TC < 10e−2
RCA	(−0)	7	7	7	7	7	7
	(0+)	7	7	7	7	7	7
	(−+)	1	1	1	1	1	1
CLA	(−0)	7	7	7	7	7	7
	(0+)	7	7	7	7	7	7
	(−+)	0	0	0	0	0	0
CKA	(−0)	16	16	16	16	16	16
	(0+)	16	16	16	16	16	16
	(−+)	3	3	3	3	3	3
CSA	(−0)	52	52	52	52	52	52
	(0+)	52	52	52	52	52	52
	(−+)	34	34	34	34	34	34
KSA	(-0)	30	30	30	30	38	39
	(0+)	30	30	30	30	38	39
	(−+)	1	1	1	2	11	13
HA	(-0)	48	48	48	48	103	105
	(0+)	41	41	41	48	103	105
	(−+)	18	18	18	21	75	80

Table 10.13. Similarly, for three different dynamic regions, we present the results in Tables 10.11, 10.12 and we observe the same result (Table 10.13).

Analytical Estimation We present the analytical and average estimation error (%) in Tables 10.14 and 10.15. For Booth multiplier, we employ fully parallel and carry-free implementation. It leads to errors of 0.27% and 0.22% for 8- and 16-bit Booth multiplier, respectively. For Vedic multiplier, we require more than two additional adders per stage to calculate the partial products in parallel. Hence, it shows error of 0.83% and 0.80% for Vedic multiplier.

10.6.3 Statistics Propagation in DSP Cores for Rare Transition Activity

We evaluate the modeling and empirical approaches on six DSP IP cores for five different input signals and three input correlations (0.1, 0.5, and 0.9). There are two outputs (`yvar2`) and (`uvar2`) associated with `Diffeq` core. For random correlation ($\rho = 0.5$), we observe lowest error which is not true for highly

Table 10.4 Comparison of switching activity (TC_{sel}) of 64-bit adder architectures under three dynamic regions

Arch. name	Dynamic region type	Rare activity nets					
		TC = 0	TC < 10e−6	TC < 10e−5	TC < 10e−4	TC < 10e−3	TC < 10e−2
RCA	(−0)	95	95	95	95	95	95
	(0+)	79	79	79	79	79	81
	(−+)	9	9	9	19	39	63
CLA	(−0)	84	84	84	84	84	84
	(0+)	70	70	70	70	70	71
	(−+)	4	4	4	12	31	53
CKA	(−0)	169	169	169	169	169	169
	(0+)	143	143	143	143	143	145
	(−+)	16	16	16	31	66	110
CSA	(−0)	336	336	336	336	336	336
	(0+)	292	292	292	292	292	296
	(−+)	81	81	81	109	167	238
KSA	(−0)	161	161	161	161	216	218
	(0+)	134	134	134	141	196	199
	(−+)	22	22	22	36	118	154;
HA	(−0)	496	496	496	496	616	624
	(0+)	397	397	397	430	562	581
	(−+)	184	184	184	235	433	537

Table 10.5 Rare nets for different activity regions (50–100%) of input bit indexes

	16-bit						32-bit						64-bit					
Adder	8	10	12	14	15	16	16	20	24	28	31	32	32	40	48	56	63	64
CLA	114	86	58	30	15	14	225	169	113	57	15	14	526	470	302	197	151	158
RCA	135	103	71	39	19	15	259	195	131	67	19	15	603	539	347	227	177	183
CKA	247	185	133	71	40	35	465	351	237	123	40	35	1074	960	618	403	313	328
CSA	412	321	228	137	95	90	834	648	462	276	143	138	1926	1740	1182	832	683	709
KSA	482	354	234	120	67	56	1118	843	565	300	119	107	3014	2710	1832	1280	1051	1077
HA	399	321	219	110	60	53	936	767	524	259	74	61	2306	2103	1324	733	318	317

correlated ($\rho = 0.9$) and anti-correlated ($\rho = 0.1$) signal, hence increases the error. This is reflected in Tables 10.16 and 10.17. We report the rare activity nets for remaining five DSP cores in Tables 10.18, 10.19, and 10.20. We observe higher error percentage for lower bit-width which is due to lack of test vector coverage of all input range. On the contrary, for higher bit-width, the word-level statistics can closely approximate multi-bit inputs. For completeness, we report the HT triggering rare nets and average estimation error (%) in Tables 10.21 and 10.22.

Table 10.6 Analytical estimation error (%) for 8-bit adder architectures under two different correlations

Input bit index	Corr (=0.5)						Corr (=0.99)					
	RCA	CLA	CKA	CSA	KSA	HA	RCA	CLA	CKA	CSA	KSA	HA
4th	0.68	0.56	0.60	0.74	0.86	0.46	0.73	0.63	0.68	0.86	0.85	0.31
5th	0.53	0.49	0.56	0.73	0.74	0.39	0.56	0.58	0.67	0.83	0.73	0.30
6th	0.49	0.44	0.55	0.58	0.66	0.39	0.51	0.52	0.59	0.72	0.63	0.28
7th	0.46	0.38	0.54	0.50	0.58	0.34	0.42	0.49	0.59	0.67	0.55	0.27

Table 10.7 Analytical estimation error (%) for 16-bit adder architectures under different correlations

Input bit index	Corr (=0.5)						Corr (=0.99)					
	RCA	CLA	CKA	CSA	KSA	HA	RCA	CLA	CKA	CSA	KSA	HA
3rd	0.73	0.66	1.30	0.83	0.92	0.90	1.36	0.64	1.67	0.72	0.82	0.86
5th	0.73	0.48	1.27	0.80	0.78	0.74	1.27	0.45	1.52	0.66	0.59	0.66
7th	0.66	0.33	1.25	0.65	0.67	0.61	1.23	0.30	1.50	0.61	0.45	0.61
9th	0.57	0.25	1.16	0.61	0.56	0.61	0.93	0.27	1.48	0.58	0.45	0.60
11th	0.50	0.23	1.02	0.56	0.46	0.59	0.68	0.26	1.43	0.58	0.44	0.59
13th	0.44	0.22	0.81	0.49	0.38	0.59	0.63	0.25	0.96	0.56	0.44	0.58
15th	0.40	0.20	0.71	0.46	0.31	0.58	0.58	0.24	0.85	0.52	0.38	0.56

Table 10.8 Analytical estimation error (%) for 32- and 64-bit adder architectures

Input bit index	32-bit (corr = 0.99)						Input bit index	64-bit (corr = 0.99)					
	RCA	CLA	CKA	CSA	KSA	HA		RCA	CLA	CKA	CSA	KSA	HA
4th	0.94	0.79	1.07	1.14	0.95	0.94	16th	1.03	0.90	0.85	0.82	0.98	0.98
8th	0.90	0.63	1.01	1.05	0.84	0.80	24th	0.93	0.66	0.63	0.60	0.84	0.92
12th	0.87	0.45	0.94	0.89	0.72	0.64	30th	0.84	0.53	0.54	0.50	0.76	0.88
16th	0.86	0.30	0.81	0.73	0.62	0.53	36th	0.69	0.41	0.53	0.49	0.69	0.84
20th	0.76	0.19	0.76	0.73	0.60	0.53	42th	0.63	0.31	0.53	0.49	0.62	0.80
24th	0.62	0.18	0.70	0.63	0.55	0.51	48th	0.49	0.23	0.51	0.45	0.59	0.55
28th	0.60	0.15	0.61	0.61	0.54	0.48	54th	0.46	0.16	0.48	0.44	0.54	0.46
31th	0.51	0.15	0.59	0.56	0.54	0.40	63th	0.42	0.12	0.46	0.34	0.50	0.42

Table 10.9 Average estimation error (%) for 8- and 16-bit adder architectures under different input bit indexes

	Arch. name	$\rho = 0.5$ 8-bit	$\rho = 0.5$ 16-bit	$\rho = 0.99$ 8-bit	$\rho = 0.99$ 16-bit
Adder	RCA	0.58	0.54	0.95	0.55
	CLA	0.47	0.34	0.56	0.35
	CKA	1.08	0.56	1.35	0.63
	CSA	0.64	0.63	0.77	0.60
	KSA	0.71	0.58	0.69	0.51
	HA	0.66	0.39	0.64	0.29

Table 10.10 Synthesis results for multiplier architectures

Multiplier architecture	8-bit Power	8-bit Area	8-bit Delay	16-bit Power	16-bit Area	16-bit Delay
Array	41.50	4885.80	18.50	130.12	16,748.30	34
Dadda	32.75	2860.16	12.91	115.09	11,545.33	23.94
Vedic	54.35	6728.80	17.72	131.43	19,131.08	25.35
Booth	38.19	3983.53	20	228.64	23,897.65	23.12

The unit of power, area, and delay is μW, μm^2, and ns, respectively

Table 10.11 Comparison of switching activity (TC_{sel}) of 8-bit multiplier architectures under three dynamic regions

Arch. name	Dynamic region type	Rare activity nets TC = 0	TC < 10e−6	TC < 10e−5	TC < 10e−4	TC < 10e−3	TC < 10e−2
Array	(−0)	92	92	92	92	92	98
	(0+)	146	146	146	146	146	176
	(−+)	71	71	71	71	71	71
Dadda	(−0)	2	2	2	2	2	2
	(0+)	4	4	4	4	4	5
	(−+)	0	0	0	0	0	0
Vedic	(−0)	386	386	386	386	387	485
	(0+)	721	721	721	721	721	916
	(−+)	367	367	367	367	379	545
Booth	(−0)	37	37	37	37	45	75
	(0+)	51	51	51	51	56	84
	(−+)	3	3	3	3	15	54

Table 10.12 Comparison of switching activity (TC_{sel}) of 16-bit multiplier architectures under three dynamic regions

Arch. name	Dynamic region type	Rare activity nets					
		TC = 0	TC < 10e−6	TC < 10e−5	TC < 10e−4	TC < 10e−3	TC < 10e−2
Array	(−0)	180	180	180	180	180	186
	(0+)	282	282	282	282	282	336
	(−+)	143	143	143	143	143	146
Dadda	(−0)	2	2	2	2	2	2
	(0+)	4	4	4	4	4	5
	(−+)	0	0	0	0	0	0
Vedic	(−0)	255	255	255	255	255	285
	(0+)	498	498	498	498	498	572
	(−+)	224	224	224	224	224	259
Booth	(−0)	127	127	127	131	136	172
	(0+)	124	124	124	124	129	165
	(−+)	105	105	105	105	113	181

Table 10.13 Rare nets for different activity regions (50–100%) of input bit indexes

	8-bit					16-bit					
Multiplier	4	5	6	7	8	8	10	12	14	15	16
Array	849	678	546	416	309	2544	1952	1450	996	792	605
Dadda	37	27	20	13	6	64	48	34	20	13	6
Vedic	3397	3076	2414	2016	1492	3697	2935	2253	1603	1269	977
Booth	1195	725	607	193	116	3208	2386	1665	876	626	378

Table 10.14 Analytical estimation error (%) for 8- and 16-bit multiplier architectures

Input bit index	8-bit (corr = 0.99)				Input bit index	16-bit (corr = 0.99)			
	Array	Vedic	Dadda	Booth		Array	Vedic	Dadda	Booth
3rd	0.91	0.87	0.82	0.73	3rd	0.96	0.96	0.91	0.41
4th	0.84	0.87	0.64	0.20	5th	0.88	0.89	0.81	0.35
5th	0.77	0.84	0.45	0.19	7th	0.83	0.83	0.69	0.27;
6th	0.72	0.79	0.44	0.15	9th	0.78	0.79	0.57	0.19
7th	0.67	0.77	0.42	0.08	11th	0.74	0.72	0.35	0.15
–	–	–	–	–	13th	0.73	0.70	0.34	0.12
–	–	–	–	–	15th	0.72	0.70	0.30	0.08

Table 10.15 Average estimation error (%) for 8- and 16-bit multiplier architectures under different input indexes

Architecture name	$\rho = 0.5$		$\rho = 0.99$	
	8-bit	16-bit	8-bit	16-bit
Array	0.76	0.72	0.81	0.78
Vedic	0.79	0.75	0.83	0.80
Dadda	0.5	0.47	0.57	0.55
Booth	0.25	0.21	0.27	0.22

Table 10.16 Analytical and average estimation error (%) for signal (yvar2) of Diffeq design under three different bit-widths

	$\rho = 0.1$			$\rho = 0.5$			$\rho = 0.9$		
Signal index	8-bit	16-bit	32-bit	8-bit	16-bit	32-bit	8-bit	16-bit	32-bit
Signal_1	0.34	0.27	0.05	0.31	0.22	0.074	0.36	0.44	0.18
Signal_2	0.37	0.29	0.18	0.28	0.25	0.01	0.41	0.36	0.23
Signal_3	0.45	0.26	0.1	0.29	0.14	0.06	0.57	0.48	0.13
Signal_4	0.37	0.38	0.12	0.3	0.02	0.01	0.25	0.54	0.15
Signal_5	0.41	0.36	0.24	0.29	0.13	0.001	0.53	0.48	0.12
Average	0.38	0.31	0.13	0.29	0.15	0.03	0.42	0.46	0.16

Table 10.17 Analytical and average estimation error (%) for signal (uvar2) of Diffeq design under three different bit-widths

	$\rho = 0.1$			$\rho = 0.5$			$\rho = 0.9$		
Signal index	8-bit	16-bit	32-bit	8-bit	16-bit	32-bit	8-bit	16-bit	32-bit
Signal_1	0.39	0.44	0.01	0.27	0.12	0.009	0.48	0.39	0.05
Signal_2	0.33	0.51	0.17	0.15	0.003	0.09	0.61	0.31	0.22
Signal_3	0.39	0.27	0.26	0.07	0.16	0.05	0.31	0.4	0.32
Signal_4	0.53	0.21	0.03	0.45	0.03	0.01	0.52	0.33	0.19
Signal_5	0.42	0.36	0.07	0.16	0.12	0.07	0.48	0.53	0.3
Average	0.41	0.35	0.11	0.22	0.08	0.04	0.48	0.39	0.21

Table 10.18 Analytical estimation error (%) of five DSP designs under three different bit-widths for ρ (=0.1)

	8-bit					16-bit					32-bit				
Signal index	Signal_1	Signal_2	Signal_3	Signal_4	Signal_5	Signal_1	Signal_2	Signal_3	Signal_4	Signal_5	Signal_1	Signal_2	Signal_3	Signal_4	Signal_5
FIR	0.057	0.05	0.21	0.0034	0.028	0.045	0.061	0.05	0.047	0.063	0.05	0.041	0.06	0.031	0.045
Latt	0.77	0.85	0.71	0.56	0.65	0.48	0.72	0.71	0.63	0.67	0.67	0.72	0.48	0.28	0.45
FFT	0.027	0.057	0.12	0.28	0.033	0.091	0.081	0.084	0.087	0.09	0.059	0.066	0.056	0.063	0.057
Paul	0.76	0.66	0.68	0.74	0.65	0.72	0.76	0.42	0.61	0.73	0.6	0.62	0.63	0.41	0.74
Wave	0.58	0.87	0.79	0.55	0.8	0.52	0.45	0.39	0.57	0.64	0.42	0.53	0.23	0.37	0.54

Table 10.19 Analytical estimation error (%) of five DSP designs under three different bit-widths for ρ (=0.5)

	8-bit					16-bit					32-bit				
Signal index	Signal_1	Signal_2	Signal_3	Signal_4	Signal_5	Signal_1	Signal_2	Signal_3	Signal_4	Signal_5	Signal_1	Signal_2	Signal_3	Signal_4	Signal_5
FIR	0.038	0.0032	0.028	0.078	0.02	0.033	0.008	0.034	0.029	0.034	0.018	0.0087	0.017	0.015	0.025
Latt	0.57	0.7	0.32	0.45	0.45	0.6	0.52	0.39	0.6	0.36	0.44	0.54	0.62	0.44	0.34
FFT	0.072	0.059	0.08	0.064	0.079	0.05	0.059	0.052	0.047	0.07	0.019	0.0057	0.014	0.022	0.019
Paul	0.72	0.58	0.62	0.52	0.64	0.42	0.57	0.44	0.68	0.61	0.41	0.5	0.35	0.42	0.38
Wave	0.63	0.66	0.67	0.72	0.58	0.27	0.66	0.47	0.37	0.16	0.25	0.26	0.11	0.48	0.18

Table 10.20 Analytical estimation error (%) of five DSP designs under three different bit-widths for ρ (=0.9)

	8-bit					16-bit					32-bit				
Signal index	Signal_1	Signal_2	Signal_3	Signal_4	Signal_5	Signal_1	Signal_2	Signal_3	Signal_4	Signal_5	Signal_1	Signal_2	Signal_3	Signal_4	Signal_5
FIR	0.27	0.079	0.069	0.2	0.013	0.11	0.085	0.036	0.044	0.13	0.075	0.085	0.082	0.078	0.066
Latt	1	0.82	0.81	1	0.87	0.89	0.87	0.94	0.75	0.89	0.79	0.74	0.65	0.94	0.72
FFT	0.27	0.063	0.28	0.32	0.031	0.006	0.0092	0.11	0.21	0.17	0.081	0.11	0.085	0.097	0.087
Paul	0.83	0.9	1.1	0.9	0.86	0.79	0.76	0.67	0.94	0.84	0.62	0.54	0.71	0.8	0.86
Wave	0.83	0.87	1.1	0.75	0.84	0.56	0.56	0.62	0.71	0.58	0.52	0.45	0.59	0.58	0.67

Table 10.21 Rare activity nets of benchmark for $\rho = 0.5$

	8-bit			16-bit)			32-bit)		
DSP cores	T.N.	R.N. (✓)	R.N. (✞)	T.N.	R.N. (✓)	R.N. (✞)	T.N.	R.N. (✓)	R.N. (✞)
Diffeq	1153	6	9	3334	173	215	6941	555	599
FIR	1319	92	95	3505	235	240	7377	472	477
Lattice	1157	12	18	2681	21	31	5471	33	49
FFT	3352	231	247	13,488	863	906	24,900	1270	1283
Paulin	1091	99	159	3579	304	468	6807	545	768
Wavelet	1032	49	81	2890	118	163	5531	183	229

T.N. = Total Nets; R.N. (✓) = Rare Net (Simulation); and R.N. (✞) = Rare Net (Analytical)

Table 10.22 Average estimation error (%) of DSP cores in rare transition activity

	8-bit			16-bit			32-bit		
DSP cores	$\rho = 0.1$	$\rho = 0.5$	$\rho = 0.9$	$\rho = 0.1$	$\rho = 0.5$	$\rho = 0.9$	$\rho = 0.1$	$\rho = 0.5$	$\rho = 0.9$
Diffeq	0.80	0.51	0.90	0.67	0.24	0.85	0.24	0.08	0.37
FIR	0.06	0.03	0.12	0.05	0.02	0.08	0.04	0.01	0.07
Lattice	0.70	0.49	0.90	0.64	0.49	0.86	0.52	0.47	0.76
FFT	0.10	0.07	0.19	0.08	0.05	0.10	0.06	0.01	0.09
Paulin	0.69	0.61	0.91	0.64	0.54	0.80	0.60	0.41	0.70
Wavelet	0.71	0.65	0.87	0.51	0.38	0.60	0.41	0.25	0.56

10.7 Conclusions

In this chapter, we provide a framework to analyze HT susceptibility empirically and analytically. Theoretically, we have shown that DBT model can provide approximation about rare activity nets in RTL IP cores. Then, for model accuracy, we provide empirical analysis to localize rare activity nets. For model validation with empirical results, we propagate the transition activity of arithmetic module architectures and find the estimation error to be around 1–2% under different bit-widths.

References

1. Deloitte: Semiconductors - the Next Wave. https://www2.deloitte.com/content/dam/Deloitte/cn/Documents/technology-media-telecommunications/deloitte-cn-tmt-semiconductors-the-next-wave-en-190422.pdf (2019)
2. Liu, B., Wang, B.: Embedded Reconfigurable Logic for ASIC Design Obfuscation Against Supply Chain Attacks. In: 2014 Design, Automation Test in Europe Conference Exhibition (DATE), pp. 1–6 (2014). https://doi.org/10.7873/DATE.2014.256
3. Guin, U., Huang, K., DiMase, D., Carulli, J.M., Tehranipoor, M., Makris, Y.: Counterfeit Integrated Circuits: A Rising Threat in the Global Semiconductor Supply Chain. Proceedings of the IEEE **102**(8), 1207–1228 (2014). https://doi.org/10.1109/JPROC.2014.2332291

4. Rostami, M., Koushanfar, F., Karri, R.: A Primer on Hardware Security: Models, Methods, and Metrics. Proceedings of the IEEE **102**(8), 1283–1295 (2014). https://doi.org/10.1109/JPROC.2014.2335155
5. Xiao, K., Forte, D., Jin, Y., Karri, R., Bhunia, S., Tehranipoor, M.: Hardware Trojans: Lessons Learned After One Decade of Research. ACM Transactions on Design Automation of Electronic Systems **22**(1), 6:1–6:23 (2016). https://doi.org/10.1145/2906147
6. Hicks, M., Finnicum, M., King, S.T., Martin, M.M.K., Smith, J.M.: Overcoming an Untrusted Computing Base: Detecting and Removing Malicious Hardware Automatically. In: 2010 IEEE Symposium on Security and Privacy, pp. 159–172 (2010). https://doi.org/10.1109/SP.2010.18
7. Waksman, A., Suozzo, M., Sethumadhavan, S.: FANCI: Identification of Stealthy Malicious Logic Using Boolean Functional Analysis. In: Proceedings of the 2013 ACM CCS, CCS '13, pp. 697–708. ACM, New York, NY, USA (2013). https://doi.org/10.1145/2508859.2516654
8. Sturton, C., Hicks, M., Wagner, D., King, S.T.: Defeating UCI: Building Stealthy and Malicious Hardware. In: Proceedings of the 2011 IEEE Security and Privacy, SP '11, pp. 64–77. IEEE Computer Society, Washington, DC, USA (2011). https://doi.org/10.1109/SP.2011.32
9. Hoque, T., Narasimhan, S., Wang, X., S, M., Bhunia, S.: Golden-Free Hardware Trojan Detection with High Sensitivity Under Process Noise. Journal of Electronic Testing **33**(1), 107–124 (2017)
10. Narasimhan, S., Du, D., Chakraborty, R.S., Paul, S., Wolff, F.G., Papachristou, C.A., Roy, K., Bhunia, S.: Hardware Trojan Detection by Multiple-Parameter Side-Channel Analysis. IEEE Transactions on Computers **62**(11), 2183–2195 (2013)
11. Narasimhan, S., Bhunia, S.: Hardware Trojan Detection, pp. 339–364. Springer New York, New York, NY (2012). https://doi.org/10.1007/978-1-4419-8080-9_15
12. Bilzor, M., Huffmire, T., Irvine, C., Levin, T.: Security Checkers: Detecting Processor Malicious Inclusions at Runtime. In: 2011 IEEE International Symposium on Hardware-Oriented Security and Trust, pp. 34–39 (2011). https://doi.org/10.1109/HST.2011.5954992
13. Waksman, A., Sethumadhavan, S.: Silencing Hardware Backdoors. In: 2011 IEEE Symposium on Security and Privacy, pp. 49–63 (2011). https://doi.org/10.1109/SP.2011.27
14. Nedospasov, D., Seifert, J., Schlosser, A., Orlic, S.: Functional Integrated Circuit Analysis. In: 2012 IEEE International Symposium on Hardware-Oriented Security and Trust, pp. 102–107 (2012). https://doi.org/10.1109/HST.2012.6224328
15. Torrance, R., James, D.: The State-of-the-Art in Semiconductor Reverse Engineering. In: 2011 48th ACM/EDAC/IEEE Design Automation Conference (DAC), pp. 333–338 (2011)
16. Chakraborty, R., Wolff, F., Paul, S., Papachristou, C., Bhunia, S.: MERO: A Statistical Approach for Hardware Trojan Detection. In: CHES 2009, pp. 396–410. Springer, Berlin, Heidelberg (2009)
17. Islam, S.A., Sah, L.K., Katkoori, S.: A Framework for Hardware Trojan Vulnerability Estimation and Localization in RTL Designs. Journal of Hardware and Systems Security **4**(3), 246–262 (2020)
18. Saha, S., Chakraborty, R.S., Nuthakki, S.S., Anshul, Mukhopadhyay, D.: Improved Test Pattern Generation for Hardware Trojan Detection Using Genetic Algorithm and Boolean Satisfiability. In: CHES 2015, pp. 577–596 (2015)
19. Huang, Y., Bhunia, S., Mishra, P.: MERS: Statistical Test Generation for Side-Channel Analysis Based Trojan Detection. In: Proceedings of the 2016 ACM SIGSAC Conference on Computer and Communications Security, CCS '16, pp. 130–141. ACM, New York, NY, USA (2016). https://doi.org/10.1145/2976749.2978396
20. Li, H., Liu, Q.: Hardware Trojan detection acceleration based on word-level statistical properties management. 2014 International Conference on Field-Programmable Technology (FPT) pp. 153–160 (2014)
21. Çakir, B., Malik, S.: Hardware Trojan Detection for Gate-level ICs Using Signal Correlation Based Clustering. In: Proceedings of the 2015 Design, Automation and Test in Europe, DATE '15, pp. 471–476. San Jose, CA, USA (2015). http://dl.acm.org/citation.cfm?id=2755753.2755860

22. Jha, S., Jha, S.K.: Randomization Based Probabilistic Approach to Detect Trojan Circuits. In: 2008 11th IEEE High Assurance Systems Engineering Symposium, pp. 117–124 (2008). https://doi.org/10.1109/HASE.2008.37
23. Salmani, H., Tehranipoor, M., Plusquellic, J.: A Novel Technique for Improving Hardware Trojan Detection and Reducing Trojan Activation Time. IEEE Transactions on Very Large Scale Integration (VLSI) Systems **20**(1), 112–125 (2012). https://doi.org/10.1109/TVLSI.2010.2093547
24. Salmani, H., Tehranipoor, M.: Analyzing Circuit Vulnerability to Hardware Trojan Insertion at the Behavioral Level. In: 2013 IEEE International Symposium on Defect and Fault Tolerance in VLSI and Nanotechnology Systems (DFTS), pp. 190–195 (2013). https://doi.org/10.1109/DFT.2013.6653605
25. Piccolboni, L., Menon, A., Pravadelli, G.: Efficient Control-Flow Subgraph Matching for Detecting Hardware Trojans in RTL Models. ACM Trans. Embedded Comput. Syst. **16**(5s), 137:1–137:19 (2017). https://doi.org/10.1145/3126552
26. Zhang, J., Yuan, F., Xu, Q.: DeTrust: Defeating Hardware Trust Verification with Stealthy Implicitly-Triggered Hardware Trojans. In: Proceedings of the 2014 ACM CCS, CCS '14, pp. 153–166. ACM, New York, NY, USA (2014). https://doi.org/10.1145/2660267.2660289
27. Nahiyan, A., Sadi, M., Vittal, R., Contreras, G., Forte, D., Tehranipoor, M.: Hardware Trojan Detection through Information Flow Security Verification. In: 2017 IEEE International Test Conference (ITC), pp. 1–10 (2017). https://doi.org/10.1109/TEST.2017.8242062
28. Ardeshiricham, A., Hu, W., Marxen, J., Kastner, R.: Register Transfer Level Information Flow Tracking for Provably Secure Hardware Design. In: Design, Automation Test in Europe Conference Exhibition (DATE), 2017, pp. 1691–1696 (2017). https://doi.org/10.23919/DATE.2017.7927266
29. Hu, W., Mao, B., Oberg, J., Kastner, R.: Detecting Hardware Trojans with Gate-Level Information-Flow Tracking. Computer **49**(8), 44–52 (2016). https://doi.org/10.1109/MC.2016.225
30. Jin, Y., Yang, B., Makris, Y.: Cycle-Accurate Information Assurance by Proof-Carrying Based Signal Sensitivity Tracing. In: 2013 IEEE International Symposium on Hardware-Oriented Security and Trust (HOST), pp. 99–106 (2013). https://doi.org/10.1109/HST.2013.6581573
31. Rajendran, J., Vedula, V., Karri, R.: Detecting Malicious Modifications of Data in Third-Party Intellectual Property Cores. In: 2015 52nd ACM/EDAC/IEEE Design Automation Conference (DAC), pp. 1–6 (2015). https://doi.org/10.1145/2744769.2744823
32. JasperGold®. https://www.cadence.com/en_US/home/tools/system-design-and-verification/formal-and-static-verification/jasper-gold-verification-platform.html
33. Ramprasad, S., Shanbha, N.R., Hajj, I.N.: Analytical Estimation of Signal Transition Activity from Word-Level Statistics. IEEE Transactions on Computer-Aided Design of Integrated Circuits and Systems **16**(7), 718–733 (1997). https://doi.org/10.1109/43.644033
34. Bobba, S., Hajj, I.N., Shanbhag, N.R.: Analytical Expressions for Average Bit Statistics of Signal Lines in DSP Architectures. In: Circuits and Systems, 1998. ISCAS '98. Proceedings of the 1998 IEEE International Symposium on, vol. 6, pp. 33–36 vol.6 (1998). https://doi.org/10.1109/ISCAS.1998.705205
35. Landman, P.E., Rabaey, J.M.: Architectural Power Analysis: The Dual Bit Type Method. IEEE Transactions on Very Large Scale Integration (VLSI) Systems **3**(2), 173–187 (1995). https://doi.org/10.1109/92.386219
36. Satyanarayana, J.H., Parhi, K.K.: Theoretical Analysis of Word-Level Switching Activity in the Presence of Glitching and Correlation. In: Proceedings Ninth Great Lakes Symposium on VLSI, pp. 46–49 (1999). https://doi.org/10.1109/GLSV.1999.757374

Chapter 11
Defense Against Hardware Trojan Collusion in MPSoCs

Chen Liu and Chengmo Yang

11.1 Threat of Hardware Trojan Collusion

As the previous chapters have mentioned, hardware Trojans come naturally as a result of the globalized IC design and production flow. To reduce manufacturing costs, many IC design houses choose to outsource circuit manufacturing to third-party foundries whose security is not always assured. In the meantime, to shorten IC development cycles, *system-on-chip* (SoC) designers tend to purchase *intellectual property* (IP) cores from third-party IP (3PIP) vendors and integrate them in their own SoC designs.

Along with the trend to integrate more 3PIPs into one MPSoC, the risk of integrating hardware Trojans into the MPSoC also increases. A rogue insider in a 3PIP house may insert hardware Trojans in 3PIPs coming out of the IP house. A hardware Trojan may have one or even multiple of the following payloads: modified chip functionality, Denial-of-Service (DoS), or a back-door to leak confidential information. Moreover, to keep the Trojan footprint small, an attacker may separate the different parts of a Trojan (e.g., trigger and payload) in multiple 3PIP cores and establish some secret communication agreement between them. This is called *hardware Trojan collusion*.

An example of hardware Trojan collusion is shown in Fig. 11.1. Cores 1, 2, and 4 have the same type of Trojan, therefore share the same trigger value. During normal execution, these cores receive different inputs and have different environmental parameters (such as temperature and power values) that diversify their triggering

C. Liu (✉)
Intel Corp., Hillsboro, OR, USA
e-mail: chen1.liu@intel.com

C. Yang
University of Delaware, Newark, DE, USA
e-mail: chengmo@udel.edu

S. Katkoori, S. A. Islam (eds.), *Behavioral Synthesis for Hardware Security*,
https://doi.org/10.1007/978-3-030-78841-4_11

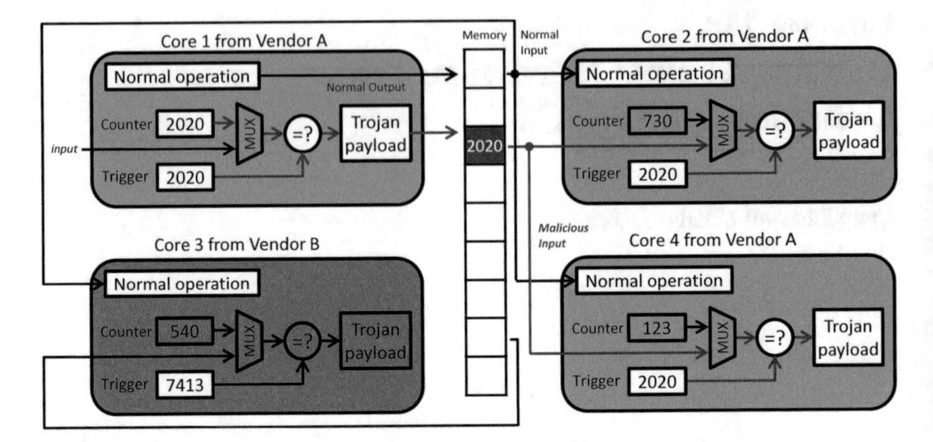

Fig. 11.1 An example of hardware Trojans in multiple cores using shared memory as a covert communication channel. The trigger value 2020 is secretly shared between three cores 1, 2, and 4, but not core 3

conditions. With the secret communication paths, however, if any core (e.g., Core 1 in Fig. 11.1) reaches the triggering condition, it will send messages to other cores (e.g., Cores 2 and 4) by secretly writing the trigger value in a pre-agreed memory location (shown in red) for the other cores to read. As a result, all Trojans of the same type become active within a short while.

11.2 Enforcing Vendor Diversity

Many existing solutions aim to capture hardware Trojans during the product testing phase. These offline techniques are limited not only by their measurement accuracy, but also by the lack of golden models [1–3] that accurately characterize the expected timing and/or power values of the design. In contrast, the technique described in this chapter integrates both *design-for-trust* (DfT) and *runtime monitoring* defenses, aiming to capture the attack patterns or mute the effects that a hardware Trojan may have.

One important constraint that can be enforced at the design stage is *vendor diversity*, an inherent property in heterogeneous systems to eliminate false negatives and improve detection accuracy. For hardware Trojans that modify chip functionality, it is highly possible that the outputs of the infected core are altered. These Trojans can be *detected* by duplicating each task on two different cores and comparing the outputs to capture any mismatch. Here, *vendor diversity* is essential to prevent two instances of the same task from producing the same incorrect outputs. This is because Trojans, unlike random faults, are inherent in all the 3PIPs of the same type. Multiple instances of the same untrusted 3PIP will contain the same Trojan and hence produce the same incorrect output under the same input. To prevent such type of false negatives, each task needs to be executed on two cores from different IP vendors.

Hardware Trojans that do not alter task outputs may escape the duplication-based detection. Nonetheless, they can still be captured by monitoring the communication paths at runtime and intercepting suspect communications, if any. Again, *vendor diversity* is essential to eliminate false negatives, that is, malicious messages camouflaged as normal messages. Normally different vendors do not share a back-door or a trigger pattern in common since to do so, a rogue element has to expose itself to other rogue elements. This is shown in Fig. 11.1, wherein the triggering condition of Core 3 is different from the other cores. Given this fact, one can enforce a constraint that *only communication channels between different vendors are considered "safe" and used to convey authorized inter-task communications*. This way, collusion between malicious 3PIPs can be *muted* or *detected* at runtime: first, if a core (e.g., Core 1 in Fig. 11.1) silently produces unexpected data at runtime, such data are only accessible to cores from different vendors (e.g., Core 3) and are ineffective for Trojan triggering; second, if any communication occurs between cores from the same vendor, it must be malicious and hence a security flag will be raised.

Same as other security features, *vendor diversity* also comes at a price. It incurs extra design cost since the designer needs to purchase licenses from multiple IP vendors, and the complexity of integrating heterogeneous cores into the MPSoC grows. Nonetheless, for applications with high security requirements, such as banking and military systems, the extra cost is acceptable since security has a higher priority than cost. In some cases, *vendor diversity* may even reduce the design cost. Previously designers tend to pay more to purchase IPs from companies with a more established reputation. Yet with *vendor diversity*, designers can use multiple cheaper IPs without worrying about their individual security problem.

To summarize, the security enhancement described in this chapter leverages vendor diversity inherent in heterogeneous MPSoCs, and adopts a design-time and runtime collaborative strategy. At design stage, security constraints are enforced. At runtime, with task duplication and communication path monitoring, potential collusion between hardware Trojans in 3PIPs can be muted or detected. This approach can be adopted by designers who have the flexibility of choosing IP cores from different vendors and binding tasks to cores, and can add glue logic to monitor inter-core communication patterns. Next section will discuss how to incorporate these security constraints into the MPSoC task scheduling step at the design stage.

11.3 Design State Enhancement: Security-Driven Task Scheduling

Task scheduling binds tasks to cores and coordinates data accesses, communication, and synchronization among the tasks [4]. It is a critical step at the MPSoC design stage that determines the performance and power consumption characteristics of an application, as well as the types and number of the 3PIP cores needed in the MPSoC. Compared to homogeneous MPSoCs, heterogeneous MPSoCs furthermore offer the flexibility of using the most suitable core for each task to improve overall performance or energy efficiency [5].

A typical MPSoC scheduler explores a three-dimensional design space of performance (modeled as schedule length), power consumption (modeled as overheads of computation and communication), and cost (modeled as types and total number of cores). To defend against potential collusion between 3PIP hardware Trojans, a set of security constraints will be added to the traditional MPSoC task scheduling process. Those security constraints will impact the three-dimensional design space. As shown in Fig. 11.2, the schedule length, communication overhead, and core count may change as a result of the extra security constraints. The design cost may increase due to the extra license cost of utilizing cores from more than one vendor in the system. Hence, the number of vendors is a critical design factor to be optimized.

The fundamental goal of the security-driven task scheduling problem is to satisfy the security constraints with minimum performance and hardware overhead. In the rest of this section, the MPSoC task scheduling problem and the security constraints will be first formulated through a *Integer Linear Programming* (ILP) description. While the problem is NP-hard, ILP solvers can achieve optimal solutions. Since

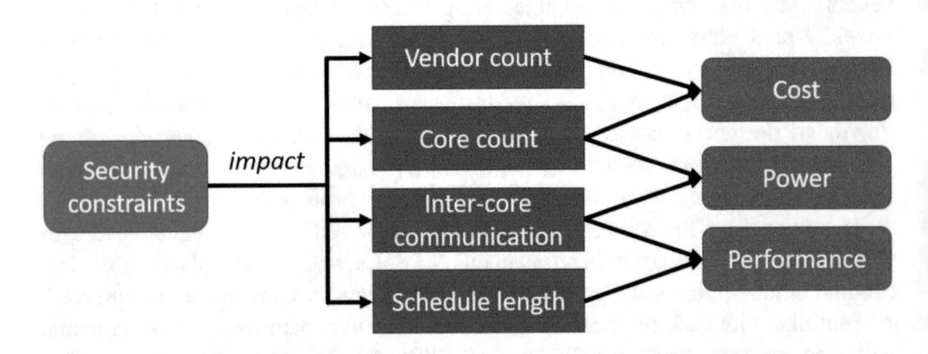

Inputs: 1) An upper bound of processing cores
2) Speed of different cores (to model heterogeneity)
3) An application represented as a DAG
- Nodes = tasks, weight = execution time
- Edges = communications, weight = communication cost

Outputs: A start time and a core assignment for every task

Goals: 1) Fulfill security constraints
2) Minimize the cost and power, and maximize the performance at a priority specified by the designer.

Fig. 11.2 Design space and definition of security-driven task scheduling

ILP solvers are time-consuming, two heuristics that exploit the flexibility in task scheduling to fulfill the security constraints will also be developed.

11.3.1 ILP Formulation of the Scheduling Problem

As Fig. 11.2 shows, security-driven task scheduling is a multi-dimensional optimization problem. Given the task graph of an application, the scheduling problem can be formalized as the association of a start time and the assignment of a core with each task. The application can be represented as a weighted *Directed Acyclic Graph* (DAG) $G = (V, E)$, which has n vertices. Each vertex $Task_i \in V$ represents a task, with a weight $exeT_i$ $(1 \leq i \leq n)$ denoting its execution time. Each edge $Dep_{i,j} \in E$ represents a communication from $Task_i$ to $Task_j$, with a weight $comT_{i,j}$ $(1 \leq i, j \leq n)$ denoting the inter-core communication overhead. The target platform is modeled as a heterogeneous MPSoC with m cores $Core_i$ $(1 \leq i \leq m)$, and each core may be produced by one of the p available vendors $Vendor_k$ $(1 \leq k \leq p)$. In a typical case, one can assume that $n \geq m \geq p$.

In the following parts, the task scheduling problem will be formulated in four steps that respectively model task bindings, task execution order, task start/finish time, and the security constraints imposed to prevent hardware Trojan collusion.

11.3.1.1 Model Task Bindings

Apparently, every task is bound to a single core, every core is bound to a single vendor, and each task should be bound to the vendor of the core it is scheduled on. To model these binding relations, three sets of binary variables $T2C_{i,j}$, $C2V_{j,k}$, and $T2V_{i,k}$ are defined:

$$T2C_{i,j} = \begin{cases} 1 \text{ if } Task_i \text{ is bound to } Core_j \\ 0 \text{ otherwise} \end{cases}$$

$$C2V_{j,k} = \begin{cases} 1 \text{ if } Core_j \text{ is bound to } Vendor_k \\ 0 \text{ otherwise} \end{cases}$$

$$T2V_{i,k} = \begin{cases} 1 \text{ if } Task_i \text{ is bound to } Vendor_k \\ 0 \text{ otherwise} \end{cases}$$

Figure 11.3 visualizes the binding relations in the three-dimensional space. $T2C_{i,j}$ represents whether $Task_i$ is bound to $Core_j$ and $C2V_{j,k}$ represents whether $Core_j$ is bound to $Vendor_k$. With these two variables, $T2V_{i,k}$, which represents whether $Task_i$ is bound to $Vendor_k$, can be further derived. With these variables, these task-to-core, core-to-vendor, and task-to-vendor binding properties can be modeled using the following three constraints:

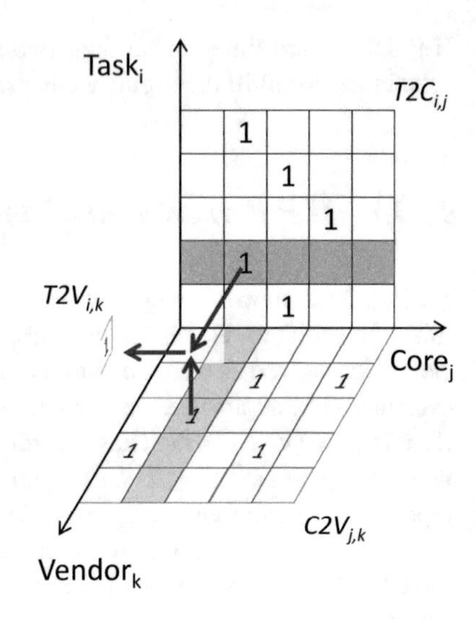

Fig. 11.3 Binding relations among tasks, cores, and vendors. The yellow row indicates that each task is mapped to one and only one core, the green column indicates that each core belongs to one and only one vendor, while the red arrows indicate the product of the two binary variables

$\forall\, Task_i$:

$$\sum_{j=1}^{m} T2C_{i,j} = 1 \tag{11.1}$$

$\forall\, Core_j$:

$$\sum_{k=1}^{p} C2V_{j,k} = 1 \tag{11.2}$$

$\forall\, Task_i$ and $\forall\, Vendor_k$:

$$T2V_{i,k} = \sum_{j=1}^{m} T2C_{i,j} * C2V_{j,k} \tag{11.3}$$

Equation (11.1) ensures that any $Task_i$ is mapped to one and only one core. It is illustrated with the rows on the $T2C_{i,j}$ plane in Fig. 11.3, as shown, each row on $T2C_{i,j}$ contains exactly a single 1. Analogously, Eq. (11.2) ensures that any $Core_j$ belongs to one and only one vendor. In Fig. 11.3, this is shown with the columns on the $C2V_{j,k}$ plane, such that each column on $C2V_{j,k}$ contains exactly a single 1. Equations (11.1) and (11.2) together ensure that there is one and only one non-zero product of the two binary variables $T2C_{i,j}$ and $C2V_{i,j}$ (indicated with the red arrows in Fig. 11.3). Therefore, Eq. (11.3) ensures that $T2V_{i,k}$ is 1 if and only if $Task_i$ is scheduled on $Core_j$ and $Core_j$ is from $Vendor_k$.

11.3.1.2 Model Task Execution Order

In addition to maintaining task binding properties, it is equally necessary to *order* the tasks, ensuring that dependency constraints are fulfilled and tasks scheduled on the same core do not overlap. We introduce another set of variables for this purpose:

$$Order_{i,j} = \begin{cases} 1 \text{ if } Task_i \text{ and } Task_j \text{ are scheduled on the same core} \\ \quad \& \; Task_i \text{ is before } Task_j \\ 0 \text{ otherwise} \end{cases}$$

Using these variables, the dependency constraint and non-overlapping property can be modeled through three constraints:

- First, a child task cannot be scheduled before any of its parent tasks:
  $\forall \; Dep_{i,j}$:

$$Order_{j,i} = 0 \tag{11.4}$$

  For a pair of dependent tasks $Task_i$ (the parent task) and $Task_j$ (the child task), if they are scheduled on the same core, Eq. (11.4) ensures that $Task_j$ cannot be scheduled before $Task_i$.
- Second, the order relation is transitive. Specifically, if $Task_i$, $Task_j$, and $Task_k$ are scheduled on the same core, with $Task_i$ before $Task_k$ and $Task_k$ before $Task_j$, then $Task_i$ must be before $Task_j$.
  $\forall \; Task_i, Task_j, Task_k$ that $i \neq j \neq k$:

$$Order_{i,j} \geq Order_{i,k} * Order_{k,j} \tag{11.5}$$

  The right-hand side of Inequality (11.5) is 1 if and only if both $Order_{i,k}$ and $Order_{k,j}$ are 1, which happens if and only if (a) $Task_i$, $Task_k$, and $Task_j$ are scheduled on the same core, (b) $Task_i$ is before $Task_k$, and (c) $Task_k$ is before $Task_j$. In this case, the left-hand side of Inequality (11.5) has to be 1, which indicates that $Task_i$ is placed before $Task_j$ on the same core.
- Third, the order relation between two tasks is mutually exclusive, meaning that $Order_{i,j}$ and $Order_{j,i}$ cannot be 1 at the same time.
  $\forall \; Task_i, Task_j$ that $i \neq j$:

$$Order_{i,j} + Order_{j,i} = \sum_{k=1}^{m} T2C_{i,k} * T2C_{j,k}$$

  If $Task_i$ and $Task_j$ are scheduled on different cores, the variables $T2C_{i,k}$ and $T2C_{j,k}$ will not be 1 for the same value of k. This means the right-hand side of Eq. (11.6) is 0, and hence both $Order_{i,j}$ and $Order_{j,i}$ are 0. On the other hand, if $Task_i$ and $Task_j$ are scheduled on the same core k, the right-hand side of Eq. (11.6) is 1. In this case, one of $Order_{i,j}$ and $Order_{j,i}$ is 1 and the other is 0.

11.3.1.3 Model Task Start and Finish Time

Once the order relationship between tasks is defined, the start and finish time of each task can be constrained. For $Task_j$, its start time ST_j is constrained by the finish time of all its predecessors and the ready time of all the incoming data. Here, the communication overhead $comT_{i,j}$ is imposed only when $Task_j$ and its predecessor $Task_i$ are scheduled on different cores, and FT_i denotes the finish time of $Task_i$.

$\forall\ Dep_{i,j}$:

$$ST_j \geq FT_i + comT_{i,j} * (1 - Order_{i,j}) \tag{11.6}$$

When $Task_j$ and $Task_i$ are scheduled on different cores, $1 - Order_{i,j}$ is 1. Hence Inequality (11.6) becomes $ST_j \geq FT_i + comT_{i,j}$, which indicates that the child $Task_j$ cannot start until (a) the parent $Task_i$ has finished, and (b) data transmission from $Task_i$ to $Task_j$ has completed. On the other hand, if $Task_j$ and its predecessor $Task_i$ are scheduled on the same core, there is no inter-core data transmission time. Therefore, inequality (11.6) becomes $ST_j \geq FT_i$.

Moreover, $Task_j$ cannot start if any $Task_i$ that starts before it on the same core (therefore $Order_{i,j} = 1$) have not finished execution:

$\forall\ Task_i, Task_j$ that $i \neq j$:

$$ST_j \geq FT_i * Order_{i,j} \tag{11.7}$$

Finally, the finish time of a task is the sum of its start time and its execution time.[1] Since heterogeneity may affect task execution time, a matrix representation is used to model different vendors. Specifically, cores from the same vendor are assumed to be the same and a task scheduled on any of them has the same execution time.[2] Therefore, in the matrix an element $exeT_{i,k}$ denotes the execution time of $Task_i$ on a core from $Vendor_k$. If $Task_i$ cannot be executed by cores from $Vendor_k$, the execution time of $Task_i$ on $Vendor_k$ will be set to a value close to infinite, enabling the scheduler to automatically avoid such binding during task assignment. Using this matrix representation of task execution time, the finish time of $Task_i$ is given by the following equation:

[1]It is assumed that the scheduling is *non-preemptive*, that is, a task in execution cannot be preempted by any other task.

[2]This assumption can be easily extended to model different cores from one vendor.

$\forall\ Task_i$:

$$FT_i = ST_i + \sum_{k=1}^{p} exeT_{i,k} * T2V_{i,k} \tag{11.8}$$

Since $T2V_{i,k} = 1$ only for a single vendor, one and only one value of $exeT_{i,k}$ will be added to ST_i, thus effectively modeling core heterogeneity and its impact on task execution time and finish time.

11.3.1.4 Model Security Constraints

As outlined in Sect. 11.2, the goals of the online detection are to (a) detect Trojans that affect task outputs, and (b) detect undesired messages between Trojans located on different cores. Two constraints are imposed to achieve these goals.

The first constraint is **duplication-with-diversity**, which detects incorrect task output (caused by a Trojan) as long as attackers in two independent 3PIP design houses do not collude to develop identical Trojans that produce identical incorrect outputs (which is highly unlikely). This constraint enforces every task to be duplicated on two 3PIP cores from different vendors, as shown below:

$\forall\ Task\ i\ and\ its\ duplicate\ Task_{i'}$ and $\forall\ Vendor_k$:

$$T2V_{i,k} + T2V_{i',k} \leq 1 \tag{11.9}$$

To accomplish task duplication, a new *DAG* is constructed combining the original *DAG* and a copy of it. Inequality (11.9) constrains that for any $Task_i$ (in the original *DAG*) and its duplicate $Task_{i'}$ (in the duplicated *DAG*), $T2V_{i,k}$ and $T2V_{i',k}$ cannot be 1 at the same time. In other words, the constraint prohibits any pair of duplicated tasks from being scheduled on the same core or cores from the same vendor.

At runtime, both $Task_i$ and its duplicate $Task_{i'}$ receive the same input, while their outputs are compared by a trusted component (e.g., designed in-house) to ensure the trustworthiness of the comparison step. Techniques to compare results from different cores, such as [6] and [7], can be adopted here. Note that to reduce overhead, the comparison is at a relatively coarse granularity: instead of performing cycle-by-cycle comparison of signals and instruction results; here we only compare the final task outcome. This relaxes the synchronization constraint, allowing tasks to be executed on cores of different processing speeds and with different instruction sets. This also minimizes the amount of data to be compared, enabling the use of low-cost comparators which can even be shared among multiple cores. The comparison latency can be further reduced if the subsequent tasks (that depend on $Task_i$) are started without waiting for the comparison output. Later, if the comparison fails, all the dependent tasks are terminated and a security flag is raised.

The second constraint, **isolation-with-diversity**, allows for quick and accurate detection of undesired and potentially malicious communications at runtime.

Specifically, the constraint confines all the legitimate inter-core communications to be between 3PIPs from different vendors. This way, at runtime, any inter-core communications between cores from the same vendor are considered malicious.

This constraint can be added to the heterogeneous MPSoC task scheduling problem at *two different granularities*. At the finest granularity, the constraint can be applied to each communication path in the *DAG*. In other words, all the communications are forced to be between 3PIP cores produced by different vendors:

$\forall\ Dep_{i,j}$ and $\forall\ Vendor_k$:

$$T2V_{i,k} + T2V_{j,k} \leq 1 \tag{11.10}$$

Inequality (11.10) constrains that for any pair of dependent tasks $Task_i$ and $Task_j$, $T2V_{i,k}$ and $T2V_{j,k}$ can never be 1 at the same time. This prohibits dependent tasks from being scheduled on the same core or cores from the same vendor. Therefore, $comT_{i,j}$ is always imposed in Equation (11.6), which in turn may delay the start and the finish time of $Task_j$ and hence increase the schedule length. Also, the energy consumption becomes higher, not only because of more inter-core communications, but also because cores need to be active for longer time due to the increased schedule length. What is more, because dependent tasks need to be bound to different vendors, the minimum number of vendors needed in the target MPSoC platform may increase. Specifically, the lowest vendor count is constrained by the minimum number of colors needed to color the task graph of the application.[3]

Given the elevated performance, energy, and vendor overhead imposed by the fine-grain Trojan isolation, the alternative is to satisfy **isolation-with-diversity** constraint at a coarser granularity, allowing dependent tasks to be scheduled on the same core (but not different cores of the same vendor) to reduce such overhead. This can be achieved by allowing both $T2V_{i,k}$ and $T2V_{j,k}$ to be 1 if $Order_{i,j}$ is 1 (implying that $Task_i$ and $Task_j$ are on the same core):

$\forall\ Dep_{i,j}$ and $\forall\ Vendor_k$:

$$T2V_{i,k} + T2V_{j,k} \leq 1 + Order_{i,j} \tag{11.11}$$

Compared to Eqs. (11.10), (11.11) also ensures that all inter-core communications are between different vendors, but does not enforce every inter-task communication to become inter-core communication. The relaxed constraint in turn provides more flexibility in scheduling, which could be exploited to reduce communication overhead and energy consumption, and to improve performance and the flexibility in coloring the task graph.

[3]Graph coloring [8] is the problem of finding the minimum number of colors that ensure adjacent vertices in a graph do not share the same color.

Table 11.1 Solution toward different Trojan categories

	Tamper with output	No change in output
Create undesired communication	Isolation w/ Diversity & Duplication w/ Diversity	Isolation w/ Diversity
No undesired communication	Duplication w/ Diversity	Not critical, left for future study

To summarize, the two security constraints together successfully mitigate the two categories of hardware Trojans discussed in Sect. 11.2. As illustrated in Table 11.1, Trojans tampering with task outputs can be detected by *duplication-with-diversity*, which exposes incorrect task outputs, preventing the incorrect data to be consumed by any dependent tasks. Meanwhile, Trojan collusion can be muted by *isolation-with-diversity*, which ensures that all the valid communication paths are between 3PIPs made by different vendors. During application execution, if one core produces unexpected data in addition to valid task outputs, the cores that execute dependent tasks (if any), as they come from different vendors and hence are unlikely to collude with this core, will not use such data for Trojan triggering. If there is any communication between two cores from the same vendor, a security flag will be raised indicating the detection of an invalid communication path.

A hardware Trojan may escape these two schemes only if it neither tampers task outputs nor sends any message through an invalid communication path. These Trojans are considered less critical and are left for future study.

11.3.1.5 Objective Functions of the ILP Model

After incorporating the security constraints, the task scheduling problem can be formulated as a multi-dimensional optimization process with two objective functions, namely, first minimizing schedule length and then minimizing vendor count.

The performance of the schedule is defined as the overall schedule length, which is computed as the finish time of the last finished task among all the tasks.

$$\textbf{Goal1}: minimize \max_{1 \le i \le n} FT_i$$

Once the minimum schedule length is determined, the next goal is to minimize the number of vendors which is crucial to the design cost of the target MPSoC. For a given schedule, the vendor count can be computed by summing up the maximum value of $C2V_{j,k}$:

$$\textbf{Goal2}: minimize \sum_{k=1}^{p} \max_{1 \le j \le m} C2V_{j,k}$$

If $Vendor_k$ is not used, then $C2V_{j,k}$ should be 0 for all cores. If $Vendor_k$ is used, then there should be at least one $C2V_{j,k}$ $(1 \leq j \leq m)$ equal to 1. Therefore, by summing up the maximum value of $C2V_{j,k}$ across all vendors, the total number of vendors used in a schedule can be obtained.

11.3.2 Scheduling Heuristics

With the ILP formulation described above, ILP solvers such as [9] can be utilized to obtain optimal solutions for the security-driven task scheduling problem. This process, however, is usually very time-consuming, especially if the search space (e.g., numbers of tasks, cores, and vendors) is large. In this section, two security-driven task scheduling heuristics will be presented, which integrate the security constraints into classic list scheduling algorithm [10] at different granularities. They can be used to obtain near-optimal solutions within reasonable computation complexity.

11.3.2.1 A Fine-Grained Scheduling Heuristic

The fine-grained heuristic satisfies both *duplication-with-diversity* and *isolation-with-diversity* at the fine-grained level. This heuristic consists of four steps, shown in Fig. 11.4. Each box represents one step, with its main purposes listed to the bottom. Each arrow represents the data transferred between steps, with the exact data type listed below the arrows.

- *Task graph coloring* enforces security constraints and determines the number of vendors needed in the target MPSoC.
- *Task scheduling* determines the total number of cores, the core assignment of each task, and its tentative start time.
- *Color to core type mapping* determines the exact type and the speed of each 3PIP core.
- *Schedule finalization* adjusts the start time and finish time of each task based on core speed.

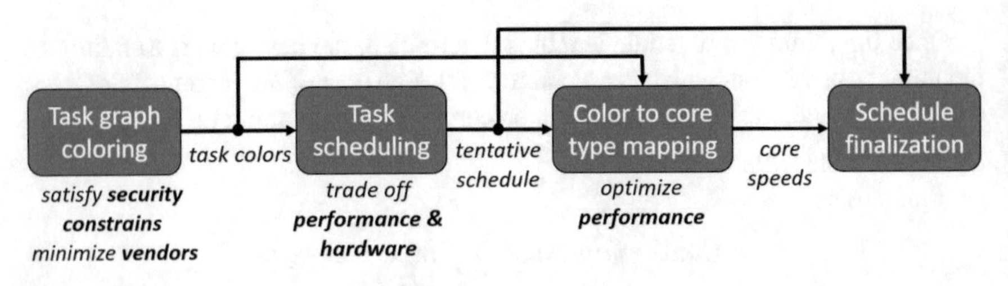

Fig. 11.4 High level flows of the fine-grained scheduling heuristic

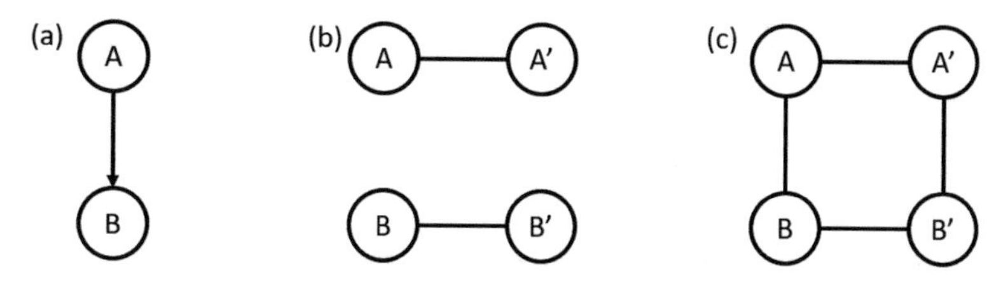

Fig. 11.5 An example showing how to construct the *task conflict graph*. (**a**) Original task graph. (**b**) Task conflict graph with duplication-with-diversity. (**c**) Task conflict graph with both security constraints

Step 1: Task Graph Coloring

The minimum number of 3PIP vendors needed to fulfill the security constraints is crucial to the design cost of the target MPSoC. This important parameter can be determined via *task graph coloring*. More specifically, to satisfy *duplication-with-diversity*, the original task graph is duplicated. Then, the duplicated task graph is transferred to a *task conflict graph* to embed task conflict relations brought by the two security constraints. Finally, the conflict graph is colored to determine the minimum number of vendors.

Figure 11.5 illustrates the process for constructing a *task conflict graph*. Vertices in the graph represent tasks, while edges represent conflicts. In Fig. 11.5b, all the vertices in the original task graph Fig. 11.5a are duplicated, and an edge is inserted between vertices corresponding to the original task and its duplication. This step embeds the *duplication-with-diversity* constraint into the conflict graph. Then, in Fig. 11.5c, an edge is inserted between any pair of dependent tasks in the duplicated task graph. This step embeds the *isolation-with-diversity* constraint.

If the original task graph contains V nodes and E edges, the task conflict graph will have $2V$ nodes and $2E + V$ edges. Standard graph vertex coloring algorithms, such as *greedy coloring* and *Brélaz algorithm* [8] can be applied to determine the minimum number of colors needed to color the graph, representing the minimum number of vendors needed in the MPSoC.

Step 2: Color-Constrained Task Scheduling

After coloring the task conflict graph constructed in Step 1, each task i and its duplicate i' has a color. This step performs color-constrained task scheduling, which is derived from traditional *list scheduling* [11]. In a typical list scheduling process, a list of tasks that are ready to be scheduled is maintained. Here, a task is considered *ready* when its predecessors are all finished. The list of ready tasks is typically sorted according to some pre-determined criteria. During every scheduling step, the task with the highest priority is picked from the list and scheduled to a core. Then the dependent tasks of the just scheduled task are checked and inserted to the list if they are ready to be scheduled. This process continues until all the tasks are scheduled.

Our task scheduling heuristic follows the list scheduling flow as well as its way of ranking the tasks in the ready list. The key difference lies in the scheduling of

Algorithm 11.1 Color-constrained task scheduling—fine-grained approach

```
Initialization:
 1: v = task to be scheduled;
 2: Corelist = ϕ;
Procedure:
 3: for all core_i do
 4:     if (core_i.color = v.color) or (core_i.color = not determined) then
 5:         Corelist ⇐ Corelist+{core_i};
 6:     end if
 7: end for
 8: for core_i ∈ Corelist do
 9:     if v.FT on core_i is the earliest till now then
10:         Schedule v on core_i;
11:     end if
12: end for
13: if core_selected.color = not determined then
14:     core_selected.color ⇐ v.color;
15: end if
```

each task. Specifically, the task color constraints need to be fulfilled: only tasks of the same color can be scheduled on the same type of cores. These constraints are fulfilled by *assigning colors to cores during the scheduling*: at the beginning, all the cores are *colorless* (i.e., without any color assigned); when a task is under scheduling, it can be placed on either a colorless core or a core with the same color as the task. If the task is assigned to a colorless core, the core immediately inherits the color of the task, and only accommodates tasks of the same color from then on.

Details of the color-constrained task scheduling algorithm are shown in Algorithm 11.1. The first loop (lines 3 to 7) identifies all the cores that a task v can be scheduled on according to its color constraint and puts them in $Corelist$. In the second loop (lines 8 to 12), all the cores in $Corelist$ are checked one by one to find the core that minimizes the finish time (FT) of V. Finally, the *if* statement (lines 13 to 15) conditionally assigns the color of task v to the selected core.

The time complexity of Algorithm 11.1 for scheduling one task is $O(C)$, with C denoting the total number of cores. Furthermore, in list scheduling, scheduling the entire task graph requires traversing all tasks and edges, leading to a time complexity of $O(V + E)$, with V representing the total number of tasks and E representing the total number of edges in the task graph.[4] Accordingly, the time complexity of color-constrained task scheduling is $O(C(V + E))$.

[4]The definitions of V, C, and E hold for the remaining algorithms as well.

Step 3: Color to Core Type Mapping

In the prior step, coloring constraints are embedded into the schedule and the color of each core is determined. Yet the color to vendor mapping is not determined, which will affect the exact speed of each core. Since tasks on critical paths have direct impact on the schedule length, it naturally motivates a heuristic that assigns a higher speed to a core with more critical tasks so as to minimize the schedule length.

The heuristic defines *critical tasks* as the tasks with 0 *timing slack* (i.e., the difference between its latest and earliest start time) in the schedule. It also defines *core priority* as the number of critical tasks assigned to a core. Critical tasks can be identified by computing the timing slack of each task in the schedule obtained in the prior step. For cores with the same number of critical tasks, higher priority is given to the one whose critical tasks either have more critical dependent tasks or the one whose critical tasks are on the upper levels of the task graph. This way, the benefit obtained by scheduling these critical tasks earlier can be maximized.

The exact procedure for ranking the cores according to their priority is shown in Algorithm 11.2. It uses the following three metrics to determine core priority:

- $metric_1(i)$ is the number of critical tasks on $core_i$.
- $metric_2(i)$ is the start time of the earliest critical task on $core_i$.
- $metric_3(i)$ is the total number of *critical dependent tasks* (i.e., the dependent tasks that are critical) of all the critical tasks on $core_i$.

In Algorithm 11.2, first the three metrics are calculated (lines 1 to 16), then the cores are ranked based on their priority (lines 17 to 30). Higher priority is given to the core with higher values of $metric_1$ and $metric_3$ and a lower value of $metric_2$. In this algorithm, the complexity of calculating the metrics is $O(V + E)$, and the complexity of ranking the cores (which is a sorting problem) is $O(ClogC)$. As a result, the overall complexity of this algorithm is $O(V + E + ClogC)$.

After determining the priority of each core, the color to core-speed mapping is performed, shown in Algorithm 11.3. During initialization, all the cores are included in *CoreList* and sorted based on the priority determined in Algorithm 11.2. At each iteration, C_{cri}, the core with the highest priority is selected from *CoreList*. This core, along with the ones that have the same color as it, is assigned the highest speed s among all the "available" speed values (lines 6 to 12). Then, these cores are removed from *CoreList* and speed s is marked as "unavailable" (line 13). This process repeats until all the cores have been assigned a specific speed. As the number of core speeds (i.e., vendors) is no more than the number of cores, the overall complexity of this algorithm is $O(C)$.

Step 4: Schedule Finalization

Since core-speed assignment in the prior step impacts task execution time, it is necessary to adjust the pre-calculated task start time (ST) and finish time (FT).

The start time of task j is constrained by either the current available time of the core (on which task j is scheduled) or the ready time of incoming data:

Algorithm 11.2 Core priority comparison

1: **for all** $core_i$ **do**
2: $metric_1(i) \Leftarrow 0,\ metric_2(i) \Leftarrow \infty,\ metric_3(i) \Leftarrow 0;$
3: **for** $T_j \in$ *all tasks scheduled on* $core_i$ **do**
4: **if** T_j *is on critical path* **then**
5: $metric_1(i) \Leftarrow metric_1(i) + 1;$
6: **if** $T_j.startTime < metric_2(i)$ **then**
7: $metric_2(i) \Leftarrow T_j.startTime;$
8: **end if**
9: **for all** $T_k \in Child\{T_j\}$ **do**
10: **if** T_k *is on critical path* **then**
11: $metric_3(i) \Leftarrow metric_3(i) + 1;$
12: **end if**
13: **end for**
14: **end if**
15: **end for**
16: **end for**
17: Sort all $core_i$ based on priority defined below:
18: **for all** $core_i$ and $core_j$ $(i \neq j)$ **do**
19: **if** $metric_1(i) > metric_1(j)$ **then**
20: $core_i.priority > core_j.priority;$
21: **else if** $metric_1(i) = metric_1(j)$ **then**
22: **if** $metric_2(i) < metric_2(j)$ **then**
23: $core_i.priority > core_j.priority;$
24: **else if** $metric_2(i) = metric_2(j)$ **then**
25: **if** $metric_3(i) > metric_3(j)$ **then**
26: $core_i.priority > core_j.priority;$
27: **end if**
28: **end if**
29: **end if**
30: **end for**

$$ST_j = max(\max_{1 \le i \le p} (FT_k,\ FT_i + comT_{i,j} * (1 - Order_{i,j}))) \tag{11.12}$$

Equation (11.12) is derived from Inequalities (11.6) and (11.7). It ensures that a task j cannot start until all the parent tasks are finished, and all the tasks scheduled before task j on the same core are finished. Here, task k is the previous task scheduled on the same core before task j, task i is one of the p parent tasks of task j, $comT_{i,j}$ is the communication time from task i to task j, $Order_{i,j}$ represents the order relation between task i and task j.

The finish time of a task is the sum of its start time and its adjusted execution time after taking core speed into consideration:

$$FT_j = ST_j + \frac{exe_j}{CoreSpeed} \tag{11.13}$$

Equation (11.13) shows that the diversity in core speed may either accelerate or decelerate the execution of task j, while Eq. (11.12) shows that this effect

Algorithm 11.3 Core-speed assignment

Initialization:
1: *CoreList* = {all cores}, sorted in descending order;
2: *SpeedList* = {all core-speed}, sorted in descending order;
Procedure:
3: **while** $CoreList \neq \phi$ **do**
4: Get the most critical core $C_{cri} \in CoreList$;
5: Get the fastest speed $s \in SpeedList$;
6: $C_{cri}.speed \Leftarrow s$;
7: $CoreList \Leftarrow CoreList - \{C_{cri}\}$;
8: **for** $C_i \in CoreList$ **do**
9: **if** $C_i.color = C_{cri}.color$ **then**
10: $C_i.speed \Leftarrow C_{cri}.speed$;
11: $CoreList \Leftarrow CoreList - \{C_i\}$;
12: **end if**
13: $SpeedList \Leftarrow SpeedList - \{s\}$;
14: **end for**
15: **end while**

propagates, impacting all the descendants of task j and all the tasks that are on the same core as j. Using these two equations, tasks are processed one by one in their scheduling order. The overall complexity of this process is $O(V + E)$.

11.3.2.2 A Coarse-Grained Scheduling Heuristic

The fine-grained scheduling heuristic described above fulfills the two security constraints at the finest granularity: *duplication-with-diversity* is added to each node in the task graph and *isolation-with-diversity* is applied to each edge in the task graph. As discussed before, one disadvantage of this approach is that all the inter-task communications are forced to be between 3PIP cores. This prohibits the scheduler from putting dependent tasks on the same core to hide communication latency, save energy, and reduce schedule length.

The coarse-grained scheduling heuristic retains the *duplication-with-diversity* constraint, but relaxes the *isolation-with-diversity* constraint to be coarse-grained, as shown in Eq. (11.11). It allows multiple dependent tasks to be scheduled on the same core. Specifically, the heuristic groups dependent tasks in the *DAG* on critical paths into a *cluster* and schedules the entire cluster to a single core. Compared to the fine-grained approach that always schedules dependent tasks across different vendors, the coarse-grained approach schedules dependent tasks either to the same core (for the intra-cluster cases) or across different vendors (for the inter-cluster cases). There is no legitimate communication between cores from the same vendor, so the security constraints are satisfied.

Clustering of critical tasks necessitates information of task criticality. Therefore, the coarse-grained scheduling heuristic first generates a performance-driven schedule and then colors the *core conflict graph* to fulfill security constraints. The entire

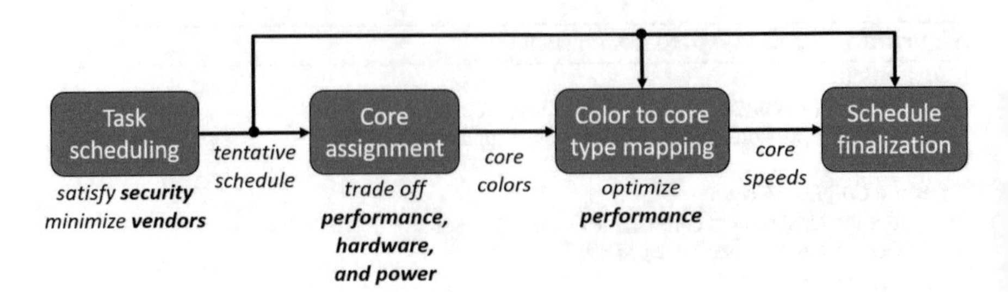

Fig. 11.6 Flows of the coarse-grained scheduling approach

scheduling process is still organized in four steps, shown in Fig. 11.6. The latter two steps (color to core-speed mapping and schedule finalization) are identical to the last two steps in the fine-grained heuristic. Here we explain the first two steps.

Step 1: Task Scheduling
This step generates a schedule and determines the total number of cores needed in the target MPSoC. Same as before, the heuristic follows a *list scheduling* process to rank ready tasks and schedule them one by one. To maximally explore the scheduler's capability in grouping critical tasks together on a single core, minimum security constraint is imposed during this step. Specifically, only the *duplication-with-diversity* constraint that precludes false negatives of Trojan detection is imposed, by preventing a task v and its duplicate v' from being scheduled on the same core. The *isolation-with-diversity* constraint is considered in the next step.

Step 2: Core Assignment
This step embeds the two security constraints into the schedule, determining the exact color of each core and the number of vendors needed in the target MPSoC. Since tasks are grouped into clusters, security constraints are imposed between *cores* instead of *tasks*. A *core conflict graph* can be constructed based on the schedule generated in the prior step. Each vertex in the graph is a core and each edge represents a conflict. An edge is inserted between cores i and j if any of the following two conditions is true:

- There is one or more pairs of duplicated tasks on i and j (to satisfy *duplication-with-diversity*).
- There is one or more communication paths between i and j (to satisfy *isolation-with-diversity*).

Standard graph coloring algorithms [8] can be used to color the core conflict graph. After that, each task inherits the color of the core it is scheduled on.

Revised Scheduling Heuristic with Vendor Count Control
While the coarse-grained approach is capable of placing critical tasks on a single core to boost performance, it may potentially increase the number of vendors needed in the MPSoC. Specifically, the number of vendors equals the number of colors needed to color the *core conflict graph*, which is no smaller than the maximum

clique size[5] of the graph. As the coarse-grained heuristic generates this graph in step 2, its maximum clique size is not considered in step 1, when the performance-driven list scheduling is performed.

To resolve this issue, we revise the task scheduling step by evaluating its impact on the maximum clique size of the *core conflict graph*. Such impact will be used to prioritize scheduling decisions. Specifically, if a task has the same earliest start time on multiple cores, a traditional scheduler would randomly assign the task to any one of them. In contrast, the new scheduling heuristic evaluates these options regarding their impact on the maximum clique size and selects the one with no or minimum impact. This way, the new heuristic is able to minimize the number of vendors without degrading schedule length.

Algorithm 11.4 Vendor-aware task scheduling—coarse-grained approach

Initialization:
1: v = task to be scheduled;
2: v' = duplicate of v;
3: N = current max-clique size;
4: $UpperBound$ = the given upper bound of clique size;
5: $Corelist_A = \phi$;
6: $Corelist_B = \phi$;
Procedure:
7: **for all** $core_i$ **do**
8:     **if** (v' not on $core_i$) **then**
9:         **if** Putting v on $core_i$ increases N **then**
10:             $Corelist_A \Leftarrow Corelist_A + \{core_i\}$;
11:         **else**
12:             $Corelist_B \Leftarrow Corelist_B + \{core_i\}$;
13:         **end if**
14:     **end if**
15: **end for**
16: **for all** core $\in Corelist_A$ **do**
17:     Find $core_A$ that allows v to have earliest start time t_A;
18: **end for**
19: **for all** core $\in Corelist_B$ **do**
20:     Find $core_B$ that allows v to have earliest start time t_B;
21: **end for**
22: **if** ($N \geq UpperBound$ or $t_A > t_B$) **then**
23:     Schedule v on $core_B$;
24: **else**
25:     schedule v on $core_A$;
26:     $N \Leftarrow N + 1$;
27: **end if**
28: Update current core conflict graph;

[5]The maximum clique of a graph is its largest complete subgraph. A graph with a maximum clique size of k needs at least k colors to color it.

This clique-size-aware scheduling heuristic is described in Algorithm 11.4. Line 8 is to filter out the cores that do not satisfy the constraint of separating a task v and its duplicate v'. Then, all the remaining cores are classified into two lists such that scheduling task v on cores in $Corelist_A$ increases the maximum clique size of the current *core conflict graph* while scheduling v on cores in $Corelist_B$ does not (lines 9–13). Given these two lists, if the current maximum clique size already reaches the upper bound, the best core from $Corelist_B$ will be selected, ensuring that scheduling of task v does not increase the maximum clique size. If the current maximum clique size is smaller than the upper bound, the best cores from $Corelist_A$ and $Corelist_B$ will be compared, and the one that allows for the earliest start time of v will be selected. This process is shown in lines 16–27. At the end, the corresponding core conflict graph and maximum clique size is updated. Overall, Algorithm 11.4 has the same time complexity of $O(C(V + E))$ as Algorithm 11.1.

Algorithm 11.5 Max-clique size estimation

Initialization:

1: v_m = *The task to be scheduled*;
2: $core_i$ = Core that v_m to be scheduled on;
3: V_{parent} = {all the parent tasks of v_m};
4: E = {all the edges in core conflict graph};
5: N = current max-clique size;
6: $count = 0$;

Procedure:

7: **for all** $v_n \in V_{parent}$ **do**
8:     $core_j \Leftarrow$ the core that v_n is scheduled on;
9:     **if** $e_{ij} \notin E$ **then**
10:         $E \Leftarrow E + \{e_{ij}\}$;
11:         $core_i.degree$++;
12:         $core_j.degree$++;
13:         **for all** $core_k$ = common adjacent node of $core_i$ and $core_j$ **do**
14:             $core_i.triangle$++;
15:             $core_j.triangle$++;
16:             $core_k.triangle$++;
17:         **end for**
18:     **end if**
19: **end for**
20: **for all** $core$ **do**
21:     **if** $(core.degree \geq N)$ and $(core.triangle \geq \frac{N(N-1)}{2})$ **then**
22:         $count$++;
23:     **end if**
24: **end for**
25: **if** $count \geq N + 1$ **then**
26:     $N \Leftarrow N + 1$;
27: **end if**

Although computing the maximum clique size of a graph is NP-complete, as the scheduler only processes one task at a time, an efficient heuristic can be developed.

The observation is that scheduling a task on core i may add one or more edge(s)[6] in the *core conflict graph*. As these edges share the same vertex i, they may increase the maximum clique size by 1 at most. Based on this observation, we develop a heuristic for estimating the maximum clique size of the core conflict graph, shown in Algorithm 11.5. The algorithm contains two major loops. The first loop (lines 7 to 19) iteratively updates two critical metrics: the *degree* (i.e., the number of edges) of each node and the *number of triangles* that each node is involved in. Every iteration inserts an edge $edge_{ij}$ into the core conflict graph and updates the degrees of $core_i$, $core_j$ as well as the triangle information of $core_i$, $core_j$, and their common neighbors $core_k$. Then, based on these updated metrics, the second loop (lines 20 to 24) computes a close estimate of the maximum clique size using a heuristic introduced in [12]. It is based on the observation of two necessary conditions for a graph to have a maximum clique size of $N + 1$. First, there should be at least $N + 1$ nodes with a degree of N. Second, each of these $N + 1$ nodes should be involved in at least $N(N - 1)/2$ triangles.

11.3.2.3 An Example Comparing the Two Heuristics

To illustrate the differences between the two scheduling heuristics, an example of applying them to the standard task graph of Gaussian Elimination is shown in Fig. 11.7. The original task graph has 10 nodes, each representing a task. Each node includes four values: Ti denoting its task ID i, Cj denoting that it is assigned to core j, and m–n denoting its start time m and finish time n. For simplicity, both the task execution time and the inter-core communication overhead are set to 10. The example also considers two vendors. Cores made by vendor B are 20% slower than the cores made by vendor A. The dark nodes are tasks assigned to the slower cores, resulting in an execution time of 12. The solid arrows represent intra-core communication (with 0 overhead), while the bold, dash arrows represent inter-core communication (with 10 overhead).

The schedule in Fig. 11.7a is the traditional schedule, which will be used as the baseline for comparison. It uses two cores ($C1$ and $C2$) from vendor A and has a schedule length of 70. Clearly, this schedule does not satisfy the security constraints: tasks are not duplicated, and there is legitimate communication between $C1$ and $C2$, which are two cores from the same vendor. The schedule in Fig. 11.7b is generated by the fine-grained heuristic. It uses 4 cores: $C1$, $C3$ from vendor A, and $C2$, $C4$ from vendor B. This schedule satisfies both security constraints at the task level: every task and its duplicate are scheduled on two cores of different types, and every communication is between two cores of different types. However, the tasks executed on the dark nodes require 20% longer execution time, leading to an overall schedule length of 108, which is 54.3% longer than the baseline. The schedule in Fig. 11.7c is generated by the coarse-grained heuristic. It is obtained by first duplicating the

[6]If no edge is added, the maximum clique size does not change.

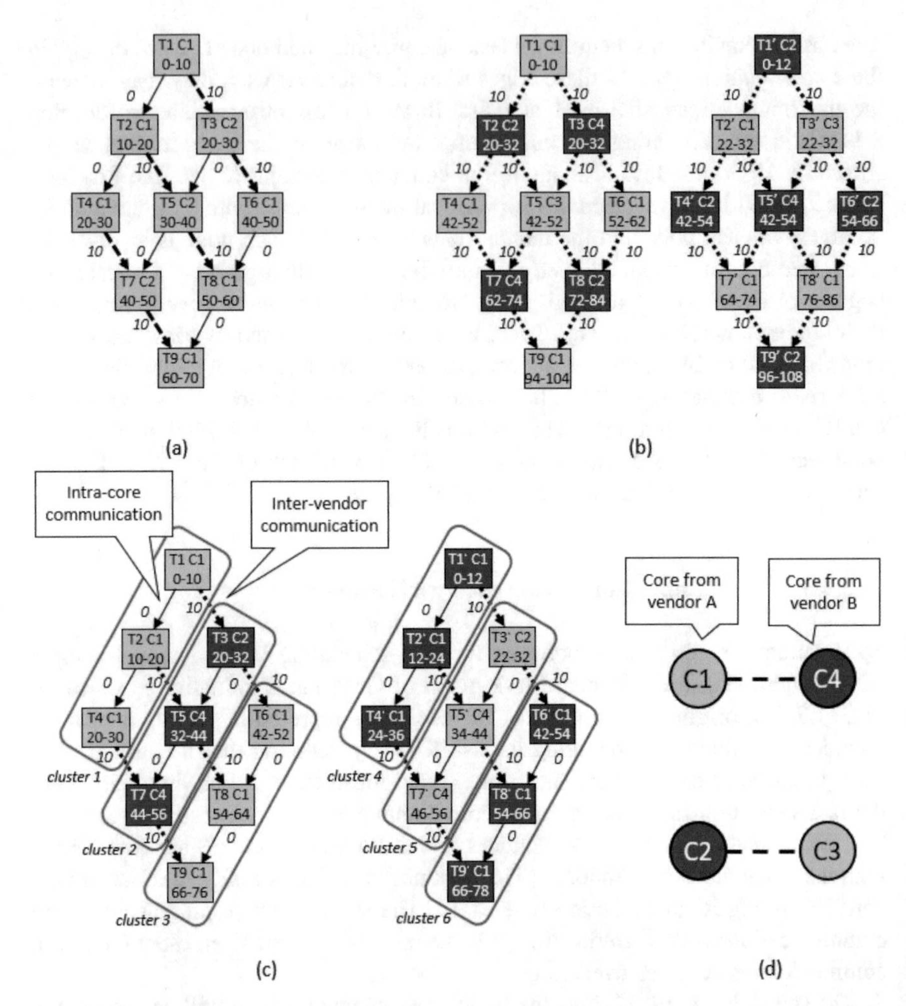

Fig. 11.7 Scheduling example of Gaussian elimination task graph with different schedulers. (**a**) Schedule w/o security constraints. (**b**) Schedule the fine-grained approach. (**c**) Schedule the coarse-grained approach. (**d**) Core conflict graph

performance-driven schedule in Fig. 11.7a and then coloring the core conflict graph (shown in Fig. 11.7d) to fulfill the security constraints. This schedule also uses 4 cores from two different vendors. It contains 6 clusters, and 50% of all the inter-task communications are intra-cluster and incur no overhead (as they are on the same core), while the remaining communications are inter-cluster and between two cores of different types. The overall schedule length is 78, which is only 11.4% longer than the baseline and 27.8% shorter than the schedule length in Fig. 11.7b.

11.4 Runtime Enhancement: Inter-IP Communication Monitoring

The security-driven task scheduling performed at the MPSoC design stage guarantees that all the legitimate inter-core communications are between 3PIPs made by different vendors. This largely simplifies the runtime communication monitoring process. In particular, there is no need to check the content of every message to distinguish legitimate and illegal communications. Instead, for every inter-core communication, only the information of the sender (i.e., data producer) and the receiver (i.e., data consumer) needs to be checked to ensure that the two involved 3PIPs are not from the same vendor. This can be easily accomplished through a lightweight extension to the underlying communication framework with only negligible hardware overhead. This section describes possible extensions to two commonly adopted communication models, namely, *shared memory* and *message passing*.

11.4.1 Monitor Communications in Shared Memory MPSoC

In a shared memory MPSoC, a communication is performed through two operations: a write to a memory location followed by a read to the same location from another core. To detect undesired communications, each data block is extended to include an *owner field* of $\lceil \log_2 m \rceil$ bits (for an MPSoC with m cores). This field uniquely specifies the owner of each data block. The memory controller is responsible for updating this field whenever a 3PIP writes the data block. With such information, whenever a 3PIP core tries to read the data block, the memory controller checks the data producer against the 3PIP requesting the data, and raises a flag if their core IDs differ but their core types are the same.

In some MPSoCs, each 3PIP core has a local and private cache, and these caches can communicate through a coherence protocol [13] without writing/reading data objects to/from the memory. To record data producer information, not only the shared memory but also the local cache of each 3PIP core should include *owner* fields for each cache line. These fields still need to be updated upon write accesses. However, they only need to be checked upon read misses, as read accesses hitting in the local cache, which are the majority cases for most workloads, are not involved in inter-core communication.

Figure 11.8 shows a detailed example, which assumes the use of write-back caches and a standard write-invalidate MSI coherence protocol.[7] In Fig. 11.8a: the owner field is *updated* upon $Core1$ issuing a *write* request to memory block A. After

[7]In MSI, each cache block can have one of three possible states: **M**odified, **S**hared, and **I**nvalid. More details about the protocol can be found in [13].

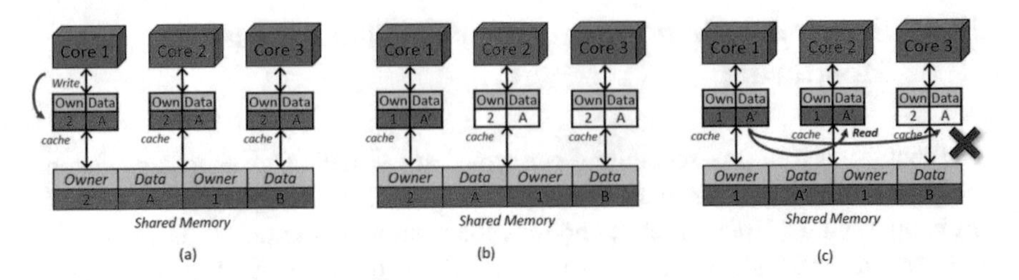

Fig. 11.8 Online monitoring example for a shared memory MPSoC with *write-invalidate* cache coherence protocol. (**a**) Core1 updates data A in its own cache. (**b**) Owner field of A is updated and copies in caches of Core2 and Core3 are invalidated. (**c**) Read attempt from Core3 to A is blocked since Core3 is from the same vendor as Core1

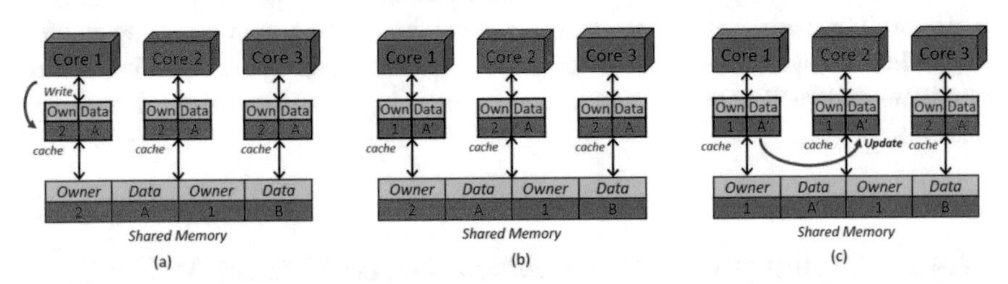

Fig. 11.9 Online monitoring example for a shared memory MPSoC with *write-update* cache coherence protocol. (**a**) Core1 update data A in its own cache. (**b**) Owner field of A is updated. (**c**) Updated data will be populated to Core2 only since Core1 and Core2 are from different vendors

this write operation, the data field and the owner field in the local cache of $Core1$ are updated, shown in Fig. 11.8b. The old copies of block A in the local caches of $Core2$ and $Core3$ are invalidated (shown in white), thus preventing incorrect data sharing. Figure 11.8c shows that upon $Core2$ and $Core3$ issuing *read* requests to block A, the owner field of A in $Core1$ is *checked*. The read issued by $Core2$ is granted since it is from a different vendor, while the read issued by $Core3$ will be denied since $Core3$ is from the same vendor as $Core1$, the owner of A. A security flag will be raised to report this unexpected (and probably malicious) read request.

Figure 11.9 shows the same example taking place in an MPSoC with write-back caches and a write-update MSI coherence protocol [13]. Same as before, upon $Core1$ issuing a write request to memory location A, as shown in Fig. 11.9a, the data field and the owner field in the local cache of $Core1$ are updated, as shown in Fig. 11.9b. As it is a write-update protocol, the updated data will then be put on the bus and sent to other cores. To prevent Trojan collusion, however, any other core from the same vendor of $Core1$, such as $Core3$, is blocked from receiving the updated data. As Fig. 11.9c shows, only the copy in the local cache of $Core2$ is updated.

11.4.2 Monitor Communications in Message Passing MPSoC

Message passing MPSoCs do not share memory between cores. Instead, inter-core communications are performed through passing messages with either an on-chip network or a centralized bus structure. In both cases, extra logic can be attached to the routers or the bus controller to act as a "firewall."

In Message Passing MPSoC, the communication pattern can be either point-to-point or broadcasting. For a point-to-point communication, both the sender and the receiver are explicit. By checking the source and the destination 3PIPs of each message, an illegal communication can be easily detected and intercepted. In comparison, for a broadcasting communication, only the sender is explicit. It is impossible to differentiate legal and illegal messages since they all share the same broadcasting channel. However, the router or the bus controller can still block, given the message sender, all the receiver cores that are from the same vendor as the sender. This way, the runtime system is able to prevent unintended message from reaching its destination.

11.5 Experimental Evaluation

In this section, we will evaluate the two security-drive task scheduling heuristics described in Sect. 11.3.2 and compare them with a classical task scheduling approach that does not consider security. The section first describes the evaluation methodology, introducing the baseline, the task graphs, and the target MPSoC, followed by a multi-dimensional evaluation of the two heuristics in terms of their cost and performance.

11.5.1 Methodology

In the experimental study, a standard list scheduling [10] algorithm is selected as the baseline. Three different scheduling approaches are implemented, including the baseline, the fine-grained, and coarse-grained scheduling heuristics. The optimal ILP solution described in Sect. 11.3.1 is not included in the comparison due to its non-polynomial complexity.

11.5.1.1 Task Graphs

The test set is composed of both standard parallel task graphs and random task graphs. Standard task graphs represent some classical algorithms or program structures, including *LU decomposition*, *Laplace equation solver*, *FFT*, *fork-join*,

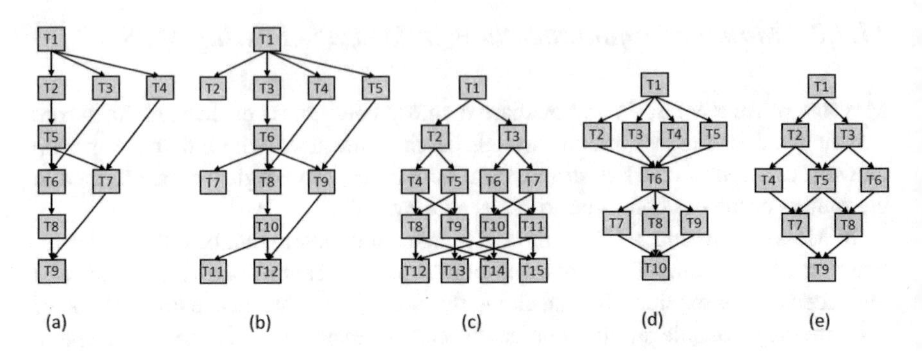

Fig. 11.10 The standard task graphs utilized in the experiment. (**a**) LU-decomposition. (**b**) Laplace equation solver. (**c**) FFT. (**d**) Fork-join. (**e**) Gaussian elimination

Table 11.2 Configurations of randomly generated task graphs

	High-comm. Task graphs	Low-comm. Task graphs
Number of tasks	50, 100, 150	50, 100, 150
Number of start nodes	1–10	1–10
Task input/output degree	4/4, 8/8	4/4, 8/8
Average of comp/comm overhead	50/50	50/5
Variation of comp/comm overhead	20/20	20/2

and *Gaussian elimination*. Their DAG representations are shown in Fig. 11.10. In addition to these task graphs, 100 random task graphs are generated with the TGFF tool [14]. Table 11.2 reports the configurations for critical parameters in TGFF. High-communication and low-communication task graphs are generated by adjusting computation and communication overhead parameters.

11.5.1.2 MPSoC Configurations

In the experiment, it is assumed that the underlying MPSoC platform can accommodate up to 16 cores. Typically the cores produced by different vendors exhibit variations in many parameters (e.g., speed, area, and power consumption). Since performance is the main concern of scheduling, we focus only on modeling speed variations across 3PIP cores, in line with the assumption made by most scheduling approaches of heterogeneous systems [15, 16]. The step of speed differences is set to 10% of the speed of the fastest core in the experiments. Here, the assumption is that a core that is $x\%$ slower than the fastest core causes the execution time of all the tasks to increase by $x\%$. Note that the core speed only affects task execution time, while the inter-core communication overhead remains intact.

11.5.2 Results

11.5.2.1 Number of Vendors Needed

The first set of experiments reports the minimum number of vendors required for incorporating the two security constraints *duplication-with-diversity* and *isolation-with-diversity*. As the number of vendors impacts both the design cost and the hardware cost, the goal is to limit this number within a reasonable range.

Figure 11.11 shows the minimum number of vendors required. It is a ratio distribution across all the task graphs. Data of both the fine-grained and coarse-grained scheduling approaches are reported, while the coarse-grained approach is scheduled with and without Algorithm 11.5 for controlling maximum clique size. The ratio is calculated using the following formula:

$$ratio_i = \frac{\#\, of\ task\ graphs\ requiring\ i\ vendors}{total\ \#\, of\ task\ graphs}$$

For the fine-grained scheduling approach, Fig. 11.11 shows that 4 vendors are sufficient for 105 out of the 107 tested task graphs (7 standard and 100 random graphs). For the coarse-grained approach, more vendors are required when the maximum clique size is not restrained during scheduling. When Algorithms 11.4 and 11.5 are applied, however, the coarse-grained approach has almost the same distribution as the fine-grained approach, with two task graphs performing even better. The study confirms the benefits delivered by the clique size minimization heuristic, which is adopted consistently in the rest of the study.

Overall, it is believed that the cost of 4 vendors is acceptable for building a trustworthy MPSoC, especially when it is used in security-critical infrastructure.

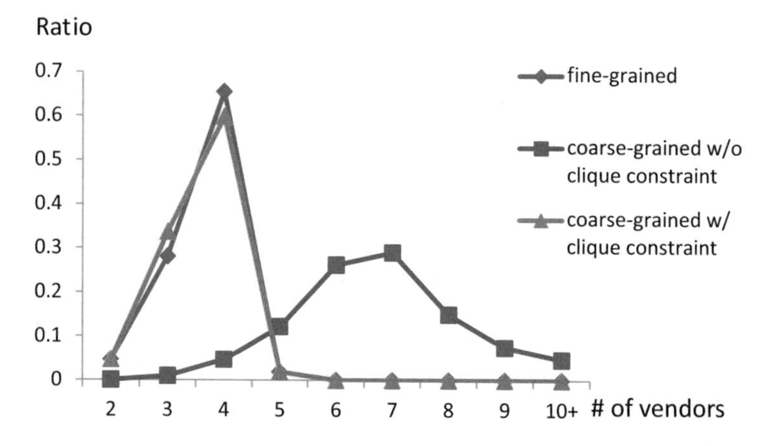

Fig. 11.11 Distribution of vendor count that fulfills security constraints with different scheduling heuristics

11.5.2.2 Lengths of Two Security-Driven Schedules

The second set of experiments evaluates the performance of the two heuristics by comparing their schedule lengths to the baseline [10]. For fair comparison, the three schedulers use the same number of cores (the one that allows the baseline scheduler to deliver best performance), and the two security-driven schedulers use the same number of vendors. Performance of the two security-driven schedulers is measured as their ratio of schedule length increase (Δ_{SL}) over the baseline, calculated using the following equation:

$$\Delta_{SL} = \frac{1}{n} \sum_{i=1}^{n} \frac{SL(i)}{SL_{baseline}(i)} - 1$$

where n is the number of task graphs and $SL(i)$ stands for the schedule length of the ith task graph. The lower the Δ_{SL}, the better the performance.

Results of schedule lengths are shown in Table 11.3. A *duplication-only* approach is tested to show the impact of the first security constraint individually. Data from both the standard and the random task graphs are reported. The random task graphs are further divided into four categories based on the number of vendors needed and the communication overhead level.

With the same number of cores as the baseline, duplicating every task increases the schedule length by 58% on average. This is expected to be lower than the overhead of the two security-driven scheduling heuristics, as both of them fulfill the *duplication-with-diversity* constraint in the same way. When both security constraints are imposed, the coarse-grained approach always outperforms the fine-grained approach. The fine-grained approach has 21–40% degradation on top of duplication only, while the coarse-grained approach only has 9–20% degradation. The average scheduling length overhead of the fine-grained and coarse-grained approaches are 103% and 81%, respectively. This confirms that by grouping critical tasks together and scheduling them to a single core, inter-core communication latency can be largely hidden, and hence schedule length can be largely reduced. This benefit is more notable for high-communication cases because the profit of hiding communication overhead is higher. Table 11.3 also shows that task graphs

Table 11.3 Increased schedule length comparison of the two security-driven approaches

	Ratio of schedule length increase		
	Duplication only	Fine-grained	Coarse-grained
Standard	0.371	0.653	0.500
Low-comm, 3-vendor	0.738	1.095	0.956
Low-comm, 4-vendor	0.726	1.196	1.079
High-comm, 3-vendor	0.497	0.979	0.667
High-comm, 4-vendor	0.573	1.205	0.857
Average	0.581	1.026	0.812

requiring 4 vendors have larger overhead in schedule length than those with 3 vendors. This is because slower cores need to be used when the number of vendors increases.

11.5.2.3 Communication Breakdown

The third set of experiments evaluates the impact of the two security-driven scheduling heuristics on inter-core communications. There are three types of communication paths: *intra-core*, *inter-core* & *intra-vendor*, and *inter-vendor*. The second type of communication violates the *isolation-with-diversity* constraint and thus should not present in any of the schedules generated. Meanwhile, a higher ratio of the third type of communication implies that more communication paths satisfy the *isolation-with-diversity*.

Ratios of the three types of communication paths over the number of all communication paths are obtained using the following formulas:

$$R_{intra-c} = \frac{1}{n}\sum_{i=1}^{n}\frac{Comm_{intra-c}(i)}{Comm_{all}(i)}$$

$$R_{inter-c\&intra-v} = \frac{1}{n}\sum_{i=1}^{n}\frac{Comm_{inter-c\&intra-v}(i)}{Comm_{all}(i)}$$

$$R_{inter-v} = \frac{1}{n}\sum_{i=1}^{n}\frac{Comm_{inter-v}(i)}{Comm_{all}(i)}$$

with $Comm_{intra-c}(i)$, $Comm_{inter-c\&intra-v}(i)$, $Comm_{inter-v}(i)$, and $Comm_{all}(i)$ respectively denoting the number of *intra-core*, *inter-core&intra-vendor*, *inter-vendor*, and all the communication paths of task graph i.

The results are shown in Table 11.4, wherein the *intra-core*, *inter-core* & *intra-vendor*, and *inter-vendor* ratios are listed from left to right for each schedule. As expected, only the baseline has the second type of communication paths, implying that security is guaranteed for the other two approaches. In the fine-grained approach, all the communication paths are forced to be inter-vendor. For the coarse-grained approach, the ratio of inter-vendor communications is very close to the ratio of inter-core & intra-vendor communications in the baseline. This is because the coarse-grained approach first generates a performance-driven schedule (in the same way as the baseline) and then imposes security constraints to protect these inter-core communication paths and achieve a trustworthy MPSoC design.

Table 11.4 Communication type breakdown of the three schedules

	Baseline			Fine-grained			Coarse-grained		
	Intra-core	Intra-vendor	Inter-vendor	Intra-core	Intra-vendor	Inter-vendor	Intra-core	Intra-vendor	Inter-vendor
Standard	0.395	0.605	0	0	0	1	0.354	0	0.646
Low-comm, 3-vendor	0.266	0.734	0	0	0	1	0.267	0	0.733
Low-comm, 4-vendor	0.236	0.764	0	0	0	1	0.240	0	0.760
High-comm, 3-vendor	0.267	0.733	0	0	0	1	0.271	0	0.729
High-comm, 4-vendor	0.258	0.742	0	0	0	1	0.253	0	0.747
Average	0.284	0.716	0	0	0	1	0.278	0	0.722

11.6 Concluding Remarks

This chapter discussed the problem of hardware Trojan collusion in MPSoCs—an increasing security threat brought by the globalized IC design flow. It also presented a countermeasure built upon a fundamental concept of *vendor diversity*. The goal was to mute or detect collusion by integrating "design-for-trust" and "runtime monitoring." At the design stage, security-driven task scheduling was utilized to schedule the application's task graph following two security constraints: *duplication-with-diversity* and *isolation-with-diversity*. At runtime, the outputs of a task and its duplicate were compared to detect Trojans that alter task outputs. Meanwhile, the source and destination of each inter-core communication were checked, and any communication between cores from the same vendor would be intercepted and reported.

The chapter described various scheduling approaches, including an optimal ILP model and two heuristics following the classical list scheduling approach. It also presented heuristics for minimizing the number of vendors required and reducing schedule length overhead. The experimental results showed that in most cases, security constraints could be fulfilled with four vendors. By scheduling dependent tasks on the same core, the schedule length overhead brought by the two security constraints could be reduced to 81% of the baseline schedule length with no extra core needed.

11.7 Further Reading

Defending MPSoCs against hardware Trojans is an evolving research topic that has attracted a lot of research attention recently. Wang et al. [17, 18] extended the work described in this chapter to minimize intra-core communication, reduce the schedule length or the number of cores needed in the MPSoC. Sun et al. [19] considered energy consumption together with security constraints in MPSoC design. Extra scheduling steps were taken to minimize energy consumption while fulfilling constraints of schedule length and security. A set of other works [20–23] tried to develop countermeasures against hardware Trojans in MPSoCs that use a Network-on-Chip (NoC). Unlike the threat model described in this chapter, those works assumed that the cores in a MPSoC are trustworthy but the NoC is provided by third-party and may have hardware Trojans. Different countermeasures including network-based, firmware-based, and software-based solutions are proposed.

Integer Linear Programming (ILP) is a common model for solving optimization problems. An ILP problem is a mathematical optimization problem in which the variables are integers and the objective function and the constraints are linear. A tutorial about ILP can be found in [24].

Scheduling is the process of allocating resources to solve tasks. It can be further classified as *static scheduling* (task graph, task execution time, and resources are pre-determined) or *dynamic scheduling* (task graph, execution time, or hardware resources may change at runtime). Static scheduling algorithms tend to be more complicated than dynamic scheduling. *List scheduling* is a type of static scheduling heuristics that schedule tasks to a system with multiple processing units (e.g., processors). A comprehensive introduction and complexity analysis of list scheduling can be found in [11].

Graph coloring is a classical problem in graph theory. It is a way of coloring the vertices of a graph such that no two adjacent vertices are of the same color; this is also called a *vertex coloring*. Refer to [25] for references of graph coloring, application to perform graph coloring, and other related information.

In a typical modern multiprocessor system, each processor owns a private cache. To ensure *cache coherency*, various cache coherence protocols can be applied. Refer to [26] for details on multiprocessor memory system architecture and cache coherence protocols.

References

1. Karri, R., Rajendran, J., Rosenfeld, K., Tehranipoor, M.: Trustworthy hardware: identifying and classifying hardware Trojans. Computer **43**(10), 39–46 (2010)
2. Tehranipoor, M., Salmani, H., Zhang, X., Wang, X., Karri, R., Rajendran, J., Rosenfeld, K.: Trustworthy hardware: Trojan detection and design-for-trust challenges. IEEE Computer **44**(7), 66–74 (2011)
3. Waksman, A., Suozzo, M., Sethumadhavan, S.: FANCI: identification of stealthy malicious logic using Boolean functional analysis. In: 2013 ACM SIGSAC Conference on Computer & Communications Security (CCS), pp. 697–708 (2013)
4. Orsila, H., Kangas, T., Hamalainen, T.D.: Hybrid algorithm for mapping static task graphs on multiprocessor SoCs. In: 2005 International Symposium on System-on-Chip (SOCC), pp. 146–150 (2005)
5. Schranzhofer, A., Chen, J.J., Thiele, L.: Dynamic power-aware mapping of applications onto heterogeneous MPSoC platforms. IEEE Trans. Ind. Inf. **6**, 692–707 (2010)

6. Foutris, N., Gizopoulos, D., Psarakis, M., Vera, X., Gonzalez, A.: Accelerating microprocessor silicon validation by exposing ISA diversity. In: 44th Annual IEEE/ACM International Symposium on Microarchitecture (MICRO), pp. 386–397 (2011)
7. Gizopoulos, D., Psarakis, M., Adve, S., Ramachandran, P., Hari, S., Sorin, D., Meixner, A., Biswas, A., Vera, X.: Architectures for online error detection and recovery in multicore processors. In: 2011 Design, Automation Test in Europe Conference Exhibition (DATE), pp. 533–538 (2011)
8. Brelaz, D.: New methods to color the vertices of a graph. Commun. ACM **22**, 251–256 (1979)
9. Lingo. http://www.lindo.com
10. Kwok, Y.K., Ahmad, I.: Static scheduling algorithms for allocating directed task graphs to multiprocessors. ACM Comput. Surv. **31**, 406–471 (1999)
11. Rădulescu, A., van Gemund, J.C.A.: On the complexity of list scheduling algorithms for distributed-memory systems. In: Proceedings of the 13th International Conference on Supercomputing (ICS), pp. 68–75 (1999)
12. Kim, K.: A method for computing upper bounds on the size of a maximum clique. Commun. Korean Math. Soc. **18**, 745–754 (2003)
13. Culler, D.E., Singh, J.P., Gupta, A.: Parallel Computer Architecture: A Hardware/Software Approach. Morgan Kaufmann, San Francisco (1999)
14. Dick, R.P., Rhodes, D.L., Wolf, W.: TGFF: task graphs for free. In: Workshop on Hardware/-Software Codesign (CODES), pp. 97–101 (1998)
15. Maheswaran, M., Ali, S., Siegel, H.J., Hensgen, D., Freund, R.F.: Dynamic matching and scheduling of a class of independent tasks onto heterogeneous computing systems. In: 8th Heterogeneous Computing Workshop (HCW), pp. 30–44 (1999)
16. Sih, G.C., Lee, E.A.: A Compile-time scheduling heuristic for interconnection-constrained heterogeneous processor architectures. IEEE Trans. Parall. Distrib. Syst. **4**, 175–187 (1993)
17. Wang, N., Yao, M., Jiang, D., Chen, S., Zhu, Y.: Security-driven task scheduling for multiprocessor system-on-chips with performance constraints. In: 2018 IEEE Computer Society Annual Symposium on VLSI (ISVLSI), pp. 545–550 (2018)
18. Wang, N., Chen, S., Ni, J., Ling, X., Zhu, Y.: Security-aware task scheduling using untrusted components in high-level synthesis. IEEE Access **6**, 15663–15678 (2018)
19. Sun, Y., Jiang, G., Lam, S., Ning, F.: Designing energy-efficient MPSoC with untrustworthy 3PIP cores. IEEE Trans. Parall. Distrib. Syst. **31**(1), 51–63 (2020)
20. Ancajas, D.M., Chakraborty, K., Roy, S.: Fort-NoCs: mitigating the threat of a compromised NoC. In: 2014 51st ACM/EDAC/IEEE Design Automation Conference (DAC), pp. 1–6 (2014)
21. JYV, M.K., Swain, A.K., Kumar, S., Sahoo, S.R., Mahapatra, K.: Run time mitigation of performance degradation hardware Trojan attacks in network on chip. In: 2018 IEEE Computer Society Annual Symposium on VLSI (ISVLSI), pp. 738–743 (2018)
22. Boraten, T., Kodi, A.K.: Mitigation of denial of service attack with hardware Trojans in NoC architectures. In: 2016 IEEE International Parallel and Distributed Processing Symposium (IPDPS), pp. 1091–1100 (2016)
23. Daoud, L., Rafla, N.: Routing aware and runtime detection for infected network-on-chip routers. In: 2018 IEEE 61st International Midwest Symposium on Circuits and Systems (MWSCAS), pp. 775–778 (2018)
24. Orlin, J.B.: Integer Programming. http://web.mit.edu/15.053/www/AMP-Chapter-09.pdf
25. Culberson, J.: Graph Coloring Page. https://webdocs.cs.ualberta.ca/~joe/Coloring/index.html
26. Solihin, Y.: Fundamentals of Parallel Multicore Architecture, 1st edn. Chapman & Hall/CRC, Boca Raton (2015)

Chapter 12
A Framework for Detecting Hardware Trojans in RTL Using Artificial Immune Systems

Farhath Zareen and Robert Karam

12.1 Introduction

One of the most challenging problems faced in hardware security is the detection of hardware Trojans, which are malicious modifications that are made to an Integrated Circuit (IC) to change its original, desired functionality. Hardware Trojans perform a wide range of undesirable or unsafe actions, such as causing system failure or leaking information to an attacker under specific conditions determined by the Trojan designer, and can be inserted at any phase of the IC design cycle, including in high-level circuit descriptions, or during fabrication in a foundry.

Since Trojans are a product of a human designer in contrast to naturally arising error caused by physical processes in fabrication, they may be viewed as intentional faults that are added specifically to induce undesirable behavior when certain rare conditions are met. For Trojans in logic or memory circuits, the *payload*, or faulty behavior, may be activated by special input combinations or sequences of input combinations to trigger this undesirable or faulty behavior [1]. These rare triggering events are designed to be stealthy and undetectable during simulation and testing.

A golden reference model that is guaranteed to be Trojan-free is typically required to detect these malicious events [2]. However, these golden reference models may not be available, since Intellectual Property (IP) cores are predominantly licensed from untrusted third-party vendors. Another drawback is that using a golden model as the basis for detection may be inconclusive or too complex for exhaustive verification, especially for larger IC designs [2]. Many techniques that use golden reference models also require direct measurement of parameters like

F. Zareen (✉) · R. Karam
University of South Florida, Tampa, FL, USA
e-mail: fzareen@mail.usf.edu; rkaram@usf.edu

S. Katkoori, S. A. Islam (eds.), *Behavioral Synthesis for Hardware Security*,
https://doi.org/10.1007/978-3-030-78841-4_12

electromagnetic (EM) emissions or power consumption and can only be effectively applied to fabricated ICs, but not at higher levels of abstraction.

In recent times, researchers have proposed detection techniques that do not require a golden reference model. The work of Narasimhan et al. [3] demonstrated an effective technique that implemented functional testing and side-channel analysis to compute transient current signature for multiple time frames under invariable state transitions and known activation times. Chakraborty et al. [4] implemented a technique based on statistical vector generation to uncover hidden Trojans in rare circuit nodes by applying test vectors to trigger rare nodes multiple times. This technique can be a valuable tool to find hidden Trojans in the circuit and can be effective at lower levels of abstraction. However, identifying unsafe behavior for designs at higher levels of abstraction requires alternative techniques.

A technique to combine multiple stages of verification comprising code-coverage, equivalence analysis, removal of redundant circuits, sequential Automatic Test Pattern Generation (ATPG), and assertion checking was proposed by Zhang and Tehranipoor to reveal existence of suspicious signals [5]. Mutation Testing is used by Fern et al. to detect Trojans in unspecified functionality such as "don't care" conditions which may go undetected by traditional verification procedures that rely on formal specification [6].

The abovementioned techniques address ways to identify hidden Trojans in rare circuit nodes or in unspecified functionality. However, the work in this chapter describes a technique to identify and classify certain *behavioral traits* in a circuit that would indicate potentially unsafe operations. A design that consists of potentially unsafe operations may have been subjected to intentional malicious modifications, such as Trojan insertions, or unintentional human error during development. Regardless, the ability to accurately classify unsafe behavior for any given design would allow IP designers to promptly verify the security of their own work, or system integrators to check third-party IP (3PIP) cores used in larger designs, during the design process, before accruing additional costs of fabrication. While other previously discussed methodologies are applied to lower levels of abstraction, analyzing a design for unsafe behavior in RTL and discovering it early in the design flow can save significant time, cost, and provide a valuable tool for hardware security research.

This chapter describes how Artificial Immune Systems [7, 8] can be used to accurately distinguish between safe and unsafe circuit behaviors by extending the concept of identifying and classifying self versus non-self traits. Some common AIS techniques are Negative Selection Algorithm, Clonal Selection Algorithm, and Immune Network Theory. Generally, AIS have been applied to adjacent security problems such as malware detection in software [9–12], where certain sequences of instructions may be considered "unsafe" and indicate the presence of malware. They have also been used for optimization and pattern recognition [13, 14].

This chapter describes how Negative Selection Algorithm (NSA) [7] and Clonal Selection Algorithm (CSA) [15] are used to identify behavioral patterns in control and data-flow graphs (CDFG) extracted from behavioral Verilog HDL circuit descriptions. The models are trained on RTL Trojan benchmarks obtained from

TrustHub [16, 17] and CDFGs are extracted using PyVerilog [18], an open-source Verilog code parser and static analyzer tool, though any CDFG extraction tool may be used. Through experiments and analysis, the described work ultimately demonstrates that AIS can effectively distinguish between circuit designs demonstrating "safe" behavior and those that show "unsafe" behavior, by recognizing patterns in Trojan-inserted designs and matching against the design under test.

The main contributions of the work described in this chapter are as follows:

1. It frames the problem of detecting unsafe behavior in RTL in terms of detecting self vs. non-self-behavior in the design's control and data flow.
2. It presents a complete software tool flow for AIS-based RTL source code analysis, including model generation and behavior classification.
3. It analyzes the efficacy of Negative Selection and Clonal Selection Algorithms on binary-encoded CDFGs for detecting unsafe behavior in RTL.
4. It demonstrates how machine learning can be used to detect unsafe behavior in Trojan-included hardware, even if the specific instance of the Trojan has not been previously encountered—similar to an immune system that responds to a foreign cell in the body, even if it has not encountered that specific cell before.

To the best of our knowledge, this is the first application of AIS to hardware Trojan detection at a high level of abstraction. This chapter is organized as follows: Sect. 12.2 introduces a formal definition of a hardware Trojan and reviews recent Trojan detection techniques based on conventional (Sect. 12.2.1.1) and machine learning approaches (Sect. 12.2.1.2). Next, an introduction to Artificial Immune System (AIS) concepts and its algorithms is provided in Sect. 12.3. Section 12.4 then delves into the AIS Trojan detection methodology. AIS training and testing procedures are then discussed in detail in Sect. 12.5. Experimental results and analysis are discussed in Sect. 12.6. Finally, Sect. 12.7 ends this chapter with conclusion and future work.

12.2 Background

This section defines the problem of hardware Trojans and discusses some traditional detection methodologies. It provides a summary of recent ML-based techniques for Trojan detection, mainly side-channel data based anomaly detectors and finally, defines Artificial Immune Systems and explores AIS algorithms that are discussed in this chapter for circuit behavior classification.

12.2.1 Hardware Trojans

Modifications made to an IC with specific malicious intents, such as to leak sensitive information or provide back-doors to carry out further malicious activity, are called hardware Trojans. These malicious modifications can be made during the design or fabrication phases by untrusted sources such as design house or foundries [2].

Trojans are classified based on physical, activation, and action attributes. Physical representation depicts the type of hardware manifestations of the Trojans in the circuit. Activation denotes different mechanisms that cause Trojan to activate (trigger), and the malicious behavior that takes place in the circuit is then defined by action (payload). The trigger periodically monitors signal activity in the circuit. In the event of a specific change in the signal, payload is activated to allow the malicious event to occur. Trigger and payload are employed to activate hardware Trojans. These trigger conditions are very rare and sometimes may only occur once in many years, making detection challenging. Trojan triggers are broadly classified as either analog or digital. Analog triggers mainly depend on activation conditions based on certain natural events such as electromagnetic emissions, temperature, and power conditions. Digital triggers, however, are comprised of sequential and combinational circuits. Trojans in combinational circuits are stateless and typically depend on rare trigger nodes to activate the payload. In sequential circuits, Trojan activation condition relies on the occurrence of a continuous sequence of rare logic conditions, making them more difficult to detect using standard verification techniques.

Due to the stealthy nature and extremely large number of activation, structure, and operation instances, a comprehensive framework that provides a systematic investigation of Trojan behavior is needed for the development of powerful and effective defenses against hardware Trojans. The first detailed taxonomy for hardware Trojans was developed by Wang, Tehranipoor, and Plusquellic [19] which was based on the three fundamental attributes of Trojans: physical, activation, and action. With the recent advancement and sophistication of hardware Trojans, new attributes were introduced to make the taxonomy more comprehensive. The most detailed taxonomy consists of six main categories that include insertion phase, abstraction level, activation mechanism, effects, and location, with each category consisting of different attributes [20–22]. In short, hardware Trojans can be better classified and identified based on these attributes to learn how to identify behaviors.

Hardware Trojan detection comprises various tools and techniques to detect malicious modifications in circuits. Since Trojan behavior depends mainly on its point of insertion, detection techniques can vary significantly. Therefore, depending on the type of Trojan to be identified, detection techniques may include logic testing [23], side-channel analysis [24], or reverse engineering [25].

12.2.1.1 Conventional Trojan Detection Approaches

Several techniques may be employed to detect hardware Trojans: (1) logic testing, (2) side-channel analysis, (3) reverse engineering via physical inspection, and (4) Design-for-Test (DfT) or Design-for-Security (DfS) based on built-in tests. Two widely used methods are logic testing and side-channel analysis. Although the other methods are still effective, they may fail to detect more stealthy and evasive Trojans [2].

Logic testing aims to uncover Trojans through a compact testing procedure to trigger as many low-probability conditions as possible within an IC; this methodology, called *Multiple Excitation of Rare Occurrences* (MERO) [4], requires finding an optimal set of vectors that activate nodes with low-probability conditions. Since certain nodes can be built to withstand random-pattern testing, toggling through multiple nodes simultaneously is better to uncover true node behavior. Since this technique narrows down node behavior to a very minute level, it may not be effective against large Trojans, as logic testing is unlikely to trigger the large number of inputs needed. Generation of large test vectors suitable for detection is also more difficult using this technique.

Detection of larger Trojans is more feasible with side-channel analysis technique, wherein physical parameters or signals based on current and delay are analyzed to investigate the presence of Trojans in a circuit. For instance, circuit delay may indicate the presence of new elements in the circuit that could have been added with malicious intent. Agarwal *et al.* used a side-channel fingerprint of the circuit using Principle Component Analysis (PCA) and compared it with a golden reference model [24]. Banga et al. introduced sustained vector technique to distinguish power consumption variations between golden reference model and infected circuits [26]. Transient current analysis, static current analysis, and path delay analysis are some other side-channel approaches. However, a golden reference model is typically required for side channel-based detection techniques. Process variation may also cause the physical parameters of a circuit under test to differ from the golden reference model.

12.2.1.2 ML-Based Hardware Trojan Detection Approaches

Aside from conventional Trojan detection techniques, several machine learning-based techniques have also been proposed for Trojan detection. However, prior research work primarily identified anomalies and classified circuits based on side-channel measurements. As a result, several of the proposed techniques require a golden reference model for HT detection. For instance, Iwase et al. [27] proposed a ML pipeline that uses Support Vector Machine (SVM) to identify Trojan-free or Trojan-inserted designs in AES circuits based on differing levels of power consumption. To do this, the authors used Discrete Fourier Transform (DFT) to convert the acquired power consumption waveform data from the time domain to frequency domain as features to train the SVM.

Lodhi et al. [28] proposed another ML classification technique that combines timing signatures and a series of classification algorithms to detect HTs. They implemented a self-learning framework that uses their proposed *macro synchronous micro asynchronous (MSMA)* signature technique for feature extraction. Here, the golden reference model is used to extract features from MSMA, which are then used to train several classifiers: k-nearest neighbors (kNN), Bayesian, and decision trees.

Bao *et al.* proposed a ML technique that does not classify circuits using side-channel measurements while requiring the use of a golden model [29]. In this technique, the IC's physical layout is extracted using reverse engineering and imaging techniques. Resulting scans of the IC's physical layout are then analyzed and used to train an SVM to characterize Trojan-inserted and Trojan-free structures in the IC. Although this RE technique does not require generating a transistor or gate netlist, it does require high resolution, low noise scans for accurate classification. This may become more challenging with future process technologies and smaller feature sizes.

Other ML techniques that do not require a golden reference model and are not based on side-channel measurement are proposed by Hasegawa et al. and Ova et al.. Hasegawa et al. [30] developed an SVM classifier that is trained to detect Trojans in a gate-level netlist using features such as flip-flop input/output and primary input/output for training. The authors found that dynamic weighting for SVM training demonstrated an accuracy of 80%.

Alternatively, Oya et al. [31] used a score-based classification technique for discerning between Trojan-free and Trojan-inserted netlists. Instead of wholly identifying a netlist as a HT-inserted circuit, the authors aimed to identify Trojan nets within these designs to classify it as a HT-inserted design; in this way, they would be able to identify what aspects of its structure exhibit unwanted behaviors. As a result, the authors observed successful Trojan detection in certain TrustHub benchmarks.

FANCI [32] and VeriTrust [33] are two other noteworthy, state-of-the-art techniques that utilize non-ML-based design stage verification on gate-level netlists. FANCI utilized "control values" that depicts how the functionality of certain wires in the IC affects others in order to identify wires that carry potential backdoor trigger signals and exhibit malicious behavior. VeriTrust used "tracers" and "checkers" to examine ICs for redundant inputs to determine if there are any signals that are carried by these inputs, which may indicate the presence of suspicious Trojan signals. A comprehensive taxonomy of Trojan characteristics and a summary of the latest Trojan detection approaches is presented in [34].

The chapter presents a detection technique based on Artificial Immune Systems (AIS), which are an alternative class of ML techniques based on biological immune systems. As opposed to the research work discussed earlier, the work discussed in this chapter does not consider side-channel data or logic functions, and does not require simulation or silicon measurements of any kind. Instead, it aims to classify *high-level behavioral traits* in a circuit based on its CDFGs, to detect unsafe behavior. Because implementation specifics are abstracted at the CDFG level and they encapsulate both sequential (control flow) and combinational (data flow) type of behavior in RTL, the Artificial Immune System can then be trained to identify a range of behaviors associated with Trojan or Trojan-like functionality.

12.3 Artificial Immune Systems

The immune system is a host defense system that encompasses various biological structures and processes to protect against diseases, wherein it fights and removes destructive pathogens such as bacteria or viruses and returns the body to a healthy state. The field of natural computing is primarily concerned with the theoretical analysis, understanding, and modeling of biological phenomena or processes in the form of algorithms that mimic processes observed in nature.

Artificial Immune Systems (AIS) are a primary example of computationally intelligent class of machine learning algorithms that are biologically inspired by the vertebrate immune systems to distinguish between negative examples (unknown) from positive examples (known). Many noteworthy algorithms have been derived from the research on AIS and their applications extend to various real world problems including anomaly detection, pattern recognition, and optimization [13, 14].

12.3.1 Overview of AIS

AIS is established on the theory that immune systems distinguish the pathogen (any malicious entity) as *non-self* and the body's cells (benign entities) as *self*. Detection of pathogens by the immune system is described in terms of discriminating "self" from "non-self," where the normal functioning of the body is characteristically classified as "self," and any anomaly or abnormality that is foreign to the body is classified as "non-self." Antigens are parts of the pathogen that alert the body to an infection. Immune cells can recognize antigens to target and remove a pathogen from the body, thereby stopping or even preventing an illness. However, not all pathogens are harmful to the body. Anything that does not harm the body is characterized as "self" and anything that endangers the normal functioning of the body is deemed as "non-self". Immune systems, therefore, primarily function to learn discriminating "self" behavior from "non-self" behavior through the process of self-replication and evolution. Hence, detection and elimination of harmful pathogens is a result of millions of immune cells circulating the body and interacting with other cells to detect and eliminate "non-self" entities [35, 36].

In biological immune systems, lymphocytes are the primary immune cells that take part in the immune response. The two dominant types of lymphocytes are B and T cells that evolve and mature in the lymphoid organ. The secondary lymphoid organs represent different locations where antigen interactions take place to stimulate an immune response. Thousands of receptors are present on the surface of lymphocytes that bind to pathogens depending on their chemical composition. The likelihood of chemical bonding increases with their affinity. When immune cells recognize a pathogen, it stimulates a response wherein the cells undergo a multiplication and separation process to produce clones or antibodies. Hence, upon exposure to a specific antigen, a large pool of antibodies are generated through a process called clonal expansion. For fast immune response, memory cells are also produced to detect similar antigens in the future.

These immune systems inspired concepts of *antibodies*, in this case self and benign entity, and *antigens*, malicious and non-self entity, are applied to the computational understanding of malicious and normal behavior of the system for anomaly detection [37, 38]. Three main immunological theories are the basis of AIS—clonal selection, negative selection, and immune networks. Research has been done on the learning and memory mechanisms of the immune system and anomalous entity detection. The immunological theories are based on adaptive and innate immune responses. By utilizing adaptive and innate immune responses, the immune system can fight or counteract pathogens. The innate immune response helps in the initialization of the immune responses to fight all kinds of pathogens in general, whereas the adaptive immune response uses lymphocytes or white blood cells, to fight against specific pathogens that invade the body [35, 36].

12.3.1.1 Motivation Behind Negative Selection Algorithm

Special white blood cells called *T cells*, which are formed from precursor cells produced in the bone marrow, are primarily responsible for immunity in the body through the elimination of pathogens that are harmful to the body [35, 36]. During their development cycle, these cells undergo a maturation process in the thymus gland referred to as *T-cell tolerance* or *negative selection* [39, 40]. This procedure is very crucial to the development of the immune system, as through this process, "faulty" or harmful cells produced through cell division are removed from the system. From a biological perspective, harmful cells are determined by those T cells that strongly bind with self-proteins, or in other words, T cells that would potentially harm self-cells that are native to the body.

This is where the recognition of *self* and *non-self* comes into play, wherein cells that are self-reactive need to be eliminated at the time of propagation. In a similar manner, the *Negative Selection Algorithm* (NSA) detects data manipulated by viruses or Trojans, whose presence demonstrates harmful or malicious behavior. After NSA is executed, the AIS acquires suitable detectors that will be trained to discriminate between manipulated and original data.

12.3.1.2 Overview of the Negative Selection Algorithm

NSA (Fig. 12.1) typically consists of two phases: the *detector generation* phase and the *detector application* phase which can also be referred to as the *censoring* and *monitoring* phases, respectively [7]. The censoring phase deals with training whereas monitoring phase tests the machine learning tasks. At the end of the censoring phase, mature detectors used to detect non-self behavior from self behavior are generated. Subsequently, the system being protected can be monitored for changes by the detectors generated in the censoring stage.

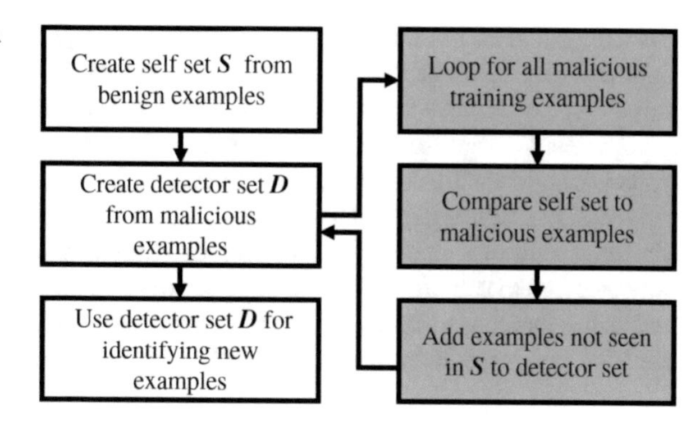

Fig. 12.1 Overview of NSA

Algorithm 12.1 Negative selection algorithm [41]

1: Let R be set of strings generated from training CDFG encodings
2: Let S_{self} be subset of strings that represents self-behavior
3: Let N be the desired number of detectors
4: Let $S_{non-self}$ be the final detector set made of non-self strings
5: **repeat**
6:     **for** strings $r \in R$ **do**
7:         **if** $r \notin S_{self}$ **then**
8:             Add r to $S_{non-self}$
9:         **else**
10:             Remove string r from R
11:         **end if**
12:     **end for**
13: **until** $size(S_{non-self}) \leq N$
14: Let T be list of strings generated from testcase CDFG encodings
15: **for** strings $t \in$ test case T **do**
16:     **if** $t \notin S_{non-self}$ **then**
17:         Label case as anomaly
18:     **end if**
19: **end for**

NSA (Algorithm 12.1), first requires a set of examples that describe typical IC behavior, denoted by S_{self}, that the AIS uses as reference for distinguishing normal self-cells from non-self cells. The data to be protected which is contained within S_{self} is represented as a binary-encoded string. The string is then split into several l-length sub-strings to make up the set of S_{self} data. Once S_{self} is populated, the algorithm then generates a set of candidate detectors R of a specified length l from the binary-encoded string of CDFGs which are obtained from the selected Trojan benchmarks. Candidate detectors, which consist of examples of malicious behavior, are obtained from the Trojan benchmarks to populate R in the censoring phase. Traditionally, random string generation is used to build the candidate detector set; however, using real examples from such benchmarks when building the detector set will significantly improve classification.

The candidate detectors in R are then matched to the entire self-set S_{self}. Any string from R that matches strings in S_{self} is eliminated from R. On the other hand, the strings that do not match any of the strings in S_{self} are kept and are added to the final detector set collection $S_{non-self}$. This process is repeated until all candidate sub-strings in R have been compared to those examples in S_{self}; after this phase is completed, a representative detector set $S_{non-self}$, comprising unsafe behavioral traits is obtained. This detector set is then used in the testing phase. During testing, binary CDF features describing the test HDL design are generated and stored in a set T; each string in T is compared to each example in the final detector set $S_{non-self}$. Any changes to self will be indicated by an overlap between T and $S_{non-self}$. If any test string matches at least one of the detector strings in $S_{non-self}$, it indicates that the NSA has identified unsafe traits that match the trained detector set.

12.3.2 *Clonal Selection Algorithm*

12.3.2.1 Motivation Behind Clonal Selection Algorithm

In vertebrate immune systems, as a response to attacks by antigens invading the body, a massive number of cells are produced by the body that are equipped to counter or combat these threats [36, 42]. This defense mechanism is based on the clonal selection principle in immunology, which describes how the B and T cells of the lymphocytes are reproduced in adaptive immune response through the process of *affinity maturation* [43], where cells are massively reproduced and repeatedly exposed to antigens to develop stronger antibodies that are resistant to antigen threats. As a result of affinity maturation, these antibodies would then be capable of eliminating the antigens by binding onto them. These ideal antibody cells are kept and could then be deployed and propagated as clones to fight against the foreign infection. Some of these cells are also retained as memory cells that will continue to circulate over a period of time, to fight any remaining antigens that are recognized in the body. The concept of clonal selection theory (CLONALG) [15] has been applied to the *Clonal Selection Algorithm*, which allows a learning model to evolve and identify key attributes that allow it to pinpoint malicious behavior while maintaining memory of positive examples for future detection. This algorithm exhibits Darwinian attributes, where only the fittest cells survive and are kept as antibodies to defend against antigen threats and that newer generations tend to vary and evolve through mutation. Through the mutation process, these newer generation cells could potentially evolve to better handle existing or future antigens.

12.3.2.2 Overview of the Clonal Selection Algorithm

The Clonal Selection Algorithm (Fig. 12.2) has traditionally been applied to pattern recognition and optimization tasks [13]. The algorithm develops a characteristic understanding of threats through the production of antibody *memory cells* that

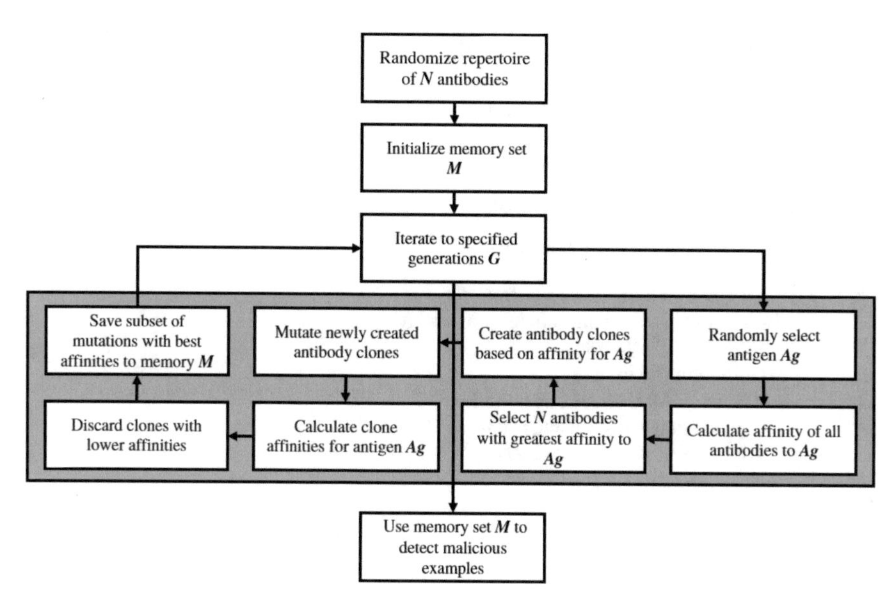

Fig. 12.2 Overview of CSA

mutate upon encountering antigens, and consequently acquire the characteristics of these antigens over time with exposure.

Affinity Maturation process [44] randomly generates a set of antibodies similar to the antigen, based on the *affinity* of the antibody to the antigen. The maturation rate is defined to reflect the amount of mutation that occurs in a particular antibody. Antibodies with a high affinity to the exposed antigens are cloned and mutated with very minimal changes as a result of affinity maturation, while antibodies with a lower affinity are typically cloned and mutated to a higher degree. Based on the maturation rate, these cells then mutate to form clones that will acquire the characteristics of the antigen. Long-term exposure of these memory cells results in a detector set of memory cells that will exhibit characteristics of the antigens that they were exposed to, resulting in an optimized pool of memory cells.

CSA (shown in Algorithm 12.2) initially generates a random set of antibodies or self examples to produce a set of detectors. These antibodies are then exposed to antigens or malicious examples to determine whether they are suitable for detection or not, which is indicated by a high degree of affinity or similarity to the antigens. With exposure to antigens, the cloned detector set acquires the features of the subjected antigens, and over a period of time, it is able to generate a single optimized clone that reflects all the characteristics of the antigens to which it has been exposed. Higher affinity clones are retained and added to a memory set, which is then applied as a detector set to identify unseen or abnormal traits as Trojan-inserted designs, after being exposed to training antigen examples over a multitude of generations.

In detail, the algorithm initially generates a set S_{Ab} containing randomized self examples or antibodies of a predetermined and fixed population size N_{pop},

Algorithm 12.2 Clonal selection algorithm [41]

1: Let G be number of generations (iterations)
2: Let M be memory set of detector clones
3: Let S_{Ab} be set of antibodies
4: Let S_{clones} be set of antibody clones
5: Let S_{Ag} be set of antigens
6: Let N_{pop} be the desired number of antibodies generated for each generation
7: Let $rate_C$ be cloning rate and $rate_M$ be mutation rate
8: **repeat**
9: randomly pick an antigen Ran_{Ag} from S_{Ag}
10: randomly generate antibodies for S_{Ab} of size N_{pop}
11: **for** antibody Ab in S_{Ab} **do**
12: calculate affinity to Ran_{Ag}
13: **if** affinity of Ab is above threshold **then**
14: Create clones of Ab using $rate_C$
15: Mutate clones of Ab using $rate_M$
16: Add clones to S_{clones}
17: **end if**
18: **end for**
19: **for** clone C in S_{clones} **do**
20: calculate affinity to Ran_{Ag}
21: **if** affinity of C is above threshold **then**
22: Add clone C to M
23: **end if**
24: **end for**
25: **until** all generations G are covered
26: Use memory set M for classifying malicious examples

and it generates a set S_{Ag} containing non-self examples or antigens, which are taken directly from Trojan-inserted benchmark designs. It then defines a subset of generated clones denoted as S_{clones} and a memory subset of clones that are selected for detection during testing, denoted as M. The process of *generation* first takes place, where each antibody in S_{Ab} is iteratively exposed to a pool of antigens S_{Ag}. For each antibody exposed to a randomly selected antigen Ran_{Ag} from S_{Ag}, its affinity value to the antigen is calculated using the Hamming distance metric; this procedure is akin to affinity maturation [44] from the clonal selection theory. Hamming distance is defined as the least number of substitutions needed to modify one string into the other [45], or it simply refers to the number of features (in this case, binary bits) that differ between Ran_{Ag} and all antibodies in S_{Ab}. The antibodies with the highest affinity to the subjected antigens are cloned at a rate reflected by their affinity and are subsequently added to S_{clones}. Specifically, the rate at which clones are produced, denoted as $rate_C$, based on a single antibody in S_{Ag} is directly proportional to its affinity. At the same time, while clones are reproduced, they will undergo mutation that may potentially increase their affinity to Ran_{Ag}. The rate at which antibody clones are mutated, denoted as $rate_M$, is inversely proportional to the affinity of a given antibody; in other words, the lower an antibody's affinity is, the more mutation occurs on each generated clone. As a result, any antibody that has a low affinity to Ran_{Ag} will undergo a greater degree of mutation compared to those with higher affinity, to potentially enable its clones to perform better than the original. The process of affinity maturation is then repeated on the cloned antibodies to determine their new affinity values. The clones with the highest degree of affinity are then retained in the detector memory subset M, while the clones with the lowest degree of affinity are replaced with other samples from R_{Ab}. After G iterations of affinity maturation, the memory set M is then ready to be used for detecting abnormalities, since it will contain entries that have exhibited Trojan behavior.

12.4 Detection Methodology

This section discusses the pipeline for Trojan detection—from CDFG generation to behavior classification. Briefly, a static analysis tool is used to generate a control and data-flow graph from which features from the RTL design are extracted and used to train the AIS. For training and testing, RTL benchmarks from TrustHub are used [16, 17], which are labeled as either Trojan-inserted (non-self) or Trojan-free (self).

12.4.1 CDFG Generation and Feature Extraction

The control and data-flow graphs of a circuit defined by HDL depict the internal sequence of operations or procedures that are performed to execute a program from beginning to end. A *control-flow graph* [46] provides a representation of all the paths that can be traversed through a program during its execution; in this graph, each node represents a basic block that expresses a sequence of consecutive statements, and edges represent any possible flow of control from one node to another. A *data-flow graph* [47] represents the data dependencies between different operations that a program performs and thus can be viewed as a fundamental expression of a computational structure. Hence, a CDFG will represent the control flow of components and any data dependencies between them. It fully encapsulates the behavior of the design, including any potentially unsafe or undesired behavior, regardless of it was integrated into the design. The underlying assumption, then, is that the CDFG is wholly and correctly extracted from the source design.

By training an AIS to recognize patterns in the CDFG as potentially unsafe, whether the intent is malicious or not, the AIS will be equipped to detect non-self behavioral traits in RTL designs. Furthermore, through the use of AIS and the evolutionary nature of the learning process, training on Trojan-included CDFGs results in the detection of similar, if not exact behavior in other designs. Hence, comprehensive training is not required to achieve high detection accuracy on unencountered Trojans.

The Python tool *PyVerilog* [18] is used to generate CDFGs for each benchmark. Initially, an abstract syntax tree representation is constructed from the behavioral Verilog code that will define the signal scope and establish all parameters and constants. Finally, an assignment tree is generated for every signal in the code to produce a thorough representation of the control and data flow.

The designs are first flattened to a single procedure since PyVerilog only generates CDFGs for single-procedure Verilog files. Then, control-flow and data-flow analysis is performed to produce CDFGs. Information regarding nodes and edges such as isolated nodes, conditional and directed edges labeled with their latencies, dependent operations etc., are extracted as features in the form of binary encoded strings by converting categorical data obtained from CDF analysis to an adjacency matrix representation. These encodings are then used as input to the AIS training procedure. The entire pipeline, from CDFG generation to feature extraction, is shown in Fig. 12.3. The format of the binary-encoded string is summarized in Fig. 12.4. Specifically, graph encoding is performed as follows:

1. Represent structure type (input, output, reg, wire) as a 2-bit *type string*.
2. Adjacency matrix appends an m-bit *edge connectivity string* obtained from a singular matrix row, where "0" suggests no existing edge between two nodes, and "1" suggests that an edge does exist.
3. This string length is equal to the number of nodes in the graph, m. Isolated nodes (nodes without outgoing edges) consist of an edge connectivity string of all 0's.
4. *Condition-based* edge connectivity string represents conditions such as "greater-than," "less-than," or "not equal to," etc. that cause a change in control flow. This string is then appended to the condition-based edge connectivity string.
5. Each condition in the condition frequency string is compressed to a count of total edges to represent a more generalized condition-based edge connectivity string.

Benign and malicious behavior is differentiated by node and edge connectivity information in the CDFGs and is essential for accurate classification of behavior. For instance, Trojan-inserted code may show isolated or atypical structures due to hidden characteristics inserted by the Trojan, whereas benign code is unambiguous with clear and typical connectivity. There may also exist specific combinations or sequences of instructions that indicate certain unspecified functionality. In such

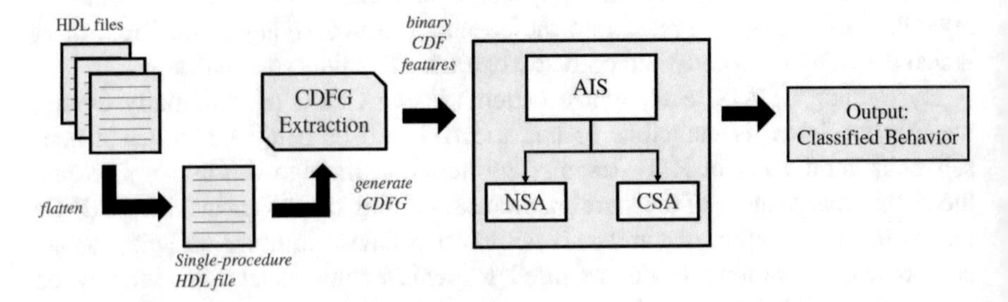

Fig. 12.3 RTL Trojan detection in CDFGs using artificial immune systems [41]

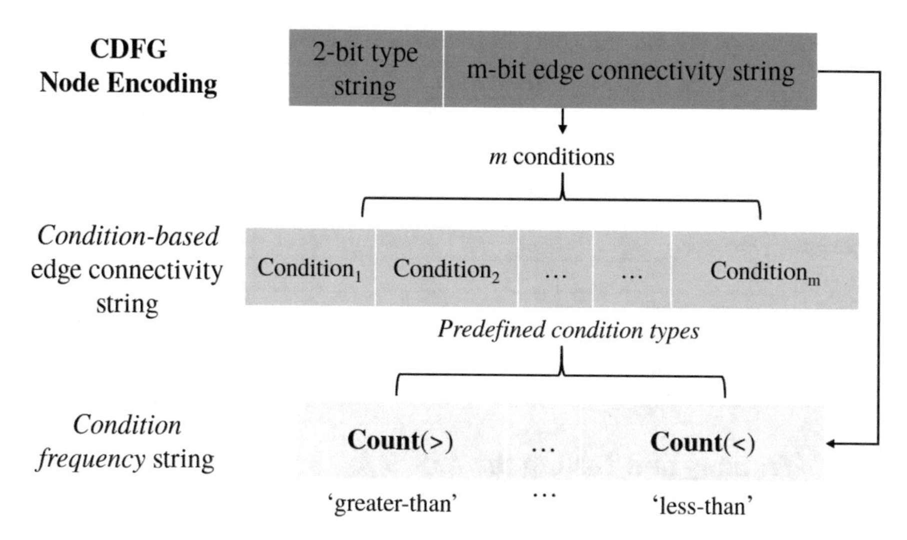

Fig. 12.4 CDFG node encoding used in training AIS [41]

Table 12.1 Hardware Trojan benchmark list obtained from TrustHub [16, 17]

Benchmark	# Trojan-free	# Trojan-inserted
AES	21	21
MC8051	7	7
PIC16F84	4	4
RS232	10	10
wb_conmax	2	2
Other	6	0

cases, frequency of conditions along with the flow of instructions of one type to another can be used to train the AIS.

12.4.2 Benchmarks

The training and testing data used in the experiments were taken from TrustHub [16, 17]. An overview of the benchmarks used can be found in Table 12.1, and a detailed description of each benchmark is shown in Table 12.2. The TrustHub repository comprises various designs categorized under different attributes such as abstraction levels, phase at which hardware Trojans were inserted, the intended location of the circuits, for instance, input/output, power, etc., and the malicious effects of those Trojans. Data leaks, manipulating instruction registers, triggering denial-of-service, or deterioration of overall performance are some of the malicious changes made by these Trojans. A total of 50 Trojan-free and 44 Trojan-inserted RTL benchmarks are used.

Table 12.2 Hardware Trojan benchmark attributes from TrustHub [16, 17]

Trojan characteristics		Benchmarks				
		AES	*MC8051*	*PIC16f84*	*RS232*	*wb_conmax*
Effect	Leak Information	✓		✓		
	Denial-of-Service	✓	✓	✓		✓
	Change functionality				✓	✓
Location	Processor	✓	✓	✓		
	I/O		✓			✓
	Power supply		✓			✓
Activation	Always On	✓				
	Trigger condition	✓	✓	✓	✓	✓

12.5 Training and Testing the AIS

AIS training and testing process is discussed in this section. The training is performed as a leave-one-out cross validation, where each classifier is trained on all examples but one type of circuit in the benchmarks, and then tested on the remaining benchmark. This approach is applied to all benchmark circuit types to have n AIS detectors for n circuit types. For instance, in the first phase, the AIS is trained on all circuit types except for AES, and then AES circuit examples are used in the testing phase. In the second phase, AIS is trained on all circuit types except MC8051 and then MC8051 circuit examples are used in the testing phase, and so on. Therefore training and testing is performed on all batches of examples in the benchmarks in a round-robin technique until there is no significant change in accuracy or loss. Leave-one-out cross validation technique enables testing on unseen benchmarks, and as a result, the AIS learns a generalization of the characteristics of a Trojan-inserted design so it can distinguish between Trojan-inserted and Trojan-free designs in unseen cases.

12.5.1 Negative Selection Algorithm

The binary-encoded CDFG features are given as input to the AIS. During the training or censoring phase, the AIS is trained on Trojan-inserted and Trojan-free examples to generate self-set and a mature detector set consisting of all the strings that were categorized as non-self. To achieve this, the AIS constructs 32-bit binary strings from the benchmark CDFGs and builds a self-set consisting of benign features obtained from Trojan-free examples. The binary string length is chosen depending on the length of the generated CDFG binaries. Larger string lengths are chosen to mitigate any feature loss. Then, the self-set is compared against the Trojan-inserted binary-encoded strings to construct the detector set consisting of strings from those examples that do not match the Trojan-free binary-encoded features. This process continues for all training examples from the Trojan-inserted benchmarks. In the testing phase, the detector set, which consists of non-self binaries, is then applied to unseen binary-encoded CDFG benchmarks to determine whether they are Trojan-inserted or Trojan-free. Whole string and partial string matching techniques are employed to distinguish self from non-self strings. Partial string matching shortens the time taken in string comparisons, but whole string matching preserves more features. Both techniques were evaluated to determine difference in performance.

12.5.2 Clonal Selection Algorithm

As from the Clonal Selection Algorithm described in Sect. 12.3.2, size of the antibody population and memory set must be defined. For the experiments, a population size of 100 and a memory-set size of 50 detector clones is used for the generation phase, and the CSA was carried out for 50 iterations. Each antibody (self example) and antigen (non-self example) is represented by 32-bit length strings. The affinity maturation process on the antibody population takes place with the antigen set comprising 22 Trojan-inserted examples. The memory set can then be used to classify unseen cases, since it holds antibodies that have a high affinity to the antigen which in this case are Trojan-inserted examples. The rate of cloning determines the rate of multiplication and reproduction of the highest affinity antibodies. The mutation rate, however, is inversely proportional to the affinity exhibited by the antibody being cloned, which means, higher the affinity, the lesser the effect of mutation on the clones. This helps in preserving the quality of the antibodies that are effective for detection. Antibodies with the highest affinity are kept in memory at the end of the generation phase to ensure that the best antibodies are used against unseen cases. During the testing phase, antigens are compared with the memory set to detect unsafe behavior. If a string shows a high affinity or overlap with the antibodies in the memory set, it is classified as consisting of unsafe behavior.

12.6 Experimental Results and Analysis

For accurate anomaly detection, it is important to correctly classify both unsafe and safe behaviors. Errors where unsafe behavior is not identified are categorized as false negatives, whereas errors where safe behavior is identified and categorized as unsafe are false positives. The terms correctly classified and incorrectly classified are used in this case to describe true positive/true negative and false positive/false negative rates, respectively. These can be described in the form of a confusion matrix as illustrated in Tables 12.5 and 12.6. An ideal system would demonstrate maximum detection accuracy for unsafe behavior (true positives), while minimizing false negatives and false positives. All experiments for the technique described in

Table 12.3 Trojan detection results for the negative selection algorithm [41]

Benchmarks	Correctly classified	Incorrectly classified
Partial string matching		
AES	90.5%	9.5%
MC8051	71.4%	28.6%
PIC16f84	75.0%	25.0%
RS232	70.0%	30.0%
wb_conmax	100%	0%
Whole string matching		
AES	90.5%	9.5%
MC8051	85.7%	14.3%
PIC16f84	100%	0%
RS232	80.0%	20.0%
wb_conmax	100%	0%

Table 12.4 Trojan detection results for the clonal selection algorithm [41]

Benchmarks	Correctly classified	Incorrectly classified
AES	90.5%	9.5%
MC8051	85.7%	14.3%
PIC16f84	75.0%	25.0%
RS232	80.0%	20.0%
wb_conmax	100%	0%

this chapter were run on 64-bit Ubuntu 16.04 with an Intel Core i7 processor and 16GB of RAM.

12.6.1 AIS Detection Accuracy

As demonstrated in Tables 12.3 and 12.4, both NSA and CSA techniques performed exceptionally well in identifying Trojan-inserted designs in the test dataset, successfully distinguishing between self and non-self behavior.

In this implementation of the NSA algorithm, two variants of string matching, partial string matching and whole string matching, were tested to determine whether there is any significant decrease in performance. For partial string matching, a 10-bit fixed length sub-string from the 32-bit length string is selected and used for censoring and monitoring. Instead of comparing a fixed range of bits from the samples of S_{self} and R, a sliding window is used to check for sub-string overlap at varying parts of the entire string. As observed in Table 12.3, Trojan detection using the partial string matching technique showed a mean detection accuracy of $81.3 \pm 13.2\%$.

For whole string matching, all of the generated 32-bit strings are used for training and testing. A mean detection accuracy $91.2 \pm 8.8\%$ was observed. Both techniques exhibit an average false negative rate of 12.6% and a false positive rate of 14.8%.

Table 12.5 Confusion matrix for NSA

Benchmarks		T′	F′
Partial string matching			
AES	**T**	19	2
	F	2	19
MC8051	**T**	5	2
	F	2	5
PIC16f84	**T**	4	0
	F	2	2
RS232	**T**	8	2
	F	4	6
wb_conmax	**T**	2	0
	F	0	2
Whole string matching			
AES	**T**	19	2
	F	2	19
MC8051	**T**	5	2
	F	0	7
PIC16f84	**T**	4	0
	F	0	4
RS232	**T**	7	3
	F	1	9
wb_conmax	**T**	2	0
	F	0	2

Based on the Student's t-test, the difference between these two methods was found to be statistically significant with at least 95% confidence and $p < 0.001$.

The detection accuracy of the Clonal Selection Algorithm is shown in Table 12.4. Across all benchmarks, a mean detection accuracy of $86.3 \pm 9.7\%$ was observed. In addition, the average false positive rate and false negative rate are 14.7% and 12.8%, respectively. The confusion matrices in Tables 12.5 and 12.6 represent true positives and true negatives which are examples correctly classified as either benign or malicious, denoted with TT' and FF', respectively. False positives and false negatives are examples that are incorrectly classified as either benign or malicious, and denoted as FT' and TF', respectively.

As seen in Tables 12.3 and 12.4, the AIS algorithms demonstrate high accuracy and efficiency in detecting unsafe behavior which represents malicious modifications made in the benchmark dataset. Larger training data and additional parameter tuning along with specialized feature extraction will help refine the AIS to further generalize and classify unsafe or Trojan-like behavior in unseen designs. For instance, exploring characteristics such as specific structure, size, and integration of the Trojan in the RTL to identify the type of threat can train the AIS model beyond a general classification of unsafe behavior to provide a more robust detection system.

Table 12.6 Confusion matrix for CSA

Benchmarks		T′	F′
AES	**T**	20	1
	F	3	18
MC8051	**T**	6	1
	F	1	6
PIC16f84	**T**	3	1
	F	1	3
RS232	**T**	8	2
	F	2	8
wb_conmax	**T**	2	0
	F	0	2

Table 12.7 Summary of CDFG generation and average AIS analysis times

			Average analysis time (s) (≈)		
			NSA		
Benchmarks	**CDFG gen. time (h)**	**File size (kb) (≈)**	**Partial str.**	**Whole str.**	**CSA**
AES	30	183.4–269.1	118	126	128
MC8051	28	108.2–126.2	110	122	124
PIC16f84	19	106.0–114.6	116	125	126
RS232	26	166.6–193.8	120	126	127
wb_conmax	15	106.3–115.7	110	118	120
		Calculated Mean	**114.8**	**123.4**	**125**

CDFG generation using PyVerilog requires designs to be flattened to single procedures, which was not the case for the benchmarks selected from TrustHub, and which will likely be the case for other real designs. Furthermore, as observed in Table 12.7, for more complex benchmarks, CDFG generation took significant time, upwards of several days for the most complex benchmarks. The average times in Table 12.7 indicate time taken on an average, to generate CDFGs for each of the respective benchmarks. On average, NSA took 123.4 ± 4.6 s and 114.8 ± 3.4 s for analysis using partial and whole string matching, respectively. These results suggest that there is a significant difference between the performance of both NSA variants (with at least 95% confidence, $p < 0.001$). In short, partial string matching, although resulting in a higher false negative rate, is significantly faster than its whole string counterpart. CSA, however, took an average of 125.0 ± 3.2 s for analysis. Improvements in CDFG generation time for high-level synthesis tools can help reduce this overhead and make the operation flow more tenable for faster classification.

Nevertheless, detecting generally unsafe behavior in CDFGs using AIS has shown to be a promising technique. One application of the proposed AIS technique is to significantly reduce the number of designs that need to be inspected for different classes of Trojans using another classifier not limited to AIS.

12.7 Conclusion and Future Work

This chapter describes a novel technique for detecting RTL Trojans using Artificial Immune Systems to identify and distinguish "safe-behavior" from "unsafe-behavior" by classifying high-level behavioral traits in HDL circuits. Control and data-flow graphs are generated from TrustHub RTL Trojan benchmarks to train and test the AIS. As demonstrated through experiments and results, Negative and Clonal Selection algorithms, with their evolutionary-like learning processes, are capable of detecting the presence of Trojans in high-level hardware descriptions. The application of AIS in HDL opens new research pathways for hardware Trojan detection and provides a valuable tool for high-level behavioral analysis to distinguishing patterns and discover anomalies early in design flow to save significant time and cost.

Future work will involve improvement in accuracy and efficiency by the process of parameter tuning. For instance, population and memory size, rate of cloning, and mutation rate, etc. can be further adjusted in the Clonal Selection Algorithm. Additionally, size of the sub-strings for partial string matching and comparison of different position of the sub-strings can be tuned in the Negative Selection Algorithm.

Building a taxonomy based on best features will provide valuable insight for classifying self and non-self behavior and improve detection of Trojan-like behavior in CDFGs. This can further lead to determining the type of hardware Trojan detected and, more importantly, the location of the hardware Trojan in the design. Reverse engineering the detection process to identify where in the original source and under what condition this unsafe behavior was triggered can provide hardware designers and system integrators with a valuable tool for improving security of hardware systems during the design stage.

References

1. Chakraborty, R.S., Narasimhan, S., Bhunia, S.: Hardware Trojan: threats and emerging solutions. In: High Level Design Validation and Test Workshop, 2009. HLDVT 2009. IEEE International, pp. 166–171. IEEE, New York (2009)
2. Bhunia, S., Hsiao, M.S., Banga, M., Narasimhan, S.: Hardware Trojan attacks: threat analysis and countermeasures. Proc. IEEE **102**(8), 1229–1247 (2014)
3. Narasimhan, S., Wang, X., Du, D., Chakraborty, R.S., Bhunia, S.: TeSR: a robust temporal self-referencing approach for hardware Trojan detection. In: 2011 IEEE International Symposium on Hardware-Oriented Security and Trust (HOST), pp. 71–74. IEEE, New York (2011)
4. Chakraborty, R.S., Wolff, F., Paul, S., Papachristou, C., Bhunia, S.: MERO: a statistical approach for hardware Trojan detection. In: Cryptographic Hardware and Embedded Systems-CHES 2009, pp. 396–410. Springer, New York (2009)
5. Zhang, X., Tehranipoor, M.: Case study: detecting hardware Trojans in third-party digital IP cores. In: 2011 IEEE International Symposium on Hardware-Oriented Security and Trust (HOST), pp. 67–70. IEEE, New York (2011)

6. Fern, N., Cheng, K.T.T.: Detecting hardware Trojans in unspecified functionality using mutation testing. In: Proceedings of the IEEE/ACM International Conference on Computer-Aided Design, pp. 560–566. IEEE Press, New York (2015)
7. Forrest, S., Perelson, A.S., Allen, L., Cherukuri, R.: Self-nonself discrimination in a computer. In: 1994 IEEE Computer Society Symposium on Research in Security and Privacy, 1994. Proceedings, pp. 202–212. IEEE, New York (1994)
8. Kephart, J.O., et al.: A biologically inspired immune system for computers. In: Artificial Life IV: Proceedings of the Fourth International Workshop on the Synthesis and Simulation of Living Systems, pp. 130–139 (1994)
9. Zhang, P.T., Wang, W., Tan, Y.: A malware detection model based on a negative selection algorithm with penalty factor. Sci. China Inf. Sci. **53**(12), 2461–2471 (2010)
10. Al Daoud, E.: Metamorphic viruses detection using artificial immune system. In: International Conference on Communication Software and Networks, 2009, ICCSN'09, pp. 168–172. IEEE, New York (2009)
11. Guo, Z., Liu, Z., Tan, Y.: An NN-based malicious executables detection algorithm based on immune principles. In: International Symposium on Neural Networks, pp. 675–680. Springer, New York (2004)
12. Al-Sheshtawi, K.A., Abdul-Kader, H.M., Ismail, N.A.: Artificial immune clonal selection classification algorithms for classifying malware and benign processes using API call sequences. Int. J. Comput. Sci. Netw. Secur. **10**(4), 31–39 (2010)
13. De Castro, L.N., Von Zuben, F.J.: Learning and optimization using the clonal selection principle. IEEE Trans. Evolut. Comput. **6**(3), 239–251 (2002)
14. Tan, K.C., Goh, C.K., Mamun, A.A., Ei, E.Z.: An evolutionary artificial immune system for multi-objective optimization. Eur. J. Oper. Res. **187**(2), 371–392 (2008)
15. De Castro, L.N., Von Zuben, F.J.: The clonal selection algorithm with engineering applications. In: Proceedings of GECCO, vol. 2000, pp. 36–39 (2000)
16. Salmani, H., Tehranipoor, M., Karri, R.: On design vulnerability analysis and trust benchmarks development. In: 2013 IEEE 31st International Conference on Computer Design (ICCD), pp. 471–474. IEEE, New York (2013)
17. Shakya, B., He, T., Salmani, H., Forte, D., Bhunia, S., Tehranipoor, M.: Benchmarking of hardware Trojans and maliciously affected circuits. J. Hardware Syst. Secur. **1**(1), 85–102 (2017)
18. Takamaeda-Yamazaki, S.: Python-based hardware design processing toolkit for verilog HDL. In: International Symposium on Applied Reconfigurable Computing, pp. 451–460. Springer, New York (2015)
19. Wang, X., Tehranipoor, M., Plusquellic, J.: Detecting malicious inclusions in secure hardware: challenges and solutions. In: IEEE International Workshop on Hardware-Oriented Security and Trust, 2008. HOST 2008, pp. 15–19. IEEE, New York (2008)
20. Tehranipoor, M., Koushanfar, F.: A survey of hardware Trojan taxonomy and detection. IEEE Des. Test Comput. **27**(1), 10–25 (2010)
21. Li, H., Liu, Q., Zhang, J.: A survey of hardware Trojan threat and defense. Integr. VLSI J. **55**, 426–437 (2016)
22. Bhunia, S., Tehranipoor, M.M.: The Hardware Trojan War: Attacks, Myths, and Defenses. Springer, New York (2017)
23. Jha, S., Jha, S.K.: Randomization based probabilistic approach to detect Trojan circuits. In: Proceedings of the 2008 11th IEEE High Assurance Systems Engineering Symposium, HASE '08, p. 117–124. IEEE Computer Society, New York (2008). https://doi.org/10.1109/HASE.2008.37
24. Agrawal, D., Baktir, S., Karakoyunlu, D., Rohatgi, P., Sunar, B.: Trojan detection using IC fingerprinting. In: Proceedings of the 2007 IEEE Symposium on Security and Privacy, SP '07, p. 296–310. IEEE Computer Society, New York (2007). https://doi.org/10.1109/SP.2007.36
25. Bloom, G., Narahari, B., Simha, R.: OS support for detecting Trojan circuit attacks. In: 2009 IEEE International Workshop on Hardware-Oriented Security and Trust, pp. 100–103 (2009)

26. Banga, M., Hsiao, M.S.: A novel sustained vector technique for the detection of hardware Trojans. In: 2009 22nd International Conference on VLSI Design, pp. 327–332. IEEE, New York (2009)
27. Iwase, T., Nozaki, Y., Yoshikawa, M., Kumaki, T.: Detection technique for hardware Trojans using machine learning in frequency domain. In: 2015 IEEE 4th Global Conference on Consumer Electronics (GCCE), pp. 185–186. IEEE, New York (2015)
28. Lodhi, F.K., Abbasi, I., Khalid, F., Hasan, O., Awwad, F., Hasan, S.R.: A self-learning framework to detect the intruded integrated circuits. In: 2016 IEEE International Symposium on Circuits and Systems (ISCAS), pp. 1702–1705. IEEE, New York (2016)
29. Bao, C., Forte, D., Srivastava, A.: On application of one-class SVM to reverse engineering-based hardware Trojan detection. In: 2014 15th International Symposium on Quality Electronic Design (ISQED), pp. 47–54. IEEE, New York (2014)
30. Hasegawa, K., Oya, M., Yanagisawa, M., Togawa, N.: Hardware Trojans classification for gate-level netlists based on machine learning. In: 2016 IEEE 22nd International Symposium on On-Line Testing and Robust System Design (IOLTS), pp. 203–206. IEEE, New York (2016)
31. Oya, M., Shi, Y., Yanagisawa, M., Togawa, N.: A score-based classification method for identifying hardware-Trojans at gate-level netlists. In: Proceedings of the 2015 Design, Automation & Test in Europe Conference & Exhibition, EDA Consortium, pp. 465–470 (2015)
32. Waksman, A., Suozzo, M., Sethumadhavan, S.: FANCI: identification of stealthy malicious logic using Boolean functional analysis. In: Proceedings of the 2013 ACM SIGSAC conference on Computer & Communications Security, pp. 697–708. ACM, New York (2013)
33. Zhang, J., Yuan, F., Wei, L., Liu, Y., Xu, Q.: VeriTrust: verification for hardware trust. IEEE Trans. Comput.-Aid. Des. Integr. Circ. Syst. **34**(7), 1148–1161 (2015)
34. Tehranipoor, M., Koushanfar, F.: A survey of hardware Trojan taxonomy and detection. IEEE Des. Test Comput. **27**(1), 10–25 (2010)
35. Alberts, B., Bray, D., Lewis, J., Raff, M., Roberts, K., Watson, J.: Molecular Biology of the Cell, 4th edn. Garland, New York (2002)
36. Murphy, K., Weaver, C.: Janeway's Immunobiology. Garland Science, New York (2016)
37. Hofmeyr, S.A., Forrest, S.: Architecture for an artificial immune system. Evolut. Comput. **8**(4), 443–473 (2000)
38. Dasgupta, D., Yu, S., Nino, F.: Recent advances in artificial immune systems: models and applications. Appl. Soft Comput. **11**(2), 1574–1587 (2011)
39. Romagnani, S.: Immunological tolerance and autoimmunity. Intern. Emerg. Med. **1**(3), 187–196 (2006)
40. Owen, J.A., Punt, J., Stranford, S.A., et al.: Kuby Immunology. WH Freeman, New York (2013)
41. Zareen, F., Karam, R.: Detecting RTL Trojans using artificial immune systems and high level behavior classification. In: 2018 Asian Hardware Oriented Security and Trust Symposium (AsianHOST), pp. 68–73 (2018). https://doi.org/10.1109/AsianHOST.2018.8607172
42. Burnet, F.M., et al.: A modification of Jerne's theory of antibody production using the concept of clonal selection. Austr. J. Sci. **20**(3), 67–9 (1957)
43. Victora, G.D., Nussenzweig, M.C.: Germinal centers. Annu. Rev. Immunol. **30**, 429–457 (2012)
44. Weinand, R.G.: Somatic mutation, affinity maturation and the antibody repertoire: a computer model. J. Theoret. Biol. **143**(3), 343–382 (1990)
45. He, M.X., Petoukhov, S.V., Ricci, P.E.: Genetic code, hamming distance and stochastic matrices. Bull. Math. Biol. **66**(5), 1405–1421 (2004)
46. Allen, F.E.: Control flow analysis. In: Proceedings of a Symposium on Compiler Optimization, pp. 1–19. ACM, New York (1970). http://doi.acm.org/10.1145/800028.808479
47. Williamson, M.C., Lee, E.A.: Synthesis of parallel hardware implementations from synchronous dataflow graph specifications. In: Conference Record of the Thirtieth Asilomar Conference on Signals, Systems and Computers, vol. 2, pp. 1340–1343 (1996). https://doi.org/10.1109/ACSSC.1996.599166

Part IV
Side-Channel Defense via Behavioral Synthesis

Chapter 13
High-Level Synthesis for Minimizing Power Side-Channel Information Leakage

S. T. Choden Konigsmark, Wei Ren, Martin D. F. Wong, and Deming Chen

13.1 Introduction

The Internet of Things (IoT)) and cloud computing have become more and more pervasive in the recent years, gradually changing the modern world by connecting a multitude of information and processing systems together. In IoT, a large number of devices, buildings, vehicles, and sensors are interconnected to build a network that gathers large amount of information and respond to it. As part of the emerging cyber-physical systems, these devices are prevalent in our daily life ranging from light control in residential buildings to smart coordination of machines in industrial production. The public and private cloud are utilized to orchestrate billions of IoT devices and perform analysis on a massive and perpetual stream of sensor data. However, IoT applications are not the only use cases for cloud computing—there is rapid development and adoption in industrial, marketing, and financial segments.

Today, security has become considerably more crucial to both IoT and cloud computing services [2]. Concerns for IoT security originate from the need for low cost, rapid development, and the lack of standardization, which inevitably increases the possibility of security vulnerabilities. The security concerns for cloud computing are different but equally important: while secure data transfer between the data source and the cloud provider is a largely solved problem, efficiently performing remote analysis over encrypted data without decryption is generally not possible or too time-consuming for meaningful real world applications. This means in practice data is still likely in the form of plaintext when being processed and analyzed remotely. Although cloud computing is a convenient and cost-effective solution,

S. T. Choden Konigsmark · W. Ren · M. D. F. Wong · D. Chen (✉)
Department of Electrical and Computer Engineering, University of Illinois at Urbana-Champaign, Champaign, IL, USA
e-mail: konigsm2@illinois.edu; weiren2@illinois.edu; mdfwong@illinois.edu; dchen@illinois.edu

S. Katkoori, S. A. Islam (eds.), *Behavioral Synthesis for Hardware Security*, https://doi.org/10.1007/978-3-030-78841-4_13

the possibility of revealing sensitive data to a remote party might limit its presence in applications with strong security requirements. Field-programmable gate arrays (FPGAs), by offering customization and complete control over data processing, could be a solution to secure data processing when provided as a service in the cloud to the users.

For both cloud and IoT), FPGAs provide excellent solutions with scalability, maintainability, low cost, and efficiency. Various proposals for FPGA-friendly protocols and security architectures have been made for these emerging fields [3, 4]. Meanwhile, high-level synthesis (HLS) provides software and hardware developers the ability to produce a low-level hardware description from a high-level system in high-level programming languages such as C and C++. Thus HLS has become a primary contributor to wider enablement of FPGA devices due to its advantages of higher design productivity, better debugging and testing capability, and faster time-to-market features.

Although basic security practices such as proper selections of encryption algorithms as well as using ciphertext in communication usually suffice to provide enough confidentiality, information can still leak through hardware implementation, which are often underestimated in a secure system design. Moreover, countermeasures for hardware attacks typically require low-level understanding and fine-grained cost balancing [5]. Thus, the hardware implementation can be a major source of information leakage which, however, is often neglected while designers are focusing more on algorithms modifications or security techniques on higher level. Such information leakage through hardware implementation can be exploited through side-channel attacks, which is exploited by performing statistical analysis to extract confidential values. For example, the presence or absence of operations in a captured power trace can disclose information about the confidential values through a simple power analysis (SPA), especially when the execution closely correlates with the confidential data. Additionally, the hardware implementation of even the simplest logic operations is typically very susceptible to information leakage, as dynamic power consumption is dependent on the number of switching bits, capacitance, and several other identifying characteristics. Consequently, more sophisticated analysis through differential power analysis (DPA) can retrieve values of potentially secret input and output. To achieve a constant and input-independent power drain, most existing techniques require deep knowledge of the lower-level implementation and they are often resource and power intensive. Without expert-level understanding of the hardware security, deep familiarity with the circuit and the implication of different techniques, such lower-level countermeasures are nearly impossible for normal user or developer to comprehend and apply in a fast-paced development.

In this work, we present the first fully automated high-level synthesis flow with the primary target focusing on minimizing the power side-channel information leakage in defense of DPA. This contribution enables developers and design engineers to efficiently mitigate side-channel leakage vulnerabilities in a practical manner: specifying confidential variables on top of the high-level specification to be synthesized is sufficient for automatic leakage analysis and scalable selection of countermeasures under various constraints. The unique contributions of this work are:

- High-level side-channel leakage characterization through derivation of per-operation confidentiality and cycle-accurate simulations.
- First side-channel leakage resistant high-level synthesis flow. Minimal annotations in addition to high-level C-code are sufficient for automated leakage analysis and insertion of countermeasures. Either resource constraints or allowable leakage threshold can be the target for this flow to optimize on.
- Automated identification and mitigation against branch imbalances that otherwise enable simple power attacks.
- Experimental evaluation with established CHStone benchmarks and a custom *IoT* benchmark. The proposed flow achieves up to 81% better leakage reduction than the reference under identical resource constraints.

13.2 Background and Related Works

13.2.1 Power Side-Channel Analysis

Side-channel information is the information exposed through unintended or unforeseen channels that are outside the designed functionality of a system. Application programs with sensitive data may use very sophisticated encryption and protection scheme. However, even with proper security review and defense against crypt-analysis on the algorithm or the program, there may still be weakness in the hardware implementation that can leak information through inadvertent means, including power consumption, time delay, and electromagnetic radiation. Such leaked information can usually provide insights on the internal operations during the computation, which may allow attackers to reveal secret data. Since different operations on sensitive data usually have different characteristics (e.g., power consumption or time delay), it is possible to extract secret information by analyzing the variations and correlations in the observed measurements.

In this work, we focus on the defense against side-channel analysis on power (or power analysis for short), which exploits power consumption information to analyze confidential values during the computation. Hardware devices such as CPU or cryptographic circuit will inherently have variations in the power consumption for different operations performed. In typical CMOS logic circuit, total power consists of both static and dynamic power [6]. Static power is consumed even when the transistors are not switching so its variation does not correlate much with data. In contrast, dynamic power is proportional to switching factor which ultimately depends on data. Figure 13.1 shows the charging and discharging behavior when output is switching in a CMOS inverter. Energy is drawn from the supply when the output changes from low to high (or $0 \rightarrow 1$), resulting in power consumed from the supply to the load capacitance. When the output switches from high to low (or

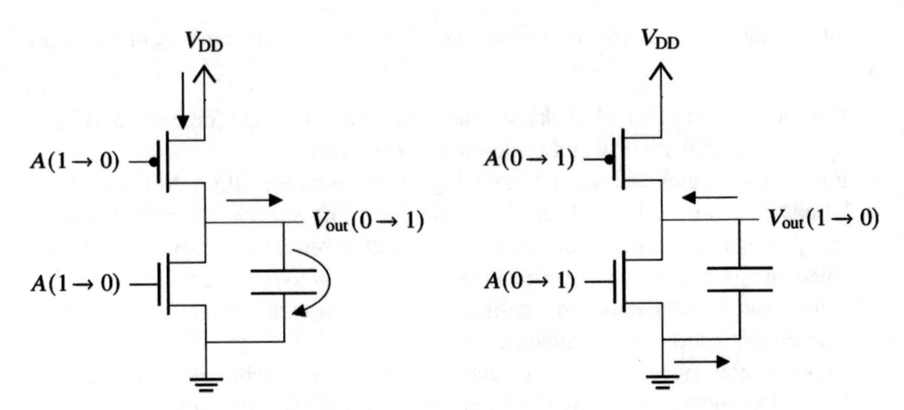

Fig. 13.1 Charging and discharging at the output of a CMOS inverter. Power is only drawn when the output transition is from low to high, resulting in dynamic power consumption.

$1 \rightarrow 0$), the energy stored in the capacitor is released (dissipated through heat in the nMOS transistor) and no energy is drawn from the supply [6]. Such fundamental asymmetry in power consumption can result in opportunities for attackers to exploit when the switching activity correlates with secret data.

Power analysis can be categorized into simple power analysis (SPA) and differential power analysis (DPA). SPA looks at the variation in one single or few power traces over time and infer sensitive data from the measurements. More advance DPA exploits sensitive information through statistical analysis and signal processing over multiple power traces.

13.2.1.1 Simple Power Analysis (SPA)

One major source of information leakage is derived from conditional operations whose execution depends on variables with confidential content. This type of leakage can be exploited without extensive computational analysis as part of a simple power analysis (SPA) attack, as the difference in operations following such a branching statement potentially reveals information about the confidential variable.

One example is the common vulnerability in early implementations of the RSA algorithm [7], which used the text-book implementation of the square-and-multiply algorithm. In this algorithm, an exponentiation of the form x^n can be restated as $x * (x^2)^{((n-1)/2)}$ when n is an odd number, or $(x^2)^{(n/2)}$ when n is even. As the operation for an odd exponent requires an additional multiplication with sufficiently different power signature, an adversary can determine each bit of the private key in a step-by-step attack. Attacks of this kind can be defeated by re-architecting the implementation of algorithms to: (1) eliminate the dependency between executed operations and confidential values or (2) decorrelate the information leaked with the secret data. One intuitive countermeasure is to "pad" a dummy multiplication when n is even to remove the difference in power signatures at the expense of larger power consumption and longer running time. More efficient countermeasure can use blinding techniques [8, 9] to make the leaked information irrelevant to secret data. In the case of RSA, the base x is randomized and only recovered after the exponentiation computation [8], which makes the leaked information unrelated to the real secret data x.

As the vulnerabilities in RSA implementations show, oversight of side-channel implication is a common problem. Especially in IoT) applications which constantly process privacy and confidentiality critical information, these countermeasures are of critical importance, yet practical considerations and the lack of automated defense mechanisms virtually guarantee that many implementations will suffer from similar vulnerabilities.

13.2.1.2 Differential Power Analysis (DPA)

The threat of side-channel leakage has received significant attention with the introduction of DPA [10] by Kocher et al. DPA is a side-channel attack that enables the extraction of secret keys through signal processing over a large number of power traces (much larger than that in SPA). In addition, sophisticated DPA often uses statistical and probabilistic models to infer secrets from power variations. It is shown that DPA can reveal the internal secrets of cryptographically secure algorithms such as the advanced encryption standard (AES) [11]. One intuitive mitigation is to insert noise (e.g., additional modules that have random power consumption) such that the signals (that could expose secrets) are well concealed. While the introduction of random noise can effectively reduce the signal-to-noise ratio, it has been shown that arbitrary noise does not provide significant security benefits and cannot efficiently hinder exploitation of side-channel leakage [7]. Such difficulty in both attack and defense makes DPA an ongoing research topic.

13.2.2 Defense Against Power Analysis

Countermeasures against power analysis have been proposed on both software level and hardware level to reduce or remove the correlation between the power consumption and the sensitive data. As summarized in [12], both kinds of countermeasures can be further categorized into two strategies: masking and elimination. Software countermeasures rely on modification in the algorithm or the software program to minimize the information leakage. The signal-to-noise ratio can be reduced by:

- Noise: As mentioned in Sect. 13.2.1.2, introduction of noise can usually hide secret by increasing the number of power traces needed for analysis. One example is the insertion of random instruction in the program [13], which leads

to a significantly higher computation complexity of correlating power traces with secret data. But it is also shown in [13] that spike reconstruction can help reduce the number of traces needed for DPA.
- Eliminating the power variation: Algorithms implementation needs to ensure that execution paths are independent of secret data. One example is to balance the sensitive branches (i.e., branches with conditions that may depend on secret data) so that power and time delay exhibits constant behavior on both sides of the branch.

Still, all the software mitigation depend on the underlying hardware implementation as well. Hardware countermeasures reduces the leakage by modifying the micro-architecture or circuit-level implementation. Like software-based approaches, hardware countermeasures can also be classified into:

- Masking: Similar to introducing noise, masking was proposed to reduce the correlation between captured power traces and the actual underlying data. Masked-AND was an early proposal for generally applicable logic design to secure AES [14].
- Eliminating (or hiding) power variation: In dynamic-CMOS logic style, dynamic differential logic (DDL) [15] is one technique with exactly one charging in precharge phase and exactly one discharging (either complemented or uncomplemented output) in evaluation phase, resulting in constant power consumption. In static-CMOS logic style, researchers have also presented various corresponding DDL (namely static-CMOS DDL) techniques to achieve the same goal by: (1) adding the complement of a circuit, and (2) mimicking the behavior of fixed output transitions in precharge and evaluation phases like dynamic-CMOS DDL. The rest of the work will refer DDL to static-CMOS DDL for simplicity unless otherwise noted. Section 13.3.1 discusses DDL selected for this work in detail.

More details on defenses against power side-channel analysis can be found in [12] and [16].

13.2.3 High-Level Synthesis

Traditionally hardware developers and designers have been using Hardware Description Language (HDL) such as Verilog to describe the functionality of circuit designs. Electronic Design Automation (EDA) tools can then translate these hardware-level descriptions into physical layouts. However, with both transistor count and the design complexity increasing over the years, it becomes much more difficult to map algorithms into circuit design in HDL. Largely because the designer has to work on Register-Transfer Level (RTL) implementation while the algorithm itself is typically expressed in high-level languages. High-Level Synthesis (HLS), also known as behavioral synthesis, helps hardware designers to automatically

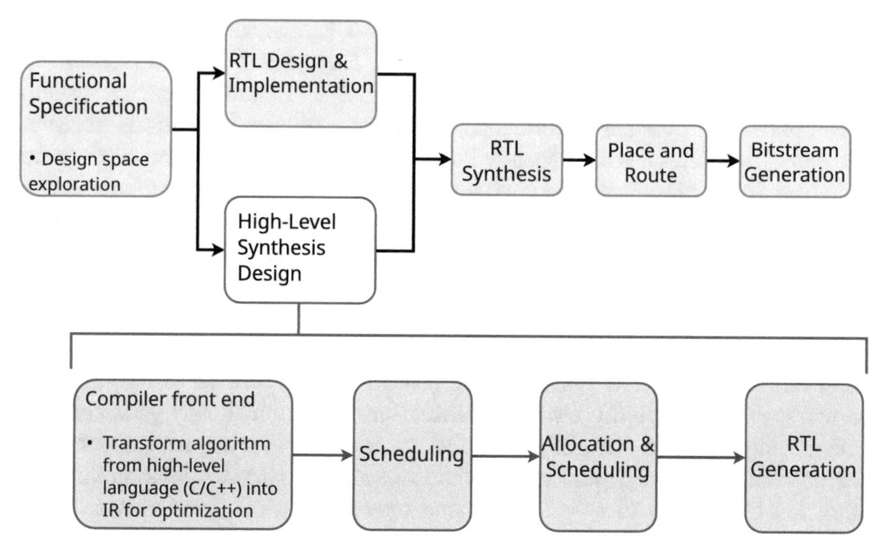

Fig. 13.2 Typical FPGA design flow with HLS. The blue blocks represent RTL design flow and the orange blocks represent HLS flow. IR here stands for intermediate representation

transform the algorithms specified in high-level languages into low-level hardware descriptions, which can be synthesized later.

Historically HLS or behavioral synthesis was debuted in EDA tools in the mid-1990s but was not widely adopted for two decades. Later works like xPilot [17] and LOPASS [18] improved scheduling and allocation/binding algorithms to enhance HLS output results. Nowadays there are many mature HLS tools, both academic and commercial: LegUp [19], Xilinx Vivado HLS, Intel HLS compiler, and many more.

Most of the HLS tools support widely known programming languages such as C and C++ as algorithm description. Designers also need to supply optimization directives and resource constraints to the HLS tool for better control of the output circuit and more efficient implementation. In typical HLS flow (refer to orange blocks in Fig. 13.2), the front end in HLS first analyzes the algorithm and exploits optimization opportunities automatically during the transformation into intermediate representation (IR). Then HLS performs scheduling, allocation, and binding to generate RTL output under design constraints while optimizing on performance, area, and/or power consumption [17, 18, 20]. The generated output of the HLS can then be integrated into system design and further synthesized in standard RTL design flow. Information in the RTL synthesis and physical design stages are also crucial to HLS. For instance, authors in [20] proposed simultaneous mapping and clustering to improve FPGA design quality. A place-and-route directed HLS flow [21] that uses time delay information after place-and-route (PAR) can be adopted to help HLS characterize timing more accurately and thus maximizes targeted clock frequency.

With HLS, designers can focus more on the architectural optimization and algorithm improvements while HLS tools take care of the detailed register and cycle-to-cycle implementation. Although manual RTL implementation usually achieves the best possible performance, it also requires very high efforts in design and debugging. HLS can significantly improve design productivity and have shorter time-to-market, while still achieving very good performance gain. Authors in [22] demonstrated that using HLS leads to a fivefold reduction in design efforts in stereo matching algorithm compared to manual RTL implementation, with significant speedup (between 3.5x and 67.9x). A case study in H.264 video decoder with HLS implementation showed that HLS is capable of synthesizing complex applications in practice [23].

Recent research found that HLS can provide unique benefits to the design of secure systems. Reliability can be increased through automatically generated On-Chip Monitors (OCMs) [24, 25]. Reliability in face of destructive hardware Trojans can be achieved through selective module selection as part of synthesis [26, 27]. HLS can be applied to detect vulnerable operations with granular insertion of hardware Trojan defenses through information dispersion [28]. Recent efforts in [29] attempt to eliminate timing side-channels with static information flow analysis in HLS. However, this chapter represents the first work to explore the benefits of HLS in securing designs against power side-channel leakage.

13.2.4 FPGA Security in Emerging Applications

With the recent advancement in machine learning and data analytics, the demand for computation power and performance has been increasing rapidly. In recent years, FPGAs have started to emerge in many applications as custom accelerators. Due to their high configurability and efficiency to perform application-specific tasks, FPGAs are well suited for both IoT and cloud computing applications, including data analytic tasks. With many applications originally developed in high-level languages, HLS can play an import role when developing applications-specific accelerators. More recently, cloud service providers such as Amazon (EC2 F1 instances [30]) and Microsoft (Brainwave project [31]) have already started to incorporate FPGAs in their cloud servers to allow users to customize the FPGA for their computation needs. However, while the benefit of accelerated computing in FPGA is obvious, the potential security risks can become a significant obstacle in both IoT and cloud computing applications.

IoT) applications often require rapid development and updates, where FPGA can be a good candidate for both prototyping and deployment. FPGAs have been shown to surpass ASICs in reliability, cost, time-to-market, and maintenance [32], especially when the deployment volume is not large. Dynamic reconfigurability of FPGAs has sparked new IoT specific architectures and applications [4]. However, the emergence of IoT also revealed numerous security problems, from plain text network traffic that can easily be intercepted and read, to an IoT-based botnet that culminated in a massive distributed denial-of-service (DDoS) attack that led to a widespread network outage on the US east coast [33].

From a security standpoint in the customer's view, and from a liability perspective from the cloud provider's view [3], cloud service providers should not be trustworthy in general. FPGAs have been proposed to offload sensitive data processing and analysis for its high degree of customization including security. These FPGAs externally consume and produce encrypted data, "such that highly sensitive plain text data is not directly revealed to the cloud provider" sounds more natural to me. Thus in such FPGA cloud environment, network-level information confidentiality can be achieved. However, the cloud service provider maintains physical access to the devices, particularly for maintenance and operation of the servers and FPGAs. Therefore, physical device access by the cloud provider for extraction of the secret keys, as well as direct extraction of the plain text, must be considered. Moreover, even if cloud service providers are trusted such that they never pry on sensitive data, the cloud environment can still be compromised by hackers or even rogue agents, giving attackers physical access to the devices. Thus, to protect sensitive information, users should always minimize the side-channel information leakage in the design.

13.3 DPA Defense in Logic Design

Several techniques in logic design have been proposed to equalize the dynamic power consumption (i.e., minimizing the variation in power) of digital circuits to reduce side-channel leakage. An introduction is given in this section to familiarize the reader with defense modules (or techniques) selected in our work.

13.3.1 Dynamic Differential Logic (DDL)

For static-CMOS circuit and FPGAs, DDLs have been proposed with the use of standard logic gates. Simple Dynamic Differential Logic (SDDL) [34, 35] produces the differential output by applying De-Morgan's law and AND-ing the pair of differential outputs with the precharge signal. SDDL aims to consistently consume power by combining complimentary standard cell gates and inputs such that both the positive and negative outputs are computed. Every input signal change requires a precharge phase and evaluation phase. In precharge phase, the precharge signal (or inverted) is set to 0 such that all outputs evaluate to 0. Thus, in the evaluation phase next, there can only be one 0-to-1 transition per output bit—either in the positive, or in the negative output. This provides for consistent dynamic power consumption regardless of any inputs. Figure 13.3 shows a 2-input AND-gate in SDDL. Truth table in Fig. 13.3 shows that only one $0 \rightarrow 1$ transition can happen during input changes after precharge signal resetting differential outputs (Z and

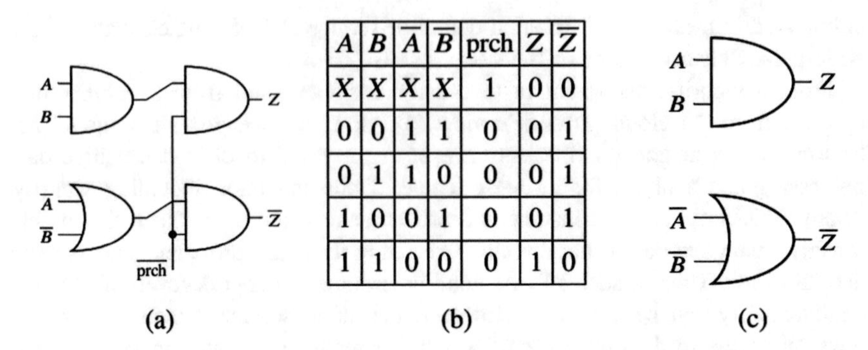

A	B	$\overline{A}$	$\overline{B}$	prch	Z	$\overline{Z}$
X	X	X	X	1	0	0
0	0	1	1	0	0	1
0	1	1	0	0	0	1
1	0	0	1	0	0	1
1	1	0	0	0	1	0

Fig. 13.3 Compund AND gate in SDDL and WDDL proposed in [34]. (**a**) A compound AND gate in SDDL. (**b**) Truth table for the AND compound gate. (**c**) A compound AND gate in WDDL

$\overline{Z}$) to 0. Compound OR gate can be constructed similarly. SDDL uses only secure compound AND and OR gates to form the circuit. NOT gate is unnecessary since all SDDL compound gates have complementary (differential) outputs.

Wave dynamic differential logic (WDDL) [34] is also proposed, which operates similarly to SDDL in terms of the complimentary part. Unlike in SDDL, WDDL does not require AND-ing zeros at the output during the precharge phase. Figure 13.3 shows an equivalent AND compound gate in WDDL. Note that it does not have AND gates at the output (AND-ing with precharge zeros only applied at input signals of the entire module). Since only AND and OR gates are present in the compound circuit, precharge inputs to low will automatically propagate zeros to all outputs (which in turn precharge zeros to inputs of other gates connected) in WDDL, eliminating the need of ANDing at all outputs.

Both SDDL and WDDL rely on standard building blocks to form secure compound gates which can be applied in a regular ASIC or FPGA design flow instead of standard cells [34]. Leakage may still occur due to timing and load capacitance variations. Due to potential glitches, SDDL cannot guarantee only one switching signal per clock cycle, and therefore is inferior to WDDL which is glitch free. However, SDDL is more flexible in FPGA implementation compared to WDDL [35]. Glitch-aware routing [36] in FPGA synthesis can also use efficient algorithms to reduce glitches and dynamic power.

However, side-channel power leakages can still be detected even when the aforementioned logic styles are utilized, due to routing and load imbalances. To mitigate such risks, Yu and Schaumont suggested a new technique named Double WDDL (DWDDL) [37], which duplicates the fully routed circuits with switched positive and negative input signals as in WDDL. Other logic styles like [38] have been proposed as well. In addition, Kaps and Velegalti suggested Partial DDL [39] to reduce resource overhead through manually applying DDL countermeasures only on a subset of the circuit while still maintaining effective leakage resistance. Compared to a full SDDL design of AES circuit, Partial SDDL reduces 24% of the total areas used.

Strong expertise and security-centric skill-set are required in all the previously described techniques and architectures to implement efficient and effective defense. Although the full adoption of advanced logic style like SDDL/WDDL may reduce power side-channel leakage significantly, it is also extremely costly in terms of resource and area usage. Therefore, we propose a resource-effective HLS flow that can automatically select and combine appropriate countermeasure techniques to different parts of the circuit under a given resources constraint or certain leakage target. More importantly, such HLS flow can provide the most benefits for the average users, while greatly lowering the need for guidance in specific security engineering expertise.

13.3.2 High-Level Leakage Characterization of Power Analysis

To fairly evaluate the effectiveness across different research works on side-channel defense, we can use the Measurements To Disclosure (MTD) [40] as the indicator for the amount of side-channel leakage, typically on circuits that are well-known and well-studied (e.g., AES). Such power measurements can be simulated with a power model, which needs to capture the relationship between dynamic power and side-channel information leakage. Menichelli et al. [41] described that the difficulty with high-level power simulation (and hence, side-channel leakage) is its focus on evaluating average power consumption. Accurate power models may not provide many useful insights, even in models with cycle-accurate simulation, due to the fact that dynamic power variation caused by signal switching behaviors correlates much more closely to the side-channel leakage variation (compared to the absolute values of power consumption). To simulate and model signal-correlated dynamic power, only the internal logic values that correspond to the signal need to be traced [41].

Fang et al. [42] showed that the Hamming distance of switching signals serves well in modeling the variations between dynamic power and side-channel leakage, and are commonly employed. Hartog and Vink have demonstrated the successful adoption of Hamming distance as the primary source to estimate and model side-channel leakage in their software tool [43] for analyzing vulnerabilities in smart card applications. This work also uses the Hamming distance of switch signals in the metric of side-channel leakage (see Sect. 13.5.2).

13.4 Threat Model

Since the major focus of this work are efficient and effective countermeasures against power side-channel analysis, the first party in the threat model of this work includes hardware designers, software developers, and end users with confidential information. The first party's primary goal in the threat model is to protect the confidential information running on the device with minimal resource overhead.

More specifically the first party will try to minimize the resource consumption for the hardware under a given side-channel leakage target, so that more resources can be utilized for normal tasks of the device (e.g., acceleration in cloud computing or data processing in IoT) applications).

In the threat model of this work, the attack goal of the second party (i.e., the adversary) is to extract the confidential information from the first party through power side-channel analysis. The attackers are assumed to have direct physical access to device and the system. In other words, adversaries can measure and collect a large number of power traces and perform sophisticated analysis to reveal hidden secrets. Additional countermeasures against physical threats or hardware Trojans are introduced in Sect. 13.3. The environment where the hardware device and applications are operated is assumed to be hostile, ranging from IoT applications in public space to data-center of an untrusted cloud service provider. However, we do not consider physical attacks (e.g., physical tampering of any type, or reprogramming with malicious Trojans) in our threat model since they are outside the scope of this work.

A neutral third party which consists of hardware vendors (i.e., FPGA manufacturers) or third-party IP providers is usually involved. However, since such third party is assumed to have no malicious intention, our threat model does not consider it. Defenses against a malicious third party are outside the scope of this work.

13.5 Methodology

We propose a new automated high-level synthesis flow for minimizing side-channel information leakage. Figure 13.4 shows an overview of the operations in the flow. It consists of three primary phases: the initial synthesis, leakage characterization, and the final security synthesis.

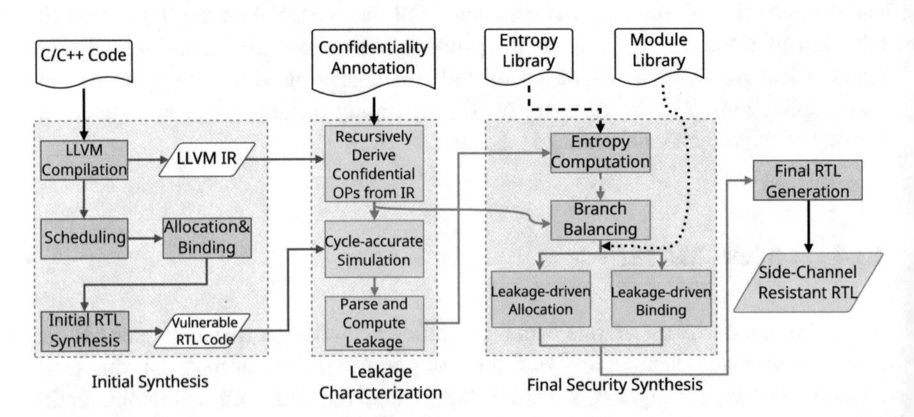

Fig. 13.4 The proposed flow extends the existing HLS flow with additional security analysis and synthesis focusing on power side-channel leakage

1. Initial synthesis: in this phase, the input source code in C/C++ is compiled into intermediate representations (IR) for optimization and security analysis. End users will also need to provide confidentiality or security annotations, which denotes the input values or secrets that should remain hidden and protected. The flow will analyze the control flow graph with IRs to derive confidential operations and outputs. Similar to traditional HLS flow, these IRs will be used to synthesize into vulnerable RTL code as the output of initial synthesis—no side-channel leakage resistance is employed in this phase.
2. Leakage characterization: in this phase, the flow simulates RTL code with confidential operations annotated in the initial synthesis. The amount of leakage is characterized and estimated for each operation, taking into account the switching behavior and degree of confidentiality.
3. Final security synthesis: in this phase, the security synthesis selects the appropriate defense modules (as mentioned in Sect. 13.3.1) for each operation based on available resource and estimated side-channel leakage. Additionally, the flow performs branch balancing to eliminate vulnerabilities in confidential branches, by removing the variation in power consumption due to conditionals that depend on confidential information.

13.5.1 Initial Synthesis

The initial synthesis flow consumes user-provided C or C++ code and compiles it into intermediate representation (IR) and synthesizes the operations into RTL code. In this work, we used LegUp HLS flow [19] which is based on the LLVM (Low-Level Virtual Machine) compiler. The LLVM IR is an assembly-like language which is machine-independent. This compilation phase already includes code optimizations provided by the LLVM compiler framework.

In addition to the source code, the proposed synthesis framework requires end users to provide high-level annotations of variables to be treated as confidential. Such variables usually refer to secret keys or authentication tokens but can also be used more generally. For example, security annotation can include the unique identifier and similar information to elevate the security of the IoT) system.

The proposed flow automatically tracks information flow starting from these user-provided high-level security annotations among all the IRs and control flows. As a result, all operations on confidential values will be identified. Here, all outputs of confidential operations are treated as confidential as well. Therefore, minimal user input is sufficient to determine a graph of confidential operations. With this graph of confidential operations, the flow can effectively reduce the simulation overhead by guiding the simulation directions specific to the confidential operations of interests.

The initial synthesis concludes with conventional steps of HLS flow, which focuses more on resource utilization and device speed. Consequently, the generated RTL code may be susceptible to side-channel attacks and does not contain any countermeasures. However, this is a valuable starting point for deeper analysis of operations which are prone to leakage.

13.5.2 Leakage Characterization and Entropy Estimation

As mentioned in Sect. 13.3.2, our proposed flow employs the Hamming distance of switching operations as the fundamental metric for side-channel leakage of a given signal. Because the variations in power consumption primarily stem from the dynamic switching behaviors in modern FPGA and integrated circuits, Hamming distance has been a widely used metric [42]. Besides the variation in dynamic power, side-channel leakage can be exploited in static power consumption as well, which the proposed flow can consider with minor modifications.

In the leakage characterization phase, the proposed flow will simulate the generated RTL code and evaluate the switching characteristics of all confidential operations. With the results from initial synthesis, the flow generates simulation scripts to be used in professional simulation tools, focusing on the confidential operations. The flow then parses the simulation outputs for leakage calculation. The flow will ignore the switching characteristics of irrelevant operations (those that are not confidential).

From the standpoint of high-level security analysis and defense, we need an overall metric on an abstracted level. For each operation, variation in dynamic power, or the leakage, can be characterized by the difference between changes in the value. We can use the Hamming distance $HD(i)$ of an operation i:

$$HD(i) = |i_{\text{out},1} - i_{\text{out},0}| \tag{13.1}$$

where $i_{\text{out},0}$ and $i_{\text{out},1}$ are the vectors of output signals before and after the operation.

The proposed HLS flow not only considers the side-channel leakage through the Hamming distance, but also takes entropy estimation into account. Here, entropy quantifies the degree of confidential information for an operation. The amount of security information contained in an operation can differ significantly depending on where it is in the control flow and its functionality. That is, as the security information flows through the circuit, the degree of confidentiality also varies in different operations. For example, the output y of the operations $y = x \geq 25$ contains a comparatively smaller degree of confidential information than its input x. Our high-level metric should capture such reduction confidential information content (i.e., entropy) for various operations in the control flow. Entropy h of each instruction i is calculated by scaling with entropy factor σ of the functional unit (FU) that it operates on and accumulating over all its parent FUs, recursively.

$$h(i) = \sigma_i \cdot \sum_{p_i \in \text{Parent}(i)} h(p_i) \tag{13.2}$$

A predetermined table or library of entropy factors can be supplied to the proposed flow to accurately model the reduction. One example can be found in Table 13.1. The overall high-level leakage γ_i of a given instruction regarding the user specified secrets is thus defined as

$$\gamma_i = h(i) \cdot HD(i) \tag{13.3}$$

13.5.3 Security Synthesis

After high-level leakage characterization and estimation, a final security synthesis will allocate and bind resources to operations under given constraints to minimize the leakage, by selecting appropriate DPA defense modules. Branch balancing is performed in addition to resource allocation and binding.

13.5.3.1 Branch Balancing

As a mitigation to SPA, branch balancing aims to eliminate variations in the dynamic power due to conditionals that depend on confidential values, as the imbalanced timing and power characteristics will expose the value. Algorithm-level modifications can solve branch balancing well while a non-trivial amount of software and hardware engineering effort is required. On the other hand, automated approaches can be faster and more comprehensive, but the resource overhead can be significantly larger than manual solutions. Therefore, the proposed flow generates detailed reports on the detected imbalanced branches to allow manual mitigation in addition to implementing an automated solution.

After the security analysis in the initial synthesis, the proposed flow can then identify all conditional statements with sensitive information in its conditions. The flow will start logging the subsequent execution or instructions of divergent paths if any sensitive information can trigger the condition. As discussed previously, any deviation between the subsequent paths, e.g., the type and number of operations, can cause considerable side-channel leakage, therefore, the number of deviating operations is computed. The flow automatically inserts dummy operations with identical or similar functional units to counterbalance the deviations. All balanced branches are recorded to allow manual adjustments.

In Fig. 13.5, an example of branch balancing is shown. In Fig. 13.5a, the divergent branches are not balanced which will cause power variation if key contains sensitive information, as the upper branch consists of additional ADD and DIV operations. In Fig. 13.5b, both branches are balanced by introducing the dummy ADD and DIV operations in the lower branch, which counteract the initial imbalance.

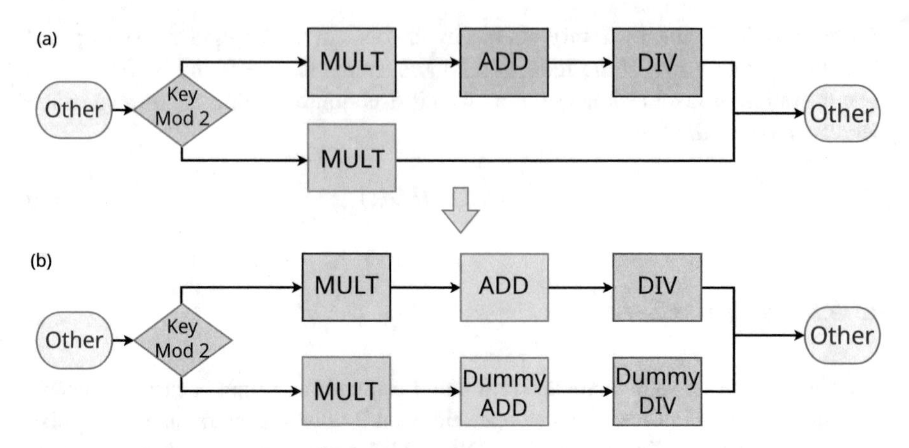

Fig. 13.5 An imbalanced branch with its condition depends on confidential value. (**a**) shows the imbalanced branch which may be exploited by SPA. (**b**) mitigates the issue by inserting dummy operations that compensate the power and timing difference in the lower branches

13.5.3.2 Leakage-Driven Allocation and Binding

In a typical HLS flow, the allocation and binding algorithms will assign hardware resources (e.g., registers and memory) to different variables and operations. Inputs and outputs are shared through multiplexer (MUX) when multiple operations are mapped to the same Functional Unit (FU). A module library is needed to target a specific hardware implementation such as FPGA or ASIC. For instance, large multiplexers are only shared for the most resource-intensive operations when necessary [19], such as division or modulo operation, due to their high resource cost in FPGAs.

Branch balancing only removes the variation due to divergence in the control flow. The final security synthesis must apply different countermeasures to minimize the security leakage, while respecting the resource constraints and security requirements. Allocation and binding algorithms are interwoven to further minimize the side-channel leakage. More specifically, the binding step will try to determine the assignment of IR operations to functional units as the basic building blocks. Compared to traditional HLS flow, the proposed flow introduces a notable security enhancement by distinguishing between high-risk operations (HROs) and low-risk operations (LROs). To minimize the possibility that the sensitive information is leaked through unwanted operations, HROs (which will be security hardened) are mapped to the same functional units while LROs are free to be assigned among the rest of the FUs. To maximize side-channel leakage reduction, the allocation of modules will also consider HRO and LROs. Figure 13.6 shows such a sharing pattern in binding. Note that such binding introduces new trade-offs between leakage minimization and resource cost: even though all operations are DIV, we need another separate FU dedicated to the operations on confidential values so that overall leakage is minimized. The binding algorithm is driven by a weighted

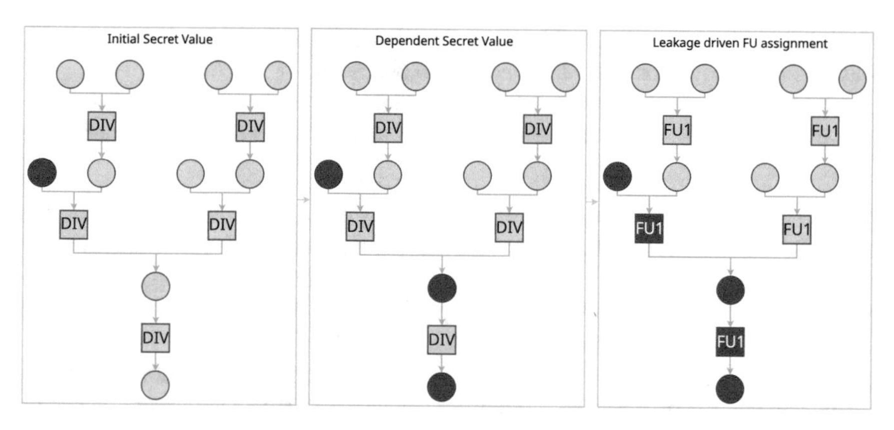

Fig. 13.6 An example of leakage-driven binding. Nodes and operations involving confidential values are marked in red. The final mapping requires two separate FUs, even all operations are of the same kind

bipartite matching algorithm, and the proposed weights between operations are determined from the estimated leakage level.

After all the operations are mapped to FUs, the proposed flow will then calculate the leakage in terms of individual leakage terms of each FU according to Eq. (13.3). The result is only a conservative approximation, as partial correlation between the underlying signals is common. With the quantified leakage, the flow can choose from a weighted module selection for each FU, thus generate a module assignment vector M. With module costs C, the flow will form the following linear programming problem to minimize the total leakage θ:

$$\text{Minimize} : \theta = \gamma \times M,$$
$$M \times C \leq b \tag{13.4}$$

Here γ is the vector of leakage estimations of all FUs and b is the resource constraint vector.

In this formulation, functional units with high leakage potential are up-sized in terms of defense mechanisms by reassigning them to more resistant modules (at the expense of higher cost), such that the overall leakage for a given resource utilization is minimized.

13.6 Experimental Evaluation

The proposed HLS flow for side-channel leakage reduction was implemented based on the LegUp HLS tool [19] and LLVM compiler infrastructure [44] (which is the basis of LegUp). Cycle-accurate simulation is performed in ModelSim. We chose a Cyclone V FPGA as the final hardware implementation platform and used Altera Quartus II to synthesize RTL-level code into bitstream.

The proposed flow is evaluated in multiple benchmarks and with regard to different characteristics. Several benchmarks are adapted from CHStone benchmarks [45]:

- AES, Blowfish, and SHA are cryptography functions commonly used in real applications as well as benchmarks. AES [46] and Blowfish [47] are two block ciphers with symmetric keys. SHA is a family of secure hashing algorithms.
- GSM benchmarks implement linear predictive coding analysis in the Global System for Mobile Communications (GSM).
- SIMON [48] is another family of lightweight block cipher algorithms proposed by the US National Security Agency's (NSA) Research Directorate, targeting at efficient hardware implementation and performance. SIMON is particularly interesting among the benchmarks since one of its purpose is to facilitate the IoT) security.
- We further developed another benchmark *IoT* which combines a mixture of general-purpose tasks such as the Fourier transformation for voice recognition, signal processing to detect human presence and temperature computation with corresponding light control. It is worth noting that all previous benchmarks implement a specific task while the *IoT* benchmark contains a group of different tasks, which can better reflect the reality of multi-task applications in the cloud or the IoT) environments.

The output of ModelSim simulation is parsed to derive the Hamming distance for leakage characterization. The total leakage (θ) for a given solution can be determined through Eqs. (13.2), (13.3), and (13.4). Table 13.1 presents entropy factors of different operations employed in the evaluation. The data were adapted from a previous work on Trojan defense [28].

13.6.1 Evaluation Reference

We presented two reference implementation for comparison. The first baseline is created using only one of the side-channel leakage defense techniques introduced in Sect. 13.3.1 for the whole design. The first baseline is named "Reference" in

Table 13.1 Empirical entropy factors for several commonly used operators

Operator	Entropy factor
Equality Comparator ('==')	0.1
Comparator (e.g. '<')	0.15
Modulo ('%')	0.3
Logical exc. XOR ('&', '|')	0.3
Other (e.g. '+', '^')	1

Table 13.2 Characteristics of different countermeasures employed in our proposed HLS flow. The resource overhead factor and leakage factor are compared to base, which does not have any countermeasures used

Logic style	Resource overhead factor	Leakage factor
Base	1	1
SDDL	4	1/7
WDDL	5.3	1/10
DWDDL	10.6	1/20
DAWDDL	13.1	1/25

later sections and tables. It mainly reflects today's design flow with largely uniform technology and IP libraries. The second reference we used is "Modular Synthesis," where the proposed side-channel synthesis is applied at modular level, as opposed to the whole design level. This design style represents a more practical approach, mainly reflecting the current modular granularity of design engineering for large ICs in enterprises. In the second reference, each module is selected to be implemented in one of the logic styles presented in Table 13.2 to achieve highest leakage resistance under the given resource constraints. We refer to Sect. 13.3.1 and references for further background on cost and leakage evaluation of SDDL [34, 35], WDDL [35, 38, 49], DWDDL [38], and DAWDDL [38]. Since the impact of resource and leakage are component-dependent, the numbers displayed in Table 13.2 are estimated and can be modeled more accurately with more empirical data.

13.6.2 Resource Targeting

The results in the following tables are presented in terms of gate-equivalent units (GEs). All the benchmarks are synthesized under a given resource target, which is selected so that it is sufficient for the baseline reference to be implemented using SDDL. This ensures that synthesized circuits generated by our proposed flow would not cost more than those using the same countermeasure for the entire design with the most resource-efficient logic style (i.e., SDDL) to reduce side-channel leakage.

Experimental results are presented in Table 13.3. The numbers in bold italics indicate that the maximum available countermeasures were applied—further reduction of total leakage (θ or θ_M) as defined in Eq. (13.4) was not possible despite available resources. Overall our proposed flow demonstrated a strong improvement: leakage is reduced by up to 72% compared to the reference, and up to 38% compared to the modular reference.

The syntheses of the GSM, *IoT*, and Blowfish benchmarks show interesting results, as both our proposed synthesis and the modular reference perform notably better in terms of side-channel leakage reduction with roughly the same amount of resources. Comparing to other benchmarks in CHStone, the GSM benchmark consists of more computation that are independent of user inputs deemed as

Table 13.3 Evaluation results of the proposed resource-targeted synthesis flow. Cost in bold italics indicates that the maximum available countermeasures were applied — further reduction of leakage (θ) was not possible despite available resources. The percentage columns ($\%_R$ and $\%_M$) denote the fraction of leakage reduction with respect to its corresponding reference (subscript R for basic reference and M for modular synthesis)

	SC resource	Balancing	Reference		Modular synthesis			Proposed SC synthesis			
Benchmark	target (GEs)	Cost (GEs)	Cost (GEs)	θ_R	Cost (GEs)	θ_M	$\%_R$	Cost (GEs)	θ	$\%_R$	$\%_M$
AES	187,000	640	186,960	394.7	186,968	323.8	18	186,856	202.0	49	38
Blowfish	88,000	0	87,360	341.3	87,556	239.3	30	87,968	231.6	32	3
SHA	62,000	0	61,440	99.4	60,176	64.6	35	61,888	49.3	50	24
SIMON	25,000	0	24,320	80.4	24,944	77.6	3	24,960	54.8	32	29
GSM	592,000	57,520	591,280	271.8	591,856	81.4	70	591,928	79.4	71	3
IoT	258,000	460	25,7900	305.3	***240,270***	85.5	72	***224,782***	85.5	72	0

secrets. Thus, both the modular reference and the proposed synthesis can utilize the resource more efficiently to reduce side-channel leakage and therefore achieve much higher leakage reduction. Similar trend can be seen in the *IoT* benchmark as enough resources were available for maximizing the countermeasures employed. Additionally, it can be observed that many of the confidential values (and their dependencies) are constrained to specific modules (basic blocks), which explains why the proposed synthesis was not significantly more efficient than the modular reference in GSM and *IoT*. In contrast, results in the Blowfish benchmark did not exhibit significant difference between the modular and our proposed synthesis because modules inside Blowfish are well-defined and nearly uniform in the degree of confidentiality.

In contrast to the GSM benchmark, the synthesis results of the SIMON benchmark shows great improvements in the proposed flow while only limited enhancement is gained in the modular synthesis reference. We believe that the underlying design is the reason for such difference—the share of confidential operands is roughly uniform across the modules in SIMON. The proposed synthesis operates on the instruction/operation level so extra leakage reduction can be further extracted. But the modular reference cannot significantly outperform the reference since it is restricted to exploit optimization opportunities across module boundaries.

To access the full capabilities of our proposed synthesis flow, we further evaluated it against the modular reference with a significantly lower resource limit. The results are listed in Table 13.4. Note that the basic reference would not yield a result in this case because the resource target was too low for it to deploy SDDL style on a full-device level. From Table 13.4, our proposed synthesis flow yields between strong reduction in side-channel leakage ranging from 22 and 40% comparing to modular reference, even under strict resource limit. Notably, the proposed flow achieved only a minor improvement of 3% compared to the modular approach under higher resource constraints for GSM but achieves a leakage reduction of 30% for the same benchmark in a more constrained environment. This signifies flexibility of the proposed flow and its suitability in embedded IoT) applications.

Table 13.4 Evaluation results of the resource-targeted synthesis flow against a modular baseline under more stringent resource constraints

		Modular synthesis		Proposed SC synthesis		
Benchmark	Resource target	Cost	θ_M	Cost	θ	$\%_M$
AES	75,000	74,980	2052.5	74,996	1221.4	40
Blowfish	35,000	34,572	2217.9	34,972	1691.8	24
SHA	23,000	18,800	592.0	22,976	361.4	39
SIMON	9500	8144	505.1	9376	395.3	22
GSM	235,000	234,992	763.0	234,952	536.0	30
IoT	75,000	74,686	1611.1	74,840	1079.2	33

13.6.3 Branch Balancing

Table 13.3 includes the resource costs of using branch balancing. The results are only presented in one column since they are identical between modular synthesis and the proposed synthesis. Not all benchmarks incur branch balancing, as some of the algorithm is already power- and time-constant by design. In other benchmarks like Blowfish and SHA, the proposed flow greatly reduces the risk of accidental branch imbalance even though the algorithm specification was not branch-balanced. The results of the GSM benchmark provide an example of the other spectrum of branch balancing: the benchmark code consists of many expensive operations that are part of the branches that (indirectly) depend on the log area ratio (LAR), which is usually specified as secret from a user's point of view. Manual branch balancing would need extensive hardware resources and human efforts to mitigate such situation. To illustrate the effectiveness of branch balancing, we consider the case that virtually the full GSM benchmark is considered to be security critical for maximum branch balancing—by specifying all the input signal to be confidential. Under this circumstance, the fully balanced design would require 184960 gate-equivalent (GE) resources just for branch balancing, 3.2 times as much as that described in Table 13.3 compared to our solution. Such difference in the resource cost demonstrates the effectiveness and convenience in our proposed HLS-based solution.

13.6.4 Leakage Targeting

We can also modify the constraints to be the total leakage instead of resource cost. This is referred to as leakage-targeted flow. The leakage-targeted flow is accessed with less severe constraints in Table 13.5 and extensively evaluated with stricter constraints in Table 13.6. With this flow, we can estimate the resource usage or mitigate overhead for a given leakage target. This flow is interesting in different practical scenarios, for example when synthesizing several sub-modules in a larger device with more resource at hand. Instead of maximizing the leakage resistance under seemingly arbitrary resource constraints, it might be more reasonable to reach for a given leakage target throughout the device.

Table 13.5 presents the results evaluated when the leakage target is modest at best. If the attackers concerned are mainly weakly motivated, then this setting is preferable since modest leakage reduction would be enough to thwart basic SPA and DPA. As shown in the results, the proposed HLS flow achieves fine-grained tuning of resource usage as they are utilized more efficiently with operation-level optimization. The resource cost is reduced by up to 67%, compared to the modular reference with the same leakage targets. Note that the achieved total leakage in modular synthesis results is much lower than the given target for the AES and Blowfish benchmark, leading to a significant reduction in leakage even

Table 13.5 Evaluation results of leakage-targeted synthesis flow under lenient leakage constraints. The cost is reported in gate-equivalent units (GE). The total leakage is denoted by θ. The percentage columns ($\%_M$) denote the fraction of resource reduction with respect to modular synthesis

		Modular synthesis		Proposed SC synthesis		
Benchmark	Leakage target	Cost	θ_M	Cost	θ	$\%_M$
AES	1000	90,120	480.2	32,160	998.2	64
Blowfish	1000	39,360	504.3	27,532	999.9	30
SHA	450	8160	398.9	5280	449.7	35
SIMON	450	8640	304.2	2880	448.4	67
GSM	1000	22,800	984.7	15,840	999.3	31
IoT	1000	21,600	802.3	12,236	999.4	43

Table 13.6 Evaluation results in leakage-target synthesis flow with more stringent leakage target. The percentage columns ($\%_R$ and $\%_M$) denote the fraction of resource reduction with respect to its corresponding reference

		Reference		Modular synthesis			Proposed SC synthesis			
Benchmark	Leakage target	Cost	θ_R	Cost	θ_M	$\%_R$	Cost	θ	$\%_R$	$\%_M$
AES	395	139,740	394.7	109,144	385.4	22	61,568	395.0	56	44
Blowfish	342	65,520	341.3	45,528	341.7	31	43,012	341.9	34	6
SHA	100	46,080	99.4	28,496	99.9	38	22,160	99.9	52	22
SIMON	81	18,240	80.4	16,320	80.5	11	12,512	81.0	31	23
GSM	272	400,320	271.8	104,624	271.6	74	93,672	271.9	77	10
IoT	306	193,080	305.3	42,930	304.3	78	36,004	305.6	81	16

below target leakage requirements. We believe such over-engineered solution can be attributed to the coarse granularity of block-based and module-level optimization. In contrast, our proposed synthesis flow can take advantage of FU-based leakage in optimization.

Table 13.6 presents the results evaluated when the leakage target is more stringent, which is best suited for scenarios to mitigate against highly motivated adversaries. Compared to the full-device baseline reference, the proposed flow reduces resource usage significantly while maintaining the side-channel leakage on the same level. As shown in Table 13.6, the reduction ranges from 31% for the highly efficient SIMON cipher to 81% for the *IoT* benchmark. The improvements over the modular reference is less significant, with reduction ranging from 6% for the Blowfish benchmark up to 44% for AES. For the AES benchmark, design synthesized by our proposed flow is implemented on a Cyclone V FPGA and the resource saving over the reference is equivalent to 12,413 Adaptive Logic Modules (ALMs). Compared to manual partitioning with a reduction of 24% presented in [39], our proposed flow achieves 56% in resource cost reduction as well as shorter development time with HLS.

13.7 Conclusion and Outlook

Due to the vast amount of information processing and the ever-increasing attention on data privacy, security has become a core requirement in cloud computing and IoT) applications. Side-channel information leakage can lead to serious data breaches if not taken seriously, since it indirectly allows malicious adversaries to access internal states in confidential applications. Detailed hardware security analysis and expertise in circuit-level countermeasures are often required to apply appropriate countermeasures defending side-channel attacks. In this work, we presented the first HLS flow that inherently maximizes the reduction of power side-channel leakage by analyzing and hardening all security critical operations given security annotation only on input variables from user. The HLS flow automatically analyzes the security information in relevant control flow and carries out simulations to perform leakage characterization based on the Hamming distance of confidential operations. The flow identifies and counteracts any imbalanced branches that might incur easy attack targets. The security-focused synthesis hardens functional units according to resource constraints and overall leakage target. From our experimental results, our proposed HLS flow can reduce side-channel leakage by up to 72% under identical resource constraints over countermeasures applied on full-device level. Additionally, resource consumption is reduced by up to 81% compared to the reference when the flow is modified to achieve a given target leakage. For future work, we can improve on the following two directions: (i) detailed evaluation and modeling of power during synthesis, as well as the impact on operation speed; and (ii) incorporation of masking to further improve the resistance against DPA.

References

1. Konigsmark, S.T.C., Chen, D., Wong, M.D.F.: High-level synthesis for side-channel defense. In: 2017 IEEE 28th International Conference on Application-Specific Systems, Architectures and Processors (ASAP), pp. 37–44. IEEE, Piscataway (2017)
2. Gubbi, J., Buyya, R., Marusic, S., Palaniswami, M.: Internet of Things (IoT): a vision, architectural elements, and future directions. Future Gener. Comput. Syst. **29**(7), 1645–1660 (2013). https://doi.org/10.1016/j.future.2013.01.010
3. Eguro, K., Venkatesan, R.: FPGAs for trusted cloud computing. In: 22nd International Conference on Field Programmable Logic and Applications (FPL), pp. 63–70 (2012)
4. Johnson, A.P., Chakraborty, R.S., Mukhopadhyay, D.: A PUF-enabled secure architecture for FPGA-based IoT applications. IEEE Trans. Multi-Scale Comput. Syst. **1**(2), 110–122 (2015)
5. Verbauwhede, I.M.R.: Secure Integrated Circuits and Systems. Springer, Berlin (2010)
6. Weste, N.H., Harris, D.: CMOS VLSI Design: a Circuits and Systems Perspective. Pearson Education, London (2015)
7. Tiri, K.: Side-channel attack pitfalls. In: 2007 44th ACM/IEEE Design Automation Conference, pp. 15–20. IEEE, Piscataway (2007). https://doi.org/10.1145/1278480.1278485
8. Kocher, P.C.: Timing attacks on implementations of diffie-hellman, RSA, DSS, and other systems. In: Proceedings of the 16th Annual International Cryptology Conference on Advances in Cryptology, CRYPTO'96, pp. 104–113. Springer, Berlin (1996)

9. Sidorov, E.: Breaking the Rabin-Williams digital signature system implementation in the crypto++ library. IACR Cryptol. ePrint Archive **2015**, 368 (2015)
10. Kocher, P., Jaffe, J., Jun, B.: Differential power analysis. In: Annual International Cryptology Conference, pp. 388–397. Springer, Berlin (1999)
11. Ors, S.B., Gurkaynak, F., Oswald, E., Preneel, B.: Power-analysis attack on an ASIC AES implementation. In: International Conference on Information Technology: Coding and Computing, Proceedings ITCC 2004, vol. 2, pp. 546–552. IEEE, Piscataway (2004)
12. Sundaresan, V., Rammohan, S., Vemuri, R.: Defense against side-channel power analysis attacks on microelectronic systems. In: 2008 IEEE National Aerospace and Electronics Conference, pp. 144–150 (2008)
13. Clavier, C., Coron, J.S., Dabbous, N.: Differential power analysis in the presence of hardware countermeasures. In: Cryptographic Hardware and Embedded Systems — CHES 2000, pp. 252–263. Springer, Berlin (2000)
14. Trichina, E.: Combinational logic design for AES SubByte transformation on masked data. IACR Cryptol. ePrint Archive **2003**, 236 (2003)
15. Tiri, K., Akmal, M., Verbauwhede, I.: A dynamic and differential CMOS logic with signal independent power consumption to withstand differential power analysis on smart cards. In: Proceedings of the 28th European Solid-State Circuits Conference, pp. 403–406. IEEE, Piscataway (2002)
16. Mangard, S., Oswald, E., Popp, T.: Power Analysis Attacks: Revealing the Secrets of Smart Cards. Advances in Information Security. Springer, New York (2008)
17. Chen, D., Cong, J., Fan, Y., Han, G., Jiang, W., Zhang, Z.: xPilot: A Platform-Based Behavioral Synthesis System. In: SRC TechCon, vol. 5 (2005)
18. Chen, D., Cong, J., Fan, Y., Wan, L.: LOPASS: a low-power architectural synthesis system for FPGAs with interconnect estimation and optimization. IEEE Trans. Very Large Scale Integr. Syst. **18**(4), 564–577 (2010)
19. Canis, A., Choi, J., Aldham, M., Zhang, V., Kammoona, A., Czajkowski, T., Brown, S.D., Anderson, J.H.: LegUp: an open-source high-level synthesis tool for FPGA-based processor/accelerator systems. ACM Trans. Embed. Comput. Syst. **13**(2), 1–27 (2013). https://doi.org/10.1145/2514740
20. Lin, J.Y., Chen, D., Cong, J.: Optimal simultaneous mapping and clustering for FPGA delay optimization. In: 2006 43rd ACM/IEEE Design Automation Conference, pp. 472–477 (2006)
21. Zheng, H., Gurumani, S.T., Rupnow, K., Chen, D.: Fast and effective placement and routing directed high-level synthesis for FPGAs. In: Proceedings of the 2014 ACM/SIGDA International Symposium on Field-Programmable Gate Arrays, FPGA '14, New York, pp. 1–10 (2014). https://doi.org/10.1145/2554688.2554775
22. Rupnow, K., Liang, Y., Li, Y., Min, D., Do, M., Chen, D.: High-level synthesis of stereo matching: productivity, performance, and software constraints. In: 2011 International Conference on Field-Programmable Technology, pp. 1–8 (2011)
23. Liu, X., Chen, Y., Nguyen, T., Gurumani, S., Rupnow, K., Chen, D.: High-level synthesis of complex applications: an H.264 video decoder. In: Proceedings of the 2016 ACM/SIGDA International Symposium on Field-Programmable Gate Arrays, FPGA '16. New York, p. 224–233 (2016). https://doi.org/10.1145/2847263.2847274
24. B. Hammouda, M., Coussy, P., Lagadec, L.: A design approach to automatically synthesize ANSI-C assertions during high-level synthesis of hardware accelerators. In: 2014 IEEE International Symposium on Circuits and Systems (ISCAS), pp. 165–168 (2014)
25. Campbell, K.A., Vissa, P., Pan, D.Z., Chen, D.: High-level synthesis of error detecting cores through low-cost modulo-3 shadow datapaths. In: 2015 52nd ACM/EDAC/IEEE Design Automation Conference (DAC), pp. 1–6 (2015)
26. Cui, X., Ma, K., Shi, L., Wu, K.: High-level synthesis for run-time hardware trojan detection and recovery. In: 2014 51st ACM/EDAC/IEEE Design Automation Conference (DAC), pp. 1–6 (2014)
27. Rajendran, J., Zhang, H., Sinanoglu, O., Karri, R.: High-level synthesis for security and trust. In: 2013 IEEE 19th International On-Line Testing Symposium (IOLTS), pp. 232–233 (2013)

28. Konigsmark, S.T.C., Chen, D., Wong, M.D.: Information dispersion for trojan defense through high-level synthesis. In: Proceedings of the 53rd Annual Design Automation Conference, pp. 1–6 (2016)
29. Jiang, Z., Dai, S., Suh, G.E., Zhang, Z.: High-level synthesis with timing-sensitive information flow enforcement. In: 2018 IEEE/ACM International Conference on Computer-Aided Design (ICCAD), pp. 1–8. IEEE, Piscataway (2018)
30. Amazon: Amazon EC2 F1 instances available in the cloud (2020). https://aws.amazon.com/ec2/instance-types/f1/
31. Fowers, J., Ovtcharov, K., Papamichael, M., Massengill, T., Liu, M., Lo, D., Alkalay, S., Haselman, M., Adams, L., Ghandi, M., et al.: A configurable cloud-scale DNN processor for real-time AI. In: Proceedings of the 45th Annual International Symposium on Computer Architecture, pp. 1–14. IEEE Press, Piscataway (2018)
32. Rao, M., Newe, T., Grout, I.: Secure hash algorithm-3 (SHA-3) implementation on Xilinx FPGAs, suitable for IoT applications. In: 8th International Conference on Sensing Technology (ICST 2014), Liverpool John Moores University, Liverpool, 2nd-4th September (2014)
33. Dobbins, R., Bjarnason, S.: Mirai IoT botnet description and DDoS attack mitigation. Arbor Threat Intell, https://www.arbornetworks.com/blog/asert/mirai-iot-botnet-description-ddos-attack-mitigation/. Accessed Jul. 2021.
34. Tiri, K., Verbauwhede, I.: A logic level design methodology for a secure DPA resistant ASIC or FPGA implementation. In: Proceedings Design, Automation and Test in Europe Conference and Exhibition, vol. 1, pp. 246–251. IEEE, Piscataway (2004)
35. Velegalati, R., Kaps, J.P.: DPA resistance for light-weight implementations of cryptographic algorithms on FPGAs. In: 2009 International Conference on Field Programmable Logic and Applications, pp. 385–390. IEEE, Piscataway (2009)
36. Dinh, Q., Chen, D., Wong, M.D.F.: A routing approach to reduce glitches in low power FPGAs. IEEE Trans. Comput.-Aid. Des. Integr. Circuits Syst. **29**(2), 235–240 (2010)
37. Yu, P., Schaumont, P.: Secure FPGA circuits using controlled placement and routing. In: Proceedings of the 5th IEEE/ACM International Conference on Hardware/Software Codesign and System Synthesis, pp. 45–50 (2007)
38. Wild, A., Moradi, A., Güneysu, T.: Evaluating the duplication of dual-rail precharge logics on FPGAs. In: International Workshop on Constructive Side-Channel Analysis and Secure Design, pp. 81–94. Springer, Berlin (2015)
39. Kaps, J.P., Velegalati, R.: DPA resistant AES on FPGA using partial DDL. In: 2010 18th IEEE Annual International Symposium on Field-Programmable Custom Computing Machines, pp. 273–280. IEEE, Piscataway (2010)
40. Tiri, K., Verbauwhede, I.: Simulation models for side-channel information leaks. In: Proceedings of the 42nd Annual Design Automation Conference, DAC'05, New York, pp. 228–233 (2005). https://doi.org/10.1145/1065579.1065640
41. Menichelli, F., Menicocci, R., Olivieri, M., Trifiletti, A.: High-level side-channel attack modeling and simulation for security-critical systems on chips. IEEE Trans. Depend. Secur. Comput. **5**(3), 164–176 (2008)
42. Fang, X., Luo, P., Fei, Y., Leeser, M.: Balance power leakage to fight against side-channel analysis at gate level in FPGAs. In: 2015 IEEE 26th International Conference on Application-Specific Systems, Architectures and Processors (ASAP), pp. 154–155. IEEE, Piscataway (2015)
43. den Hartog, J., de Vink, E.: Virtual analysis and reduction of side-channel vulnerabilities of smartcards. In: IFIP World Computer Congress, TC 1, pp. 85–98. Springer, Berlin (2004)
44. Lattner, C., Adve, V.: LLVM: a compilation framework for lifelong program analysis & transformation. In: Proceedings of the 2004 International Symposium on Code Generation and Optimization (CGO'04), Palo Alto (2004)
45. Hara, Y., Tomiyama, H., Honda, S., Takada, H.: Proposal and quantitative analysis of the CHStone benchmark program suite for practical C-based high-level synthesis. J. Inf. Process. **17**, 242–254 (2009)

46. Standard, N.F.: Announcing the advanced encryption standard (AES). Federal Inf. Process. Standards Publ. **197**(1–51), 3–3 (2001)
47. Schneier, B.: Description of a new variable-length key, 64-bit block cipher (Blowfish). In: International Workshop on Fast Software Encryption, pp. 191–204. Springer, Berlin (1993)
48. Beaulieu, R., Shors, D., Smith, J., Treatman-Clark, S., Weeks, B., Wingers, L.: The SIMON and SPECK families of lightweight block ciphers. IACR Cryptol. ePrint Archive **2013**(1), 404–449 (2013)
49. Moradi, A.: Side-channel leakage through static power. In: International Workshop on Cryptographic Hardware and Embedded Systems, pp. 562–579. Springer, Berlin (2014)

Chapter 14
S*FSMs for Reduced Information Leakage: Power Side Channel Protection Through Secure Encoding

Mike Borowczak and Ranga Vemuri

14.1 Introduction

The world of electronic communication and thereby the transfer of information and control of digital components is often designed, modeled, and then implemented with the assistance of Finite State Machines (FSM). From integrated circuit communication buses to larger control systems, our world operates and functions because these systems work as designed. These communication mediums, however, are also prime targets for statistically driven data exfitration attacks which, unfortunately, become easier to perform if we continue to focus solely on optimization and minimization of FSM structures and encodings.

These statistically driven attacks utilize side channels of information, in conjunction with known operating characteristics of a device, to exfiltrate the internal state(s) of a system. While side channels can take many forms, from power consumption to timing information to glitching characteristics, they all have some relationship to the internal operation of a device and attackers fuse this information with other available data—from device input/output to specific knowledge of a devices' implementation to reduce the search space of discovering some internal secrets such as secret keys, early stage boot signals, or activation and verification sequences for custom IP-blocks. Leveraging decades of work in creating optimized Finite State Machines (FSMs) for a large variety of target applications, from communication protocols to transmitter/receiver pairs to message encoders and more, we propose the application of a two-part process to harden the security of

M. Borowczak (✉)
University of Wyoming, Laramie, WY, USA
e-mail: mborowcz@uwyo.edu

R. Vemuri
University of Cincinnati, Cincinnati, OH, USA
e-mail: vemurir@ucmail.uc.edu

S. Katkoori, S. A. Islam (eds.), *Behavioral Synthesis for Hardware Security*,
https://doi.org/10.1007/978-3-030-78841-4_14

FSMs from statistically based power side channel attacks. The resulting FSMs are named S*FSMs. While the development of Secure Finite State Machines (S*FSMs) does not ignore advances in FSM optimization [1], it does not require an optimized machine as an initial starting point. It is important to note that the techniques for creating S*FSMs are based on an imposition of a set of additional constraints that result in an increase in both state space and encoding length. As with any design decision, this security trade-off is more easily accepted if the FSM that you are trying to secure is already minimized. For the purposes of this chapter, however, we utilize FSMs that may not always be fully optimized, for space and/or encoding, in order to provide clear examples of the potential vulnerabilities and the steps required to create S*FSMs. This chapter contains motivating examples of insecure FSMs. These two examples are revisited throughout the chapter to provide the reader tangible examples of the steps required to create S*FSMs.

The remaining of this chapter provides context for both FSMs in system design as well as threats and vulnerabilities, and it continues with the general approach for securing FSMs. It then addresses some of the complexities when attempting to develop S*FSMs from creating secure optimal encodings to dealing with synthesis tools that "optimize away" the required structural and/or encoding changes.

14.2 Motivating Examples

14.2.1 Finite State Machines

A Finite State Machine, or FSM, is a mathematical computational model comprised of one or more states. An FSM can only be in one state of a finite set of states at any given time and thus transitions between states, based on some set of inputs, are also included with the model. FSMs can be defined as the quintuple $(\Sigma, S, s_0, \delta, F)$ where:

- Σ is the finite, non-empty, set of symbols.
- S is the finite, non-empty, set of states.
- $s_0 \in S$ is the initial state.
- $\delta : S \times \Sigma \rightarrow S$ is the state-transition function.
- F is the set of final states.

When implemented as digital logic circuits, the FSM states are encoded as Boolean vectors and usually stored in flip-flops. State encoding has a significant impact on the size power and security of the implementation. As shown in the following two examples (see Figs. 14.1 and 14.2), FSMs are easily expressed using graphs where FSM states are nodes and FSM transitions are directed edges between nodes. A directed graph, formally defined as $G = (V, E)$, contains:

- A set of vertices V, or nodes, and
- A set of ordered vertex 2-tuples E, or edges.

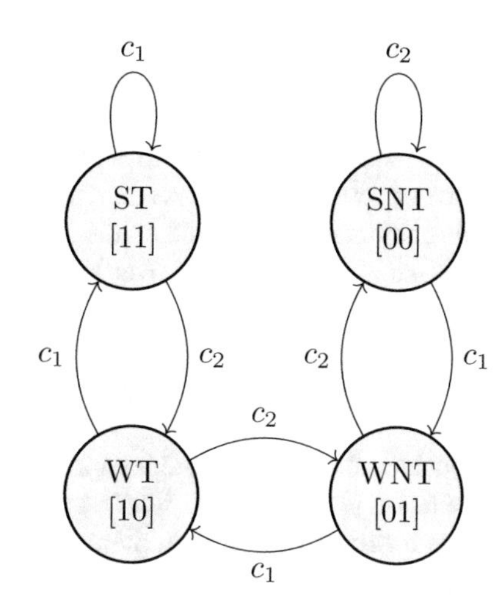

Fig. 14.1 A Branch Predictor with states Strongly Taken (ST), Weakly Taken (WT), Weakly Not Taken (WNT), and Strongly Not Taken (SNT). Transitions occur based on Branch Taken (BT) input. The c_1 transitions represent BT = True, while the inner counter-clockwise path, labeled c_2, represents BT = False. The self-loops on ST and SNT are associated with transitions c_1 and c_2, respectively

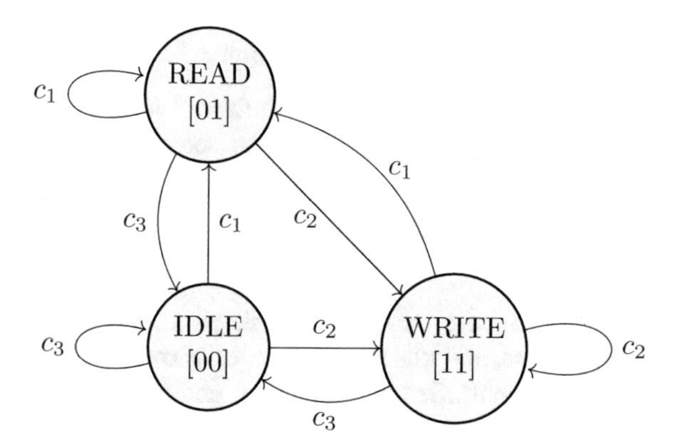

Fig. 14.2 A generic SRAM FSM with three states Read, Write, and Idle and nine unique transitions. The nine transitions are controlled by three conditions (c_1,c_2,c_3), where c_1 and c_2 utilize three active low signals ChipSelect (CS), WriteEnabled (WE), and OutputEnabled (OE) and c_3 represents all other transition. Specifically, $c_1 = \overline{CS}$ & $\overline{OE}$ & WE, $c_2 = \overline{CS}$ & $\overline{WE}$, and $c_3 = \overline{c_1}$ & $\overline{c_2}$. Each state has two inbound and two outbound transitions not including one self-loop

When mapping a FSM onto a graph, the set of vertices consists of a direct mapping of finite state space of the FSM $V = S$ while the edges form a similar mapping between edges and state transitions $E = \delta$. In addition, the nodes and edges of a graph, when labeled and annotated, can represent the remaining components of the formal FSM definition. For the purposes of our discussion, we label edges with a generic condition label and nodes with a name and, when appropriate, a binary encoding.

In the context of creating secure FSMs (S*FSMs), our discussion uses the simplified graph representation which focuses on set of states (S) and the transitions

(δ) between them. Those wanting more details on FSMs are directed to the large body of literature on the subject including these two resources covering theory and application [2, 3].

For the sake of clarity, while the examples that follow are small FSMs with only three or four states and less than ten transitions each, the processes presented within this chapter are extensible to any size FSM and have been used on larger FSMs with over a hundred states and over two hundred transitions.

14.2.1.1 Branch Predictor

Consider the classical computer architecture problem of designing a dynamic branch predictor [4, p. 82–83]. While several schemes exist, Fig. 14.1 shows a modified branch predicting FSM implementation that utilizes a 2-bit saturating counter. In this branch predictor, when a branch is *taken* (BT=T) it moves the predictor toward the strongly taken (ST) state, while if the branch is *not take*n (BT = F) the predictor moves toward strongly not taken (SNT) state. Each state of the counter is assigned a binary encoding to satisfy the following conditions:

1. The overall fixed encoding bit-length is minimized ($l = 2$).
2. The encoding of states SNT<WNT<WT<ST are sequential ordered from 0 to 3.

14.2.1.2 SRAM FSM

While many memory controllers exist, Fig. 14.2 shows a simple three-state machine controlling the Read, Write, and Idle states of a SRAM module. The SRAM FSM's transitions (c_1,c_2,c_3) are controlled by Boolean functions that utilize three active low inputs, namely ChipSelect (CS), WriteEnabled (WE), and OutputEnabled (OE). Each state has each of the three unique transitions and thus given the proper conditions each state can transition to every other state as well as itself. Though the Boolean functions that make up the three conditions are not required to understand the security methods presented in this chapter, they are explained in the Figure for completeness.

14.2.2 Side Channels

Side Channels are non-functional outputs of a system that are somehow related to processing of information. A side channel is unlikely to have complete fidelity into the underlying data processed—but it could. As an example, consider the amount of time it takes to perform the following operation $\mathbf{a} \times \mathbf{b}$. The time to compute $\mathbf{7} \times \mathbf{3}$

should differ from the time it takes to compute **37x42**. Even if we could find a way to collect the specific **a** $\times$ **b** timing information, this would still not be enough to guess the actual numbers multiplied—but it would reduce the possible search space.

Side channels exist the moment a design is realized—every action taken within a system requires some amount of energy and time. The goal of a side channel is to measure these byproducts and associate the result with some model of the functionality of the system. In some cases, the model is based on knowledge of only the function being performed, while in other cases, an input or output might also be known.

14.2.2.1 Attack and Protection Fundamentals

Side channel attacks are attack vectors that exploit the relationships between logical/functional operations of a device and their physical byproducts. These statically driven attacks can be used to determine information within a device such as a hidden functional input (e.g., secret key), or the state/values on a communication bus.

There are two basic requirements in order to perform side channel attacks—the first requires a side channel that is related to the (sub)function while the second requires a modeled sub-function to be proportional to the side channel and intrinsically the underlying (sub)function. Given a relationship between a side channel ($f(sc)$) and model m, each containing j aligned data-points, the correlation, $C(m, sc)$, is defined by Eq. 14.1, where m_j are individual model outputs, sc_j are individual side channel output, and $\bar{m}$ and $\bar{sc}$ represent the average model output and side channel output, respectively.

$$C(m, sc) = \frac{\sum \left(\left(m_j - \overline{m}\right) \times \left(sc_j - \overline{sc}\right)\right)}{\sqrt{\sum \left(m_j - \overline{m}\right)^2 \times \sum \left(sc_j - \overline{sc}\right)^2}} \tag{14.1}$$

14.2.2.2 The Power Side Channel: Source and Models

The power side channel is one of the most prevalent side channels due to its accessibility and low hardware cost to measure power. These attacks target devices by exploiting information embedded within the data-dependent power consumption [5].

Modeling of the power side channel is relatively straightforward for most CMOS technologies. This is due to the asymmetric nature and associated power consumption in CMOS logic styles and is predominately a result of static and dynamic power dissipation.

Given the difference in the sources of power consumption models of the power side channel, a needed component for attacks typically relies on hamming-based measures. The measures include the Hamming Weight (HW) and the Hamming Distance (HD) which are formally defined in Eqs. 14.2 and 14.3, respectively. These models focus on the current or changing state of set of bits within a device (e.g., register, bus) by modeling either the total number of bits "on," the Hamming Weight, or the total number of bits that are switching, or Hamming Distance [6]. Thus, Hamming Weight for register, or bus, or any multi-bit segment S is the number of the "1"-bits. The Hamming Distance between two consecutive states at times t and $t+1$ of any multi-bit segment (S) is the number of bits that changed from time t to $t+1$. When unguarded against, side channel attacks present an extremely powerful attack vector, especially in low-power and resource constrained hardware, as variations in hamming models are highly related to variations in the power consumed [7].

$$\mathrm{HW}(S[x..0]) = \sum_{i=0}^{x} S[i] \tag{14.2}$$

$$\mathrm{HD}(S_t[x..0], S_{t+1}[x..0]) = \sum_{i=0}^{x} S_t[i] \oplus S_{t+1}[i] \tag{14.3}$$

14.2.2.3 Modeling of the Branch Predictor FSM

The Hamming Weights for our small FSM examples are easily computed as the number of the "on"-bits while the Hamming Distance is dependant on the transition between the current and next state. Table 14.1 shows the summary of both hamming models for the branch prediction FSM seen in Fig. 14.1. Interestingly, we can already see potential sources of information leakage from both models, specifically that ST and SNT are immediately distinguishable via the hamming weight alone, and the transition between the WT and WNT state is also immediately apparent. When combining the information extracted from both hamming models, the original states of the machine can be immediately extracted as seen in Fig. 14.3.

Table 14.1 Hamming Weight for the current state and the Hamming Distance to each of the two, condition-dependant NextStates transitions of the branch prediction FSM from Fig. 14.1

			Transition c_1			Transition c_2		
Current state	[Enc]	HW	NextState	[Enc]	HD	NextState	[Enc]	HD
ST	[11]	2	ST	[11]	0	WT	[10]	1
WT	[10]	1	ST	[11]	1	WNT	[01]	2
WNT	[01]	1	WT	[10]	2	SNT	[00]	1
SNT	[00]	0	WNT	[01]	1	SNT	[00]	0

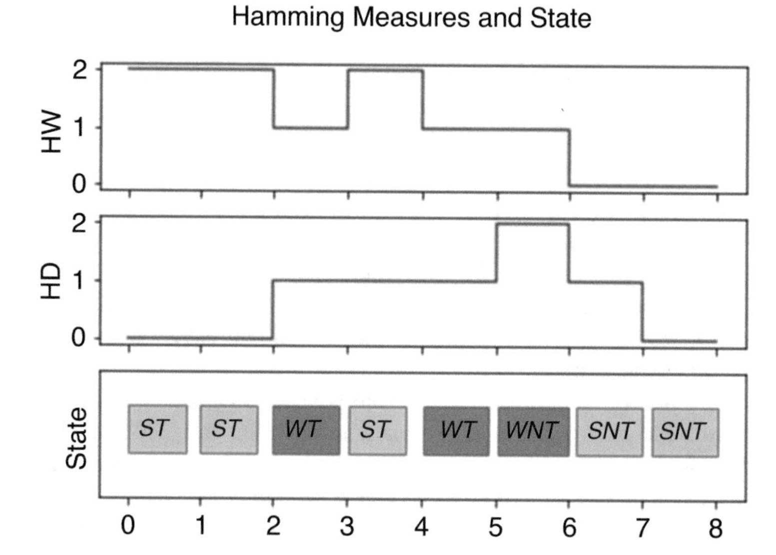

Fig. 14.3 The combination of the Hamming models allows for a perfect recovery (or extraction) of the branch predictor (Fig. 14.1) states. The WT and WNT states share a common Hamming Weight and thus extraction also requires the Hamming Distance

Table 14.2 Hamming Weight for the current state and the Hamming Distance for each of the possible transitions derived from the SRAM FSM in Fig. 14.2

Current state	[Enc]	HW	Transition c_1 Next St.	[Enc]	HD	Transition c_2 Next St.	[Enc]	HD	Transition c_3 Next st.	[Enc]	HD
READ	[01]	1	READ	[01]	0	WRITE	[11]	1	IDLE	[00]	1
WRITE	[11]	1	READ	[01]	1	WRITE	[11]	0	IDLE	[00]	2
IDLE	[00]	2	READ	[01]	2	WRITE	[11]	2	IDLE	[00]	0

14.2.2.4 Modeling of the SRAM FSM

Table 14.2 shows the summary of both hamming models for the branch prediction FSM seen in Fig. 14.2. Exact state information is directly extracted from the Hamming Weight model of this minimally encoded FSM.

While not as informative, the Hamming Distance model for this FSM can be used to reduce the possible state space that the machine was in. For example, a Hamming Distance of 2 occurs twice as frequently for outbound transitions from the IDLE state than any other scenario. Similarly, a Hamming Distance of 1 occurs twice as frequently on outbound transitions from the READ state than all others. Side channel attacks leverage these imbalances through statistically driven attacks that aim to reduce the overall search space of a problem. Since each additionally known bit (0/1) of information from system reduces the search space of an embedded secret by a factor of two, significant research and effort has gone into protecting sensitive hardware components.

14.2.2.5 Existing Approaches

The two broad approaches to mitigating vulnerabilities associated with the power side channel have focused on either masking or mitigating the channel. Masking strategies, which can include physical or electronic means, aim to lower the signal to noise ratio between information being processed and the overall operation of a device. Unfortunately, this strategy often takes the form of some security through obscurity mechanism. Unless the signal can be completely eliminated information can theoretically be extracted. Of more practical interest to hardware designers are the mitigation approaches which modify the power consumption of critical components, via either low-level logic modification and circuit designs, or high-level custom restructuring.

The protection of electronic devices from power side channel attacks has been well documented at low levels of abstraction since the early 2000s. A majority of existing low-level solutions focus on methods that reduce intra-cycle current variations through custom logic cells which generally contain complementary logic paths. These specialized logic styles include, among many others, Dynamic Differential Logic (DDL) [8] and Wave Dynamic Differential Logic (WDDL) [9], Reduced Complementary Dynamic and Differential Logic (RCDDL) [10], and Secure Differential Multiplexer Logic using paththrough transistors(SDMLp) [11].While these and other low-level methods have been shown effective against power side channel attacks, they come at a significant design cost and increased complexity in downstream design and implementation [12].

Other approaches include special asynchronous designs methods [13] or the redesign of secure individual cryptographic algorithms in order to prevent information leakage [14–16]. These approaches target specific algorithms (e.g., AES, DES), or algorithmic components (e.g., S-Boxes), and typically deal with masking specific computations. These techniques while effective when tuned and adapted for a specific algorithm have limited widespread reusability and may suffer from higher-order side channel attacks.

14.2.2.6 Security Metrics

Quantifying the relationship between the power side channel and side channel hamming models provides a baseline of comparison for different designs. In previous, cell-level work many authors claimed that reduced variation in the power signal should reduce correlation—since variation still existed in the signal correlation metric was still defined. Using correlation, especially as a security metric, involves making some strong assumptions about the underlying relation of the two data sources. In addition, in order to assess the security of a system using correlation analysis alone would require access to both data sources. A high-level hardware designer is unlikely to have details surrounding specific power consumption of a device. This work focuses on a method that, as we will see later, does not require access to both sources as it attempts to mitigate the amount of mutual information between two sources—a task which can be measured and tracked given only one of the two sources—in this case the model.

Consider the two data sources used in a side channel attack—mainly power P and a Hamming-based power model M. Each source contains some amount of information, which can be defined as entropy (H) as shown in Eq. 14.4 for a variable A. Entropy is a function of the probability of the possible values $p(a)$ of a source A. The entropy of the power and model can be computed independently of one another. The joint entropy, seen in Eq. 14.5, describes the total amount of entropy associated with two related variables, A and B. The mutual information between these two sources ($I(A, B)$), as defined in Eq. 14.6, quantifies how much entropy (or information) is shared or can be removed in one source by knowing the other. Replacing A and B with a device's power (P) and model (M) in Eq. 14.6 enables a metric to quantify the amount of information captured in a particular implementation. The goal of creating a secure FSM is to minimize or even better eliminate the mutual information between the power consumption and its side channel models.

$$H(A) = -\sum_{a \in A} p(a) \times \log_2(p(a)) \tag{14.4}$$

$$\begin{aligned} H(A, B) &= -\sum_{a \in A; b \in B} p(a, b) \times \log_2(p(a, b)) \\ &= -\sum_{a \in A; b \in B} p(b|a) \times p(a) \log_2(p(b|a) \times p(a)) \end{aligned} \tag{14.5}$$

$$I(A; B) = H(A) + H(B) - H(A, B) \tag{14.6}$$

Attacks that leverage power-based models can do so because of two properties: first there is a non-zero amount of information contained within the model M and second there is an inherent probabilistic relationship between the model M and the power consumed P and thus Mutual Information exists.

14.3 The Two-Part S*FSM Approach

A S*FSM eliminates information leakage from the two common power side channel attack models. It is important to note that when approaching side-channel security from the attack-model perspective, should fundamental power consumption characteristics change in devices, the techniques presented will need to be adjusted.

Given the two information-based models, Hamming Weight (HW) and Hamming Distance (HD), the S*FSM approach consists of two primary objectives [17]. These objectives are to eliminate mutual information between the FSMs:

1. State encoding and the HW model, and
2. State transitions and the HD model.

Achieving these two objectives would remove the ability for an attacker to correlate (and extract) the current state of a machine when applying Hamming-based models to extract information from power side channels. In order to remove the mutual information between the model and side channel, their data must be independent or one of the two must have zero entropy. Since, at a high-level we have no control over whether or not a relationship will exist between a model and the power side channel, our approach focuses on minimizing the entropy of the Hamming models. Intuitively, as we decrease the entropy of the Hamming models, the potential mutual information that exists between the models and the power side channel decreases.

In the best case, we would eliminate the entropy of the model entirely. Zero entropy occurs when there is "no information" in the signal; in this case the signal remains constant. This implies the probability of occurrence of the one event is 1. Equations 14.7 and 14.8, refined from [18], show the impact on the joint entropy and mutual information when one of the two data sources, B in this case, is held constant. This approach achieves zero mutual information between the model and the power side channel, irrespective of characteristics of the power side channel.

$$\begin{aligned} H(A, b) &= -\sum_{a \in A} p(b|a) \times p(a) \log_2 (p(b|a) \times p(a)) \\ &= -\sum_{a \in A} 1 \times p(a) \log_2 (1 \times p(a)) \\ &= H(A) \end{aligned} \tag{14.7}$$

$$\begin{aligned} I(A; b) &= H(A) + H(b) - H(A, b) \\ &= H(A) + 0 - H(A) \\ &= 0 \end{aligned} \tag{14.8}$$

We define two conditions that FSMs are required to maintain in order to remove the entropy from their associated Hamming models. First, as shown in Eq. 14.9, all states (or vertices) within a S*FSM must have the same constant Hamming Weight. Second, as shown in Eq. 14.10, the transitions or edges between connected states (s and s') must have the same Hamming Distance.

$$\forall s \in S : \mathrm{HW}(s) = c_1 \tag{14.9}$$

$$\forall s, s' \in S \mid \exists \alpha \in \Sigma : s' = \delta(s, \alpha) \rightarrow \mathrm{HD}(s, s') = c_2 \tag{14.10}$$

The second condition, maintaining a constant HD, poses considerable challenges for many FSMs. Looking back at the two example FSMs there is no encoding strategy that can eliminate the variability of the HD model. Specifically, any time a state transitions to itself, the HD will be zero while for all other transitions, the HD is non-zero.[1]

[1] Designers looking to control power characteristics may tune the values of c_1 and c_2 to suit their needs [19] or add additional constraints for power-security trade-offs [20].

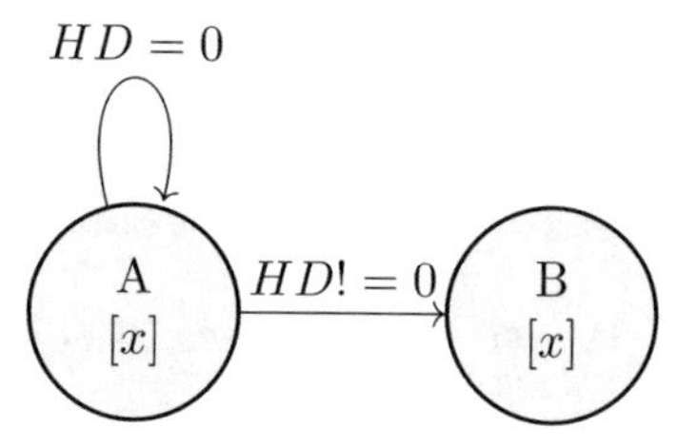

Fig. 14.4 A two state FSM motivating the need for structural modification. Irrespective of the encoding length or scheme, the HD between all states can never be constant

This enforces a structure change in any FSMs that contain states with self-loops. Thus, in order to create S*FSMs, a two-part FSM hardening process includes (1) structural modifications and (2) a re-encoding of states.

14.3.1 Structure

The need for this structural modification in S*FSM is demonstrated when considering the need for a constant HD between all state transitions. Consider that the two-node FSM in Fig. 14.4, regardless of the encoding selected, can never satisfy the second condition (Eq. 14.10).

The requirement for all Hamming Distances to be the same across all transitions is impossible in the presence of self-looping transitions. There is only one mechanism that can resolve this inconsistency for any FSM presented. Loop-unrolling, though counter to the end goal of state collapsing and FSM minimization to reduce complexity [21], it is a cost trade-off required for enhanced hardware security [17]. Thus, in order to satisfy the second condition, **a multi-state, hardened FSM cannot contain any self-loops**.

Algorithm 14.1, *Direct Loop Remove*, is an effective method for removing self-loops in any FSMs and is presented assuming a graph representation of an FSM. The algorithm examines each node, checks for the existence of a self-looping transition (or edge), and upon locating such an edge removes it from the transition list. A new state is added along with two corresponding edges, each with the self-looping edge's original transition condition. Finally all out-going edges are replicated from the original state that had the self-loop to the newly created state, maintaining the original functionality of the FSM.

The result of applying the Loop Unroll Algorithm (Algorithm 14.1) to our two example FSMs can be seen in Fig. 14.5. The initial FSMs both expand to six-node FSMs, with the restructured Branch Prediction FSM, Fig. 14.5a, increasing by two nodes or 50% and the restructured SRAM FSM, Fig. 14.5b, increasing by three nodes or 100%. The net change in transitions also increases at a predictable rate equal to all of the outbound transitions from each looping state.[2] The net new computation takes into account the removal of one looping transition and the

[2]For the purposes of this discussion a self-loop is considered an outbound transition.

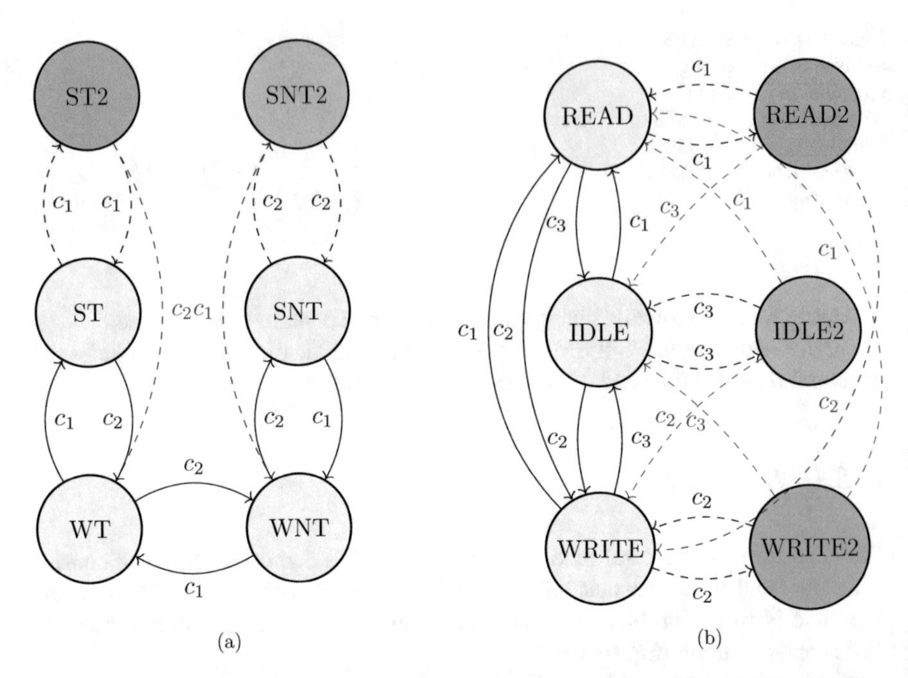

Fig. 14.5 Two examples FSMs after the application of the structural modification algorithm. New states, generated in line 6 of the algorithm, are shaded in color. The two transitions that replace the loop, generated by line 7 of the algorithm, are dashed and in black. Finally, the new outbound transitions of the new state, generated in lines 8-10 of the algorithm, are dashed in the color of the new node. (**a**) Unrolled branch predictor from Fig. 14.1 +(2s,4t). (**b**) Unrolled SRAM from Fig. 14.2 +(3s,9t)

insertion of two transitions between the existing and newly created state. Thus, the restructured Branch Prediction FSM has an increase of two transitions per new node for a total of four net new transitions, similarly the restructured SRAM FSM has three new transitions per new node for a total of nine net new transitions.

Algorithm 14.1 Direct loop remove

Procedure:

1: **for** $v \in V$ **do**
2:     **if** $\exists e : e(v, v, c) \in E$ **then**
3:         $E \leftarrow E - e$ *// Remove Self-Loop Edge*
4:         $v' \leftarrow v$ *// Create NewState node*
5:         $V \leftarrow V \cup \{v'\}$ *// Add NewState to set of states*
6:         $E \leftarrow E \cup \{\{v, v', c\}, \{v', v, c\}\}$ *// Add edges to/from new state w/cond. c*
7:         **for** $u \in V | u \neq v, e(v, u, c) \in E$ **do**
8:             $E \leftarrow E \cup e(v', u, c)$ *// Add Outbound Edge to NewState*
9:         **end for**
10:     **end if**
11: **end for**

14.3.2 Encoding

The two conditions required to remove all information entropy from the Hamming models require maintaining two constraints, one for each model. The first constraint (Eq. 14.9) requires little effort, but does have implications on total number of states (S) that can be encoded with a given bit-length (n). In a typical minimal binary encoding the total number of bits required is the ceiling of the $\log_2$ of the number of states ($\lceil \log_2 |S| \rceil$). In the constrained encoding space, the value of the Hamming Weight constant c is flexible and could range from 1 to $n - 1$ so long as Eq. 14.11 is satisfied. The total number of possible encodings of an $n - bit$ encoding is maximized when $c = \frac{n}{2}$. S*FSMs are limited to any binary constant weight, or $m - of - n$, code [22]; these can include standard encoding strategies such as one-hot-encodings ($1 - of - S$) or balanced encodings ($\frac{n}{2} - of - n$) where the number of "on"-bits equals the number of "off"-bits [23]. Preferably, however, the S*FSM encoding should have a tuned Hamming Weight constraint c resulting in per-FSM $c - of - n$ encoding that defines the lower bound encoding length prior to attempting to solve the Hamming Distance encoding constraint.

$$\binom{n}{c} \geq s \tag{14.11}$$

While defining a base encoding scheme that satisfies the Hamming Weight constraint was straightforward, solving the Hamming Distance constraint for S*FSMs, shown in Eq. 14.10, for all but one-hot encodings ($(n, c) = (S, 1)$) is a hard problem. Akin to many other constraint satisfaction problems and mostly closely resembling the k-coloring of non-planar graphs, satisfying the Hamming Distance constraint is an NP-complete problem [24]. While other methods are possible, the most straightforward path to determining valid S*FSM state encodings is through constraint programming.

14.4 Encoding With Constraint Programming

Constraint programming (CP) is a declarative programming paradigm often used to solve combinatorial problems. Constraint programming consists of the two-part process of modeling a problem as a set of variables and constraints on those variables, and then applying one of several search or dynamic programming techniques to solve the constraint satisfaction problem. While several languages are specifically designed for CP (e.g. CLP(R) [25], CHIP [26], Kaleidoscope [27]) and others have built in support for constraint programming (e.g., Oz), many imperative languages have external constraint solving libraries to enable CP functionality.

Due to its power and broad set of available bindings (interfaces) to common programming languages including .NET, C, C++, Java, Python, and OCaml, we utilize the Z3 satisfiability modulo theories (SMT) solver [28] to identify valid constraint-satisfying encodings for S*FSMs. Of particular interest is Z3's ability to generate and solve models involving arbitrary bit-vectors along with manipulations and calculations across those vectors. While technically not necessary, we utilize several of the equations detailed earlier to limit the overall search space of constraint solving.

The Z3 solver and constraint programming can be used to find a set of possible encodings, if they exist, that satisfy the two Hamming-based constraints for specific bit-length vectors. Explicitly mapping this challenge to a model suitable for CP is straightforward: the variables consist of a $S\ n{-}length$ bit-vectors, each representing an individual $n - bit$ state encoding.[3] Constraints applied to these variables map directly to the constraints in Eqs. 14.9 and 14.10. The overall program structure is seen in the following program listing.

Python + Z3: Constraint Programming Structure for S*FSM Encoding

```
from z3 import *
from math import *
for bits in range( MAX,  MIN, -1):
    //DEFINE VARIABLES
    st_1,...,st_n = BitVecs('st_1 ... st_n',bits)
    s = Solver();
    //ADD GENERAL FSM CONSTRAINT
    s.add(Distinct(st_1,...,st_n)
    //ADD S*FSM CONSTRAINTS
    ...
     //RUN & CHECK SAT SOLVER
    if(s.check() == sat):
        print("Satisfied in %d bits" % (bits))
        m = s.model()
        for d in m.decls():  //OUTPUT MODEL SOLUTION
            print ("%s->%s" % (d.name(),bin(m[d].as_long())))
    else:
        print("Not Satisfied In %d bits" % (bits))
```

14.4.1 Hamming Constraint Functions

There are multiple ways to define the constant Hamming Weight constraint for all nodes. First, however, a procedure to compute the Hamming Weight of a single bit-vector must be defined. The procedure iterates through the bit-vector one bit at a time and sum all of the bit values individually:

[3] Alternatively, this problem could be approached as a constraint minimization on the bit-length n.

```
Sum([( (bitVec & (2**(i)))/(2**(i)) ) for i in range(bits)])
```

There happens to be a quicker way to extract bits from bit-vectors using `Extract(start,end,bit-vector)` which produces a bit-vector of length one. Since it needs to be summed, we extend the length of the bit-vector using `ZeroExt(size,value)` where size is the length of the bit-vector required to store maximum number of 1's we could ever expect in a hamming weight. This is directly computed by our overall bit-length as $\lceil \log_2(bits) \rceil$. The following program listing defines an efficient computation of the Hamming Weight of a Z3 Bit-Vector and utilizes a helper function to return the integer ceiling of the $\log_2$ of a number:

Hamming Weight Function and IntCeilLog2 Helper

```
def IntCeilLog2(val):
    return int(ceil(log2(val)))

def HW(bv):
    return Sum([
        ZeroExt(IntCeilLog2(bv.size()),Extract(i,i,bv))
        for i in range(bv.size())
    ])
```

A similar function is defined to compute the Hamming Distance between two Z3 Bit-Vectors, note the only difference is the bitwise exclusive-or of the two vectors ($bv1$ ^ $bv2$) within the `Extract` function call:

Hamming Weight Function

```
def HD(bv1,bv2):
    return Sum([
        ZeroExt(IntCeilLog2(bv1.size()),Extract(i,i,bv1^bv2))
        for i in range(bv1.size())
    ])
```

With efficiently defined Hamming functions, all that remains is to add the constraints to the targeted FSM. When it comes to the Hamming Weight, a straightforward approach is to perform $s - 1$ comparisons between an arbitrary state in the FSM and every other state. Alternatively, every state could be compared to a specific constant value (cv), though that it would require s total comparisons of the form $s.add(\mathrm{HW}(st_i) == cv)$.

Table 14.3 One potential 4-bit encoding of the restructured Branch Predictor in Fig. 14.5a that has both constant Hamming Weight ($HW = 2$) and Distance ($HD = 2$)

HD	ST2 [0101]	ST [0110]	WT [0011]	WNT [1001]	SNT [1010]	SNT2 [1100]
ST2 [0101]	–	2	2	–	–	–
ST [0110]	2	–	2	–	–	–
WT [0011]	–	2	–	2	–	–
WNT [1001]	–	–	2	–	2	–
SNT [1010]	–	–	–	2	–	2
SNT2 [1100]	–	–	–	2	2	–

Hamming Weight Constraints

```
s.add(HW(st_1) == HW(st_2))
..
s.add(HW(st_1) == HW(st_i))
...
s.add(HW(st_1) == HW(st_n))
```

Unlike the Hamming Weight constraints, the addition of Hamming Distance constraints requires knowledge of the underlying transitions. Consider that unlike the Hamming Weight, the Hamming Distance check cannot simply occur between one state and every other state in the FSM. Just because the Hamming Distance between one state and every other state is a value, it does not guarantee that the Hamming Distance between those other states is that same value.

Take the original SRAM FSM in Fig. 14.2 and for a moment ignore the self-loops. The Hamming Distance between the READ state [01] and the two other states, IDLE[00] and WRITE [11], is $HD = 1$—the distance between IDLE and WRITE, however, is $HD = 2$.

Perhaps another approach would be to add constraints between each state and every other state, creating a fully connected check that ensures that the Hamming Distance is the same between all states.[4] This approach is unfortunately an overly restrictive constraint. Take the restructured Branch Predictor in Fig. 14.5a, if fully connected the shortest valid encoding is a 6-bit one-hot encoding (or its dual 5-of-6 encoding), resulting in a Hamming Distance of 2 between all states. An optimal-length encoding, however, only requires 4-bits as shown in Table 14.3.

[4]Since the Hamming measures are transitive, the total number of checks can be minimized to the number of edges in a fully connected graph: $(n \times (n - 1))/2$.

To link constraints without needing to specify a target Hamming Distance, a check between each outbound transition of every state is utilized. The following listing shows the constraints for the Restructured Branch Predictor from Fig. 14.5a.[5]

All Constraints for Restructured Branch Predictor (Fig. 14.5a)

```
for bits in range(6, IntCeilLog2(6)-1, -1):
    st_st, st_st2, st_wt, st_wnt, st_snt, st_snt2 =\
    BitVecs('st_st st_st2 st_wt st_wnt st_snt st_snt2',bits)
    s = Solver();
    s.add(Distinct(st_st,st_st2,st_wt,st_wnt,st_snt,st_snt2))
    s.add(HW(st_st) == HW(st_st2))
    s.add(HW(st_st) == HW(st_wt))
    s.add(HW(st_st) == HW(st_wt))
    s.add(HW(st_st) == HW(st_wnt))
    s.add(HW(st_st) == HW(st_snt))
    s.add(HW(st_st) == HW(st_snt2))
    s.add(HD(st_st2,st_st)==HD(st_st2,st_wt))
    s.add(HD(st_st,st_st2)==HD(st_st,st_wt))
    s.add(HD(st_wt,st_st)==HD(st_wt,st_wnt))
    s.add(HD(st_snt2,st_snt)==HD(st_snt2,st_wnt))
    s.add(HD(st_snt,st_snt2)==HD(st_snt,st_wnt))
    s.add(HD(st_wnt,st_snt)==HD(st_wnt,st_wt))
    if(s.check() == sat):
        print("Sat In %d Bits" %(bits)),
        m = s.model()
        for d in m.decls():
            print ("%s->%s" % (d.name(),bin(m[d].as_long()))),
    else:
        print("NotSat in %d Bits" %(bits))
```

14.5 Results

In order to quantify the impacts of this approach two sets of real-world FSM benchmarks were selected and utilized. The goal is to quantify, using these real-world FSM benchmarks, the effectiveness and cost of implementing S*FSMs in hardware. These results capture the impacts of the techniques described above on 150 FSM benchmarks that were either generated for boundary condition validation or acquired from the authors of BenGen ($n = 7$) [29] and MCNC ($n = 10$) [30]. The FSMs ranged in size from four to 60 states and contained between four and 216 transitions. It is important to note that while the BENGEN and custom FSMs benchmarks are synthetic, the MCNC benchmark suite represents more typical real-world FSMs.

[5]Readers may access python code and web-enabled notebooks from the following source code repository: https://gitlab.ssc.dev/community/sfms/.

14.5.1 Restructuring Impact on States and Transitions

In order to visualize the effect of the secure FSM strategy on several selected benchmarks, FSMs from the BenGen and MCNC benchmark suites are shown in Fig. 14.6 along with the restructured versions of those benchmarks in Fig. 14.7. The restructured FSMs show newly added states shaded and labeled by using the name of state it duplicates followed by "_n."

14.5.1.1 Quantifying State/Transition Count Changes

While the overall change in S*FSMs state space for all benchmarks ranged between 0 and 100%, this directly corresponds to the two bounding conditions of FSM containing no self-loops and FSM containing only states with self-loops. The BenGen and MCNC FSM benchmarks, containing on average 36.3 and 31.8 states, saw their average size increase to 68.7 and 42.6, respectively. Thus, the observed net change in FSM size was +72% for the BenGen FSMs and +58% for the MCNC FSMs. Unsurprisingly, the changes in transition counts map closely back to the observed changes in state space. The BenGen FSM benchmarks, which had an average over 119 transitions saw a 77% increase in transition counts to an average of 231 transitions per restructured FSM. The MCNC benchmarks, which had an average of 103 transitions per FSM had a 68% increase in their average transition count to an average of 161 transitions per restructured FSM. These upfront increases in state space and transition counts form a lower bound that quantifies the impact of securing FSMs, the impact of encoding these states informs a more accurate view.

14.5.2 Impact of Re-encoding on Bit-Length Requirements

As described in the approach earlier, the encoding of states is critical in preventing information leakage from both hamming models. In order to quantify the impact of the encoding strategy independently of the restructuring, the minimal binary encoding requirements for the original and restructured FSM are presented alongside the optimal secured encoding as determined through the methods presented.

The select BenGen FSM benchmarks required an average of 4.6 bits to encode prior to restructuring. After restructuring, these same benchmarks required 5.3 bits to encode using a minimal binary encoding. This 15% increase on the binary encoding requirement would serve as the minimum possible impact for restructuring an FSM. After application of the encoding method, the average encoding length increased 72% from 4.6 to 7.9 bits.

As with the state space and transitions requirements, the MCNC benchmarks provide a more representative baseline for bit-length encoding impact in real FSMs (See Table 14.4), though the difference between benchmark suites is less apparent due to the logarithmic nature of bit encoding requirements.

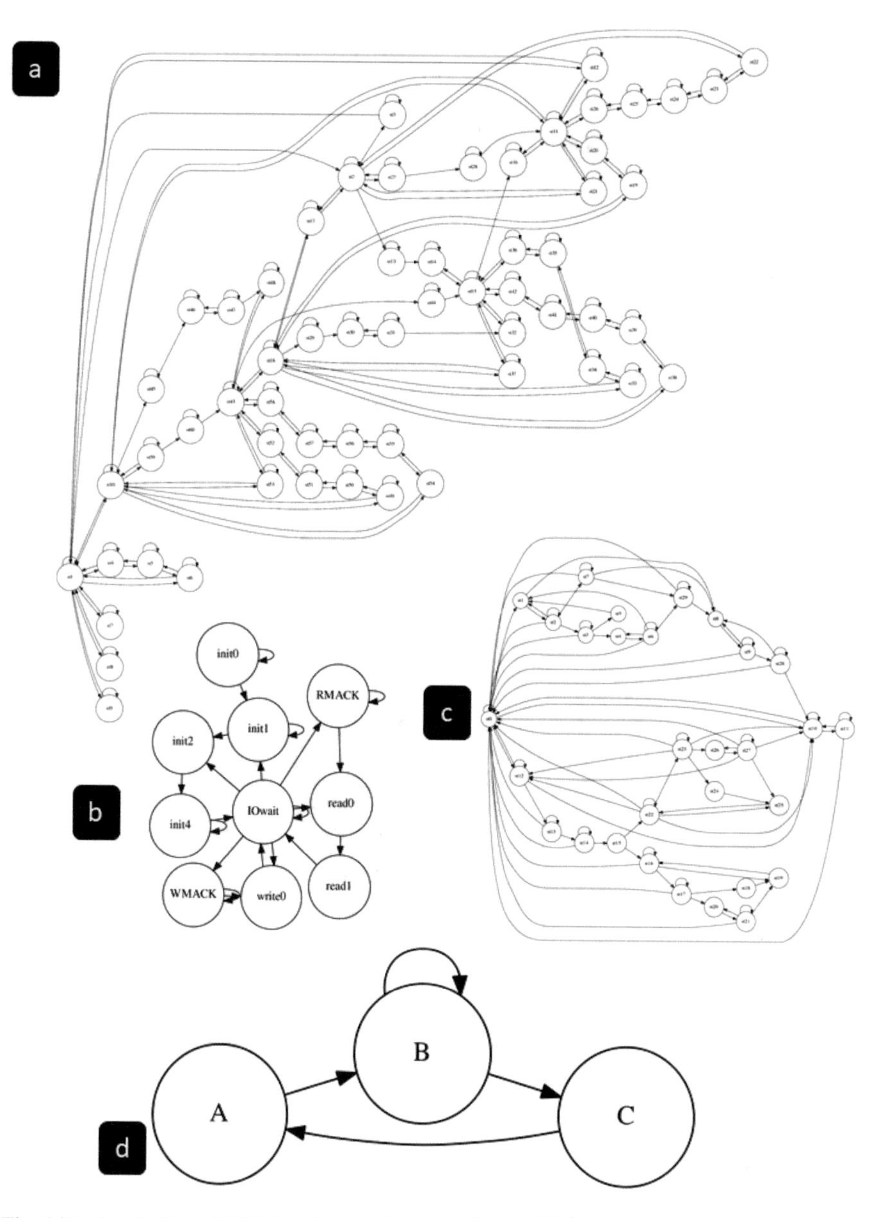

Fig. 14.6 A selection of FSM benchmarks from BenGen (**a**: FSM 108, **c**: FSM 147), and MCNC (**b**: opus, **d**: MotEx) prior to restructuring

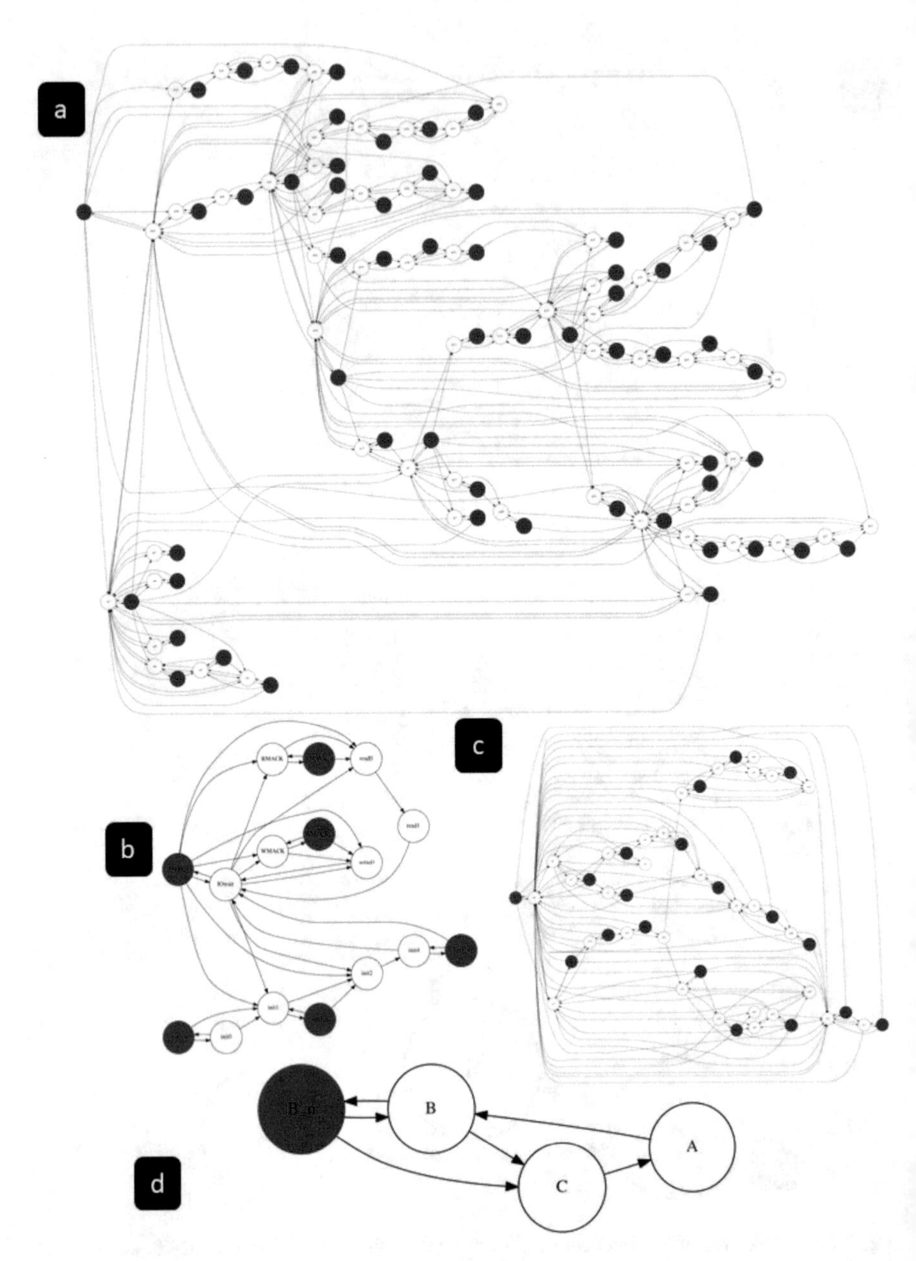

Fig. 14.7 Restructured versions of the select benchmarks shown in Fig. 14.6 with new states highlighted to include the original label and appended "_n"

Table 14.4 Encoding requirements for MCNC FSMs (binary encoding) and their restructured counterparts (binary encoding and secure optimal)

	Bits needed				
FSM	MCNC-BE	S*FSM-BE	% increase	S*FSM-S*O	% increase
Modulo12	4	5	25	7	75
Opus	4	5	25	7	75
Bbara	4	5	25	8	100
Bbsse	4	5	25	8	100
Sse	4	5	25	8	100
Sand	5	6	20	8	60
Planet	6	6	0	8	33
Ex1	5	6	20	9	80
Styr	5	6	20	9	80
Scf	7	7	0	9	29
Avg	4.8	5.6	19	8.1	73

Table 14.5 Correlation between HW/HD Model and state/transition for common encoding strategies for FSMs and their restructured S*FSM counterparts

		FSM		S*FSM		
	# cycles	BE	OH	BE	OH	S*O
State,HW	10	0.71	–	0.45	–	–
	500	0.62	–	0.59	–	–
	5000	0.63	–	0.59	–	–
	50,000	0.63	–	0.60	–	–
Tran,HD	10	0.61	0.61	0.01	–	–
	500	0.17	0.17	0.30	–	–
	5000	0.17	0.17	0.32	–	–
	50,000	0.18	0.18	0.32	–	–

14.5.3 Validation of Mutual Information Removal

An event driven FSM simulator was used to instrument the variability of the HW and HD models during the normal operation of any FSM. The simulator applied randomized transition vectors, ranging from 10–50,000 transitions to the FSM benchmarks. During each transition, the state and transitions were recorded along with the HW and HD values. This experimentally verified if and when relationships between the recorded state and transition events and the two hamming models occurred.

Table 14.5 summarizes the simulation-based maximum theoretical correlation for the standard FSM as well as the restructured S*FSM for three unique encoding schemes including: Minimal Binary Encoding (BE), One-Hot Encoding (OH), and a Constraint Optimized Secure Optimal Encoding (S*O). The correlation values reported are an average over 100 unique simulation runs at given cycle lengths. The unique set of 100 input vectors was utilized for each encoding to eliminate random bias.

Table 14.6 Average layout area (nm^2) in 90 nm SAED

	BenGen ($n = 7$)	MCNC ($n = 5$)
FSM	14.6 K	2.9 K
S*FSM	25.2 K	5.4 K
Overall increase	96%	85%

The results highlight first and foremost that a proper encoding strategy is key: FSMs, regardless of structure, are insecure when encoded using a space-optimal binary encoding scheme. Secondly, the only way to eliminate variability of both the HD and HW models is through the combination of restructuring and use of a secure encoding scheme (S*O or OH).

14.5.4 Implications on Physical Layout

The original and secured-restructured benchmarks were synthesized *without any optimizations* using the Synopsys 90 nm standard cell library. The resulting layout areas are summarized in Table 14.6. Note that while the increase in size varied from a 50 to 160% (for the smallest circuit due to overhead) the average increase, even without optimizations, is 4% smaller than 1-to-1 low-level duplication methods. Extending the analysis to include a larger subset ($n = 25$) of the original benchmarks shows an average total area increase closer to 75%. As with the previous metric, the MCNC benchmarks provide a more accurate baseline for layout area in real FSMs. Overall increases in area, when including small FSM outliers, tend toward a 100% layout increase—without these the average tends closer to a 65% increase.

14.6 Discussion and Summary

A historical need for reducing circuit sizes, power, and costs has enabled hardware implementations that are susceptible to side channel attacks. Given advances in technology and reduced feature sizes, many custom circuits are no longer bound by prior constraints. Unfortunately, these types of devices are also prime targets for side channel attacks. This chapter describes an automated high-level solution that, through constraint programming, can automatically convert any Finite State Machine circuit into a restructured and securely encoded version. While the trade-off costs of securing FSMs vary, the methods presented here require less overhead when applied to real-world FSMs. The methods presented in this chapter are only a starting point and are based on present-day side channel attack models. Modifications to these strategies may be required for new logic styles and device structures that depart from traditional CMOS operating characteristics or if new non-hamming based power models emerge.

References

1. Kohavi, Z., Jha, N.K.: Switching and Finite Automata Theory. Cambridge University Press, Cambridge (2009)
2. Pedroni, V.A.: Finite State Machines in Hardware: Theory and Design (with VHDL and SystemVerilog). MIT Press, Cambridge (2013)
3. Wagner, F., Schmuki, R., Wagner, T., Wolstenholme, P.: Modeling Software with Finite State Machines: A Practical Approach. CRC Press, Boca Raton (2006)
4. Hennessy, J., Patterson, D., Arpaci-Dusseau, A.: Computer Architecture: A Quantitative Approach. No. v. 1 in The Morgan Kaufmann Series in Computer Architecture and Design. Morgan Kaufmann, Burlington (2007)
5. Kocher, P., Jaffe, J., Jun, B.: Differential power analysis. In: Wiener, M. (ed.) Advances in Cryptology – CRYPTO' 99. Lecture Notes in Computer Science, vol. 1666, pp. 789–789. Springer, Berlin (1999)
6. Verbauwhede, I.: Secure Integrated Circuits and Systems. Integrated Circuits and Systems. Springer, London (2010)
7. Mangard, S., Oswald, E., Popp, T.: Power Analysis Attacks: Revealing the Secrets of Smart Cards (Advances in Information Security). Springer, New York (2007)
8. Tiri, K., Akmal, M., Verbauwhede, I.: A dynamic and differential CMOS logic with signal independent power consumption to withstand differential power analysis on smart cards. In: Proceedings of the 28th European Solid-State Circuits Conference, 2002. ESSCIRC 2002, pp. 403–406 (2002)
9. Tiri, K., Verbauwhede, I.: A logic level design methodology for a secure DPA resistant ASIC or FPGA implementation. In: Proceedings Design, Automation and Test in Europe Conference and Exhibition, pp. 246–251 (2004)
10. Sundaresan, V., Rammohan, S., Vemuri, R.: Power invariant secure IC design methodology using reduced complementary dynamic and differential logic. In: IFIP International Conference on Very Large Scale Integration, 2007. VLSI - SoC 2007, pp. 1–6 (2007)
11. Ramakrishnan, L.N., Chakkaravarthy, M., Manchanda, A.S., Borowczak, M., Vemuri, R.: SDMLp: On the use of complementary pass transistor logic for design of DPA resistant circuits. In: 2012 IEEE International Symposium on Hardware-Oriented Security and Trust (HOST) (2012)
12. Schaumont, P., Tiri, K.: Masking and dual-rail logic don't add up. In: Paillier, P., Verbauwhede, I. (eds.) Cryptographic Hardware and Embedded Systems—CHES 2007. Lecture Notes in Computer Science, vol. 4727, pp. 95–106. Springer, Berlin (2007)
13. Kulikowski, K., Smirnov, A., Taubin, A.: Automated design of cryptographic devices resistant to multiple side-channel attacks. In: In Workshop on Cryptographic Hardware and Embedded Systems, pp. 339–413 (2006)
14. Golić, J., Tymen, C.: Multiplicative masking and power analysis of AES. In: Kaliski, B., Koç, K., Paar, C. (eds.) Cryptographic Hardware and Embedded Systems—CHES 2002. Lecture Notes in Computer Science, vol. 2523, pp. 31–47. Springer, Berlin (2003). http://doi.org/10.1007/3-540-36400-5_16
15. Oswald, E., Mangard, S., Pramstaller, N., Rijmen, V.: A side-channel analysis resistant description of the AES S-box. In: Fast Software Encryption, pp. 413–423 (2005). https://doi.org/10.1007/11502760_28
16. Oswald, E., Mangard, S., Pramstaller, N., Rijmen, V.: A Side-Channel Analysis Resistant Description of the AES S-Box. Springer, Berlin (2005)
17. Borowczak, M., Vemuri, R.: S*FSM: An paradigm shift for attack resistant FSM designs and encodings. In: Redefining and Integrating Security Engineering, 2012. RISE 2012. ASE International Conference on Cyber Security, pp. 651–655 (2012)
18. Borowczak, M.: Side Channel Attack Resistance: Migrating Towards High Level Methods. Ph.D. Thesis, University of Cincinnati (2013)

19. Borowczak, M., Vemuri, R.: Mitigating information leakage during critical communication using S*FSM. IET Comput. Digital Techni. **13**(4), 292–301 (2019)
20. Agrawal, R., Borowczak, M., Vemuri, R.: A state encoding methodology for side-channel security vs. power trade-off exploration. In: 2019 32nd International Conference on VLSI Design and 2019 18th International Conference on Embedded Systems (VLSID), pp. 70–75. IEEE, Piscataway (2019)
21. Grune, D., Jacobs, C.: Parsing Techniques: A Practical Guide. Monographs in Computer Science. Springer, Berlin (2008)
22. Brouwer, A.E., Shearer, J.B., Sloane, N.J.A., Smith, W.D.: A new table of constant weight codes. IEEE Trans. Inf. Theory **36**(6), 1334–1380 (1990)
23. Knuth, D.: Efficient balanced codes. IEEE Transact. Inf. Theory **32**(1), 51–53 (1986)
24. Garey, M.R., Johnson, D.S., Stockmeyer, L.: Some simplified NP-complete problems. In: Proceedings of the Sixth Annual ACM Symposium on Theory of Computing, STOC '74, p. 47–63. Association for Computing Machinery, New York (1974). https://doi.org/10.1145/800119.803884
25. Jaffar, J., Michaylov, S., Stuckey, P.J., Yap, R.H.C.: The CLP(R) language and system. ACM Trans. Program. Lang. Syst. **14**(3), 339–395 (1992). https://doi.org/10.1145/129393.129398
26. Dincbas, M., Hentenryck, P.V., Simonis, H., Aggoun, A., Graf, T., Berthier, F.: The constraint logic programming language CHIP. In: Proceedings of the International Conference on Fifth Generation Computer Systems, FGCS 1988, Tokyo, November 28-December 2, 1988, pp. 693–702. OHMSHA Ltd. Tokyo and Springer, Berlin (1988)
27. Lopez, G., Freeman-Benson, B., Borning, A.: Kaleidoscope: A constraint imperative programming language. In: Mayoh, B., Tyugu, E., Penjam, J. (eds.) Constraint Programming, pp. 313–329. Springer, Berlin (1994)
28. De Moura, L., Bjørner, N.: Z3: An efficient SMT solver. In: Tools and Algorithms for the Construction and Analysis of Systems, pp. 337–340 (2008)
29. Jozwiak, L., Gawlowski, D., Slusarczyk, A.: An effective solution of benchmarking problem FSM benchmark generator and its application to analysis of state assignment methods. In: Euromicro Symposium on Digital System Design, 2004. DSD 2004, pp. 160–167 (2004)
30. Yang, S.: Logic Synthesis and Optimization Benchmarks User Guide Version 3.0 (1991)

Chapter 15
Generation and Verification of Timing Attack Resilient Schedules During the High-Level Synthesis of Integrated Circuits

Steffen Peter and Tony Givargis

15.1 Introduction

Side-channel attacks are malicious attempts at obtaining secret data from a computing system. A particular category of side-channel attacks, called timing attacks, work by exploiting the variations in execution time of processes under different load and data input scenarios [1]. Despite considerable research and development in the fight against timing attacks, existing system continues to be vulnerable. Examples of state-of-the-art counter measures include the implementation of cryptographic protocols such as RSA [2] and Elliptic Curve Cryptography (ECC) [3]. Timing attacks predominantly aim to steal private cryptographic encryption keys, but they can also be used to extract intellectual property, algorithm innovations, and implementation details of deployed systems. While timing attacks apply to both software and hardware, in this chapter we focus on vulnerabilities of ICs. Our goal in the design of the approach presented in this chapter is to provide a degree of guarantee that variabilities in execution time as a function of input data do not result in exploitable timing variances, that is, that the number of clock cycles, needed to process the data, does not leak sensitive and private information.

Several existing countermeasures against timing attacks work by tackling low-level implementation up to higher-level algorithm design. Examples of such low-level approaches include path and timing obfuscation methods [4], the addition of temporal monitors [5] or manual balancing of the HDL code. However, the application of such approaches is error-prone and complicated since it contains many manual design steps [6]. On a higher-level, timing invariance has been addressed at the algorithm design, so that operations in different execution paths are

S. Peter · T. Givargis (✉)
Center for Embedded Systems and Cyber-Physical Systems (CECS), University of California, Irvine, CA, USA
e-mail: st.peter@uci.edu; givargis@uci.edu

S. Katkoori, S. A. Islam (eds.), *Behavioral Synthesis for Hardware Security*,
https://doi.org/10.1007/978-3-030-78841-4_15

balanced. Examples of such approach include the Lopez-Dahab ECC algorithm [7], or the aimed addition of dummy operations [8]. The problem is that implementations of balanced algorithms still might leak timing information on lower levels because synthesis tools attempt to optimize register transfers and usage patterns of functional units.

In this chapter we look at methods to guard against timing attacks and address effective methodologies that may be used in state-of-the-art high-level synthesis (HLS) tool chains. HLS transform an abstract high-level description (e.g., in C or System C) of a system to low-level implementable netlist of circuit components intended for hardware design. In the process, HLS solves the problem of allocating, scheduling, and binding of processing elements needed to achieve the high-level behavioral description [9]. Applications of HLS include derived industrial/academic tools and systems [10, 11]. These HLS tools improve the design practice of ICs with a heavy and specific emphasis on performance and power, rather than system security. Timing is only supported to determine the upper bound or the worst-case execution time. In this work we want to answer two questions:

1. Can timing invariance be integrated into an HLS design flow to prevent timing side channels?
2. Can we verify that the generated schedule is timing invariant with regard to the information that should be protected?

In response to these types of issues, we propose the design flow shown in Fig. 15.1, as an extension of the work in [12]. Briefly stated, the primary input of the HLS flow is the system specification captured as plain C-code, while the output is the implementation model in a hardware description language (HDL) containing security features. In this methodology, a list of protected variables (LPV) and appropriate Timing Annotation (TA) for the input C-code is created. The TA provides constraints for the scheduling phase of HLS in order to balance the execution paths of the overall circuit. The result of the modified schedule is an operation schedule in which all data and control dependencies originating from the LPV are time invariant. The timing invariance is verified during a final validation step, based on the generated schedule and the LPV. Hence, the contributions of our work include:

1. A methodology to add time-balancing constraints to the schedule generator of an HLS tool. The scheduling constraints are based on timing annotations in the system specification.
2. An algorithm to identify and annotate paths with security-related data processing.
3. Automated verification of the timing invariance based on the generated HLS scheduling information. The verification applies a Satisfiability Modulo Theory (SMT) solver and Bounded Model Checking (BMC).

We have validated our approach using an implementation based on LegUp (an open source HLS tool) [10], using the LLVM compiler back-end, and the SDC scheduler [13]. We have tested our proposed solution with a range of benchmarks, including applications of cryptographic standards ECC and RSA. Our approach has the benefit

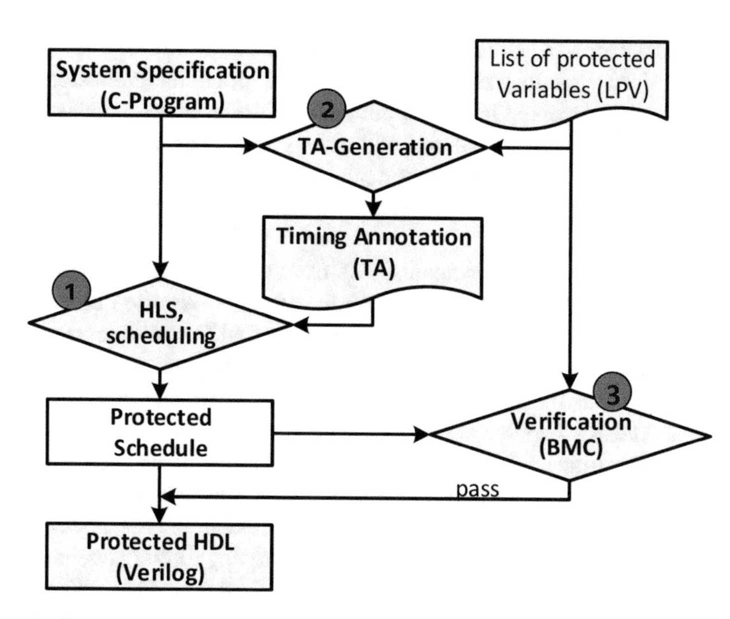

Fig. 15.1 Proposed design flow for a secure HLS that balances and verifies the execution time of sensitive paths. The chapter addresses (1) the addition of scheduling constraints, (2) the TA-generation, and (3) Verification of the generated schedules

of being fully automated while having marginal and negligible overhead in terms of chip area, critical paths, or worst-case execution time.

The organization of the rest of this chapter is as follows: After a review of related work in Sects. 15.2 and 15.3 provides an overview on preliminaries and the problem formulation. In Sect. 15.4 we present our approach for timing annotation and their support in the HLS scheduling process. Section 15.5 discusses the verification, and Sect. 15.6 reports our experimental results.

15.2 Related Work

In this section we provide a brief overview on related work on side-channel attacks and countermeasures, followed by an overview on high-level synthesis of systems and timing annotations. Timing attacks are only one of many possible attacks that exploit side channels of ICs. Other attacks exploit electro-magnetic side channels, power consumption, or faults [14]. Many side-channel resistant design flows [6] consider timing- and other side-channel attacks. However, these design flows address lower implementation levels and their application still requires significant manual efforts. An automated toolchain to prevent power attacks and fault attacks is presented in [15]. The approach converts a synchronous to an asynchronous design with dynamic balanced gates.

Other mechanisms to address timing attacks include the integration temporal monitors [5], timing randomization methods [4], or clock randomization in GALS design [3]. These approaches have in common that they require significant modifications of the design, are implementation technology dependent and require additional logic. Randomization in general might negatively affect real-time applications and production tests. The approach presented in this chapter provides deterministic timing.

Few works tackled protection against side-channel attacks already on the system level. The idea is to balance the operations in different execution paths, either by diligent algorithm design [7] or by the placement of additional dummy operations [8]. While our approach does not rely on a time-balanced system model, a balanced input might improve the system utilization, since fewer idle cycles need to be inserted by the HLS scheduler.

Balanced algorithms still are vulnerable to timing attacks, due to possible compiler, pipeline, and loop-optimizations in the synthesis process. The extent of such effects for the LegUp [10] HLS tool has been studied in [16], however, mostly aiming for performance improvement instead of obtaining timing invariance. LegUp utilizes SDC (system of difference constraints) scheduling [13] which supports a range timing, latency, and cycle time constraints, which are the basis for the implementation in our work. While a range of works exist to use HLS to improve the system's performance or its robustness [17], so far, no approach is known that prevents leakage of information by timing.

In this chapter we discuss C as high-level system description language and annotation in an external file. Several other system description languages have been proposed with optional timing-related annotations. For SpecC [18] and SystemC exist concepts like waiting states [19] that facilitate the specification of the timing of states. However, the timing annotations are not applied to steer the synthesis process. A designer must address the timing properties at lower abstraction levels to meet the constraints.

15.3 Preliminaries

In this section we provide the preliminaries for the HLS scheduling algorithm and its variables, as well as motivational examples we use throughout this chapter.

15.3.1 High-Level Synthesis and Scheduling

As shown in Fig. 15.1, the HLS process synthesizes an implementation model in a hardware description language (HDL) from a system description in a high-level programming language. Specifically in this chapter, we generate a Verilog program from a C program. To synthesize a system, the HLS needs to allocate the functional

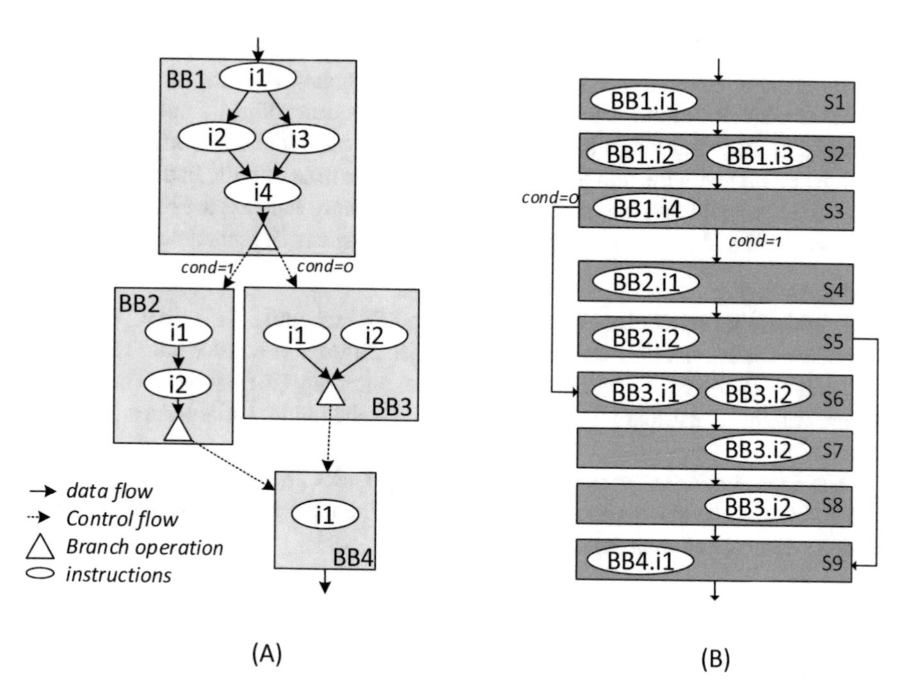

Fig. 15.2 An example for (**a**) a CDFG, and (**b**) a possible FSMD. In the example it is assumed that instruction i2 of BB3 requires 3 clock cycles, while all other instruction complete within 1 cycle

units (FUs), bind operations to FUs, and schedule when the operations on the FUs are executed.

The schedule is generated from the control and data flow graph (CDFG). The CDFG contains basic blocks (BB), operations (Op), data dependencies ($Op \times Op$) and control flows ($BB \times BB$). Each BB contains sequences of operations with data dependencies without loops and branches. Loops and branches are indicated as possible control paths between BBs. CDFGs do not contain scheduling information, neither for the operations in a BB nor for the sequence in which BBs are executed. As an example, Fig. 15.2a shows a CDFG with four BBs and one condition at the end of BB1.

Operations to states are assigned by the scheduler. The state diagram can be expressed as a finite state machine with data (FSMD), as shown in Fig. 15.2b. Each state can execute more than one operation (see S2 and S6), whenever data dependencies are resolved and no resource conflicts are present. Operations might require more than one state if an operation cannot be executed in one cycle. In Fig. 15.2, operation BB2.i2 requires three cycles (from S6 to S8).

While the FSMD assigns operations to states, the FSMD does not present a static global schedule, because typically several execution paths exist. In the example of Fig. 15.2, the program requires 6 cycles if $cond = 1$, and 7 cycles if $cond = 0$. That is an example for a timing variance that might leak information about $cond$, simply by monitoring the time needed to execute the program. The aim of the approach discussed in this chapter is to balance possible execution paths, if, and only if $cond$ is based on data that needs to be protected. In that case, idle states will be inserted into the FSMD with the aim to balance the execution time of BBs. In the example of Fig. 15.2 we could add one state between S4 and S5 to balance the execution time for $cond = 0$ and $cond = 1$. We will address this issue in the scheduler of the HLS.

SDC Scheduling
The underlying scheduler of our work is the SDC scheduler [13]. SDC expresses the scheduling constraints as a set of integer difference constraints. The set of constraints is solved by an integer linear programming (ILP) solver. The general format of the constraints is $\sum_i sv_i \leq b$, for scheduling variables sv_i, and the constraint b.

The most important variables we need to consider are the start ($sv_{beg}(i)$) and the end ($sv_{end}(i)$) states of operations $i \in Op$. The difference constraints allow to express control and data dependencies, such as

$$sv_{beg}(\text{BB1}.i2) - sv_{end}(\text{BB1}.i1) \geq 1,$$

latencies, such as

$$sv_{end}(\text{BB3}.i2) - sv_{beg}(\text{BB3}.i2) \geq 3,$$

as well as resource and timing constraints. The typical optimization criterion of the SDC scheduler is to minimize the total execution time, i.e., minimize the sum of the timings of the scheduled operations:

$$\min \sum_{v \in V_op} sv_{beg}(v).$$

The result is the FSMD in which the scheduling constraints are satisfied and all operations are assigned to states. In Sect. 15.4, we describe how to add new scheduling constraints and variables that help to balance the execution time of different paths in the FSMD.

15.3.2 Examples

This subsection introduces examples, used throughout this chapter, and additionally motivate the need for timing invariance.

Max
The *Max* example computes the maximum of two numbers if the enable signal is set.

This pseudo-code for the example is:

```
if (en)
  if (a<b)
    a=b;
  return a
```

Assumed variable b is a secret and a can be freely set, an attacker can discover b by systematically trying different a.

Algorithm 15.1 Inner loop of the Montgomery kP multiplication [7]

Input: $k = (k_{t-1}, ..., k_1, k_0)_2$ with $k_{t-1} = 1$, $P = (x, y) \in E(F_{2^m})$

Procedure:

1: **for** $i = t - 2$ downto 0 **do**
2: **if** $k_i = 1$ **then**
3: $T \leftarrow Z_1$; $Z_1 \leftarrow (X_1Z_2 + X_2Z_1)^2$; $X_1 \leftarrow xZ_1 + X_1X_2TZ_2$
4: $T \leftarrow X_2$; $X_2 \leftarrow X_2^4 + bZ_2^4$; $Z_2 \leftarrow T^2Z_2^2$
5: **else**
6: $T \leftarrow Z_2$; $Z_2 \leftarrow (X_2Z_1 + X_1Z_2)^2$; $X_2 \leftarrow xZ_2 + X_1X_2TZ_1$
7: $T \leftarrow X_1$; $X_1 \leftarrow X_1^4 + bZ_1^4$; $Z_1 \leftarrow T^2Z_1^2$
8: **end if**
9: **end for**

ECC

Algorithm 15.1 shows the inner loop of the Lopez-Dahab Elliptic Curve Cryptography (ECC) point multiplication [7]. The algorithm is already balanced and seems resistant against timing attacks. For each bit of the key k, one of two execution paths are followed. Both paths contain 6 multiplications, 4 additions, and 7 square operations. However, the symmetric ECC design that implements the algorithm discussed in [3] is still vulnerable against a timing attack. In that implementation $k_i = 1$ requires 54 clock cycles, and $k_i = 0$ requires 56 clock cycles. The reason can be found in the last cycle of the loop.

There the two operations are:
for $k_i = 1: \ Z_2 \leftarrow T^2Z_2^2$, and for $k_i = 0: \ Z_1 \leftarrow T^2Z_1^2$.
In the first case, the synthesis tool uses the accumulator register to store the Z_2. In the alternative case, the value is copied from the accumulator to Z_1. In order to reduce the number of needed registers and to improve the overall performance, a timing side-channel was inserted during the synthesis.

RSA

RSA is a cryptographic public key protocol that relies on modular exponentiation (ModExp) in a large finite field. RSA is susceptible to side-channel attacks [20] due to the structure of the inner loop of the ModExp, which performs for each bit of the key:

```
    if (bit is set)
      result = result * base mod modulus
    else
      result = result;
```

This difference is easily exploitable since a modular multiplication of typical RSA values requires several hundreds to thousands clock cycles.

We implemented the examples and applied the HLS design flow to generate the HDL model. We discuss the results further in the evaluation in Sect. 15.6.

15.4 Balancing Execution Paths

In this section we discuss methods to balance the execution paths of the generated FSMD. We first discuss new scheduling constraints that facilitate path balancing. Then we show how timing annotations in the system description can be supported in the HLS process. In the second part of this section we discuss how execution paths can be balanced automatically.

To demonstrate the practical feasibility of our approach, we extend the toolchain of the LegUp open source HLS tool and its SDC scheduler. As highlighted in Fig. 15.3, which outlines the LegUp toolchain, we add a Path *Balancer* module parallel to the SDC module. Like the SDC module, the *Balancer* uses the intermediate code representation of the C program, generated by LLVM, and adds scheduling constraints that are solved by an ILP solver to generate a preferable schedule. In this section we discuss the *Balancer* and the required annotations in the C-code and the external configuration file `config.tcl` that helps to parametrize the synthesis process.

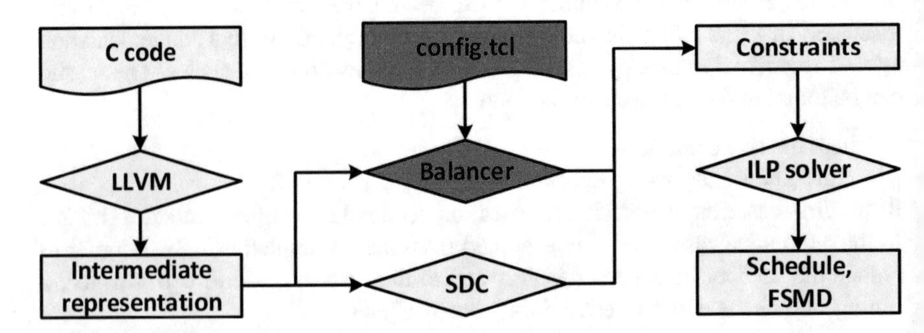

Fig. 15.3 The LegUp HLS scheduler toolchain: The *Intermediate Representation* of the C-code is translated to ILP constraints in the *SDC* scheduler. We add the *Balancer* module and annotations into the LegUp *config.tcl* to add constraints to balance the timing of the generated schedule

15.4.1 Scheduling Variables for BBs

As introduced in Sect. 15.3.1, we apply the SDC scheduler that assigns operations to states. The SDC scheduler performs the scheduling for each BB separately, so that timings for operations within a BB are fixed. However, BBs might be executed earlier or later depending on the run-time branch decisions. To balance executions paths in this environment, it is our aim to instruct the scheduler to add idle states into shorter BBs, so that all possible execution paths that follow a security-related branch require an equal amount of clock cycles. To balance the execution time of BBs, variables for the relative start and end time of BBs are required. Therefore, we add new scheduling variables:

- $sv_{beg}(BB_i)$ is the relative start time of basic block i, which is bound to the earliest start state $sv_{beg}(op)$ of any operation $op \in BB_i$
- $sv_{end}(BB_i)$ is the relative end time of basic block i, which is bound to the latest end state $sv_{end}(op)$ of any operation $op \in BB_i$
- $latency(BB_i)$ is the total execution time of basic block i, which is difference between end and start time of BB_i: $latency(BB_i) = sv_{end}(BB_i) - sv_{beg}(BB_i)$.

The three new variables facilitate a range of options to constrain the execution time. For instance, by constraining $latency(BB_i)$ to a certain value, we enforce the relative difference of start and end state of the BB, which in turn constrains the assigned state for the first and the last operation. Therefore, the scheduler has to insert idle states, if $latency$ exceeds the number of actual operations.

In the following subsections we discuss how the described basic concept can be applied in practice. We start with a simple manual annotation in Sect. 15.4.2 and present an extension that automatically generates the annotations in Sect. 15.4.3.

15.4.2 Static BB Latency Constraints

Assumed the designer knows the identifier of the BBs and the required states of each BB, the latency of the BBs could simply be constrained by defining a fixed $latency$ variable, as discussed in the previous subsection. Practically that can be achieved by adding an annotation in the `config.tcl` configuration file. `config.tcl` is evaluated in the balance, so that the parameter

```
set_parameter Cycles_BB_i n
```

defines $latency(BB_i) = n$, i.e., the latency of BB_i is constrained to n. For the example in Fig. 15.2, we could define

```
set_parameter Cycles_BB_2 3,
```

to constrain the latency of BB2–3 cycles. The resulting schedule for the instructions of BB2 is shown in Fig. 15.4. In Fig. 15.4 a shaded cell indicates that the operation (row) is scheduled for the specific state (column). We see that instr1 is assigned to state S4 and instr2 is assigned to state S6. The final branch operation in S6 jumps

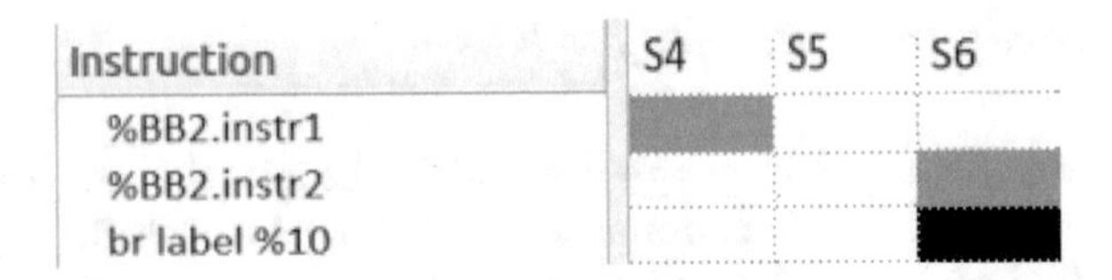

Fig. 15.4 Example schedule for BB2 of Fig. 15.2. Instruction1 is scheduled in state S4, while instruction2 and the exit branch are scheduled for S6. State S5 is idle

unconditionally to S10, which is the start of BB4. As a result, the execution times for BB2 and BB3 are both 3 cycles, so that regardless of the run-time conditions, the number of cycles, following BB1 is invariant.

One advantage of the direct BB annotations is that it does not require a modification of the C-code. However, the identification of BB-id in practice is non-trivial as it requires insight into enumerations of the intermediate LLVM representation. In addition, the BB-ids might change with any modification of the input C-code, which renders the applicability of the approach questionable.

BB Annotations To improve the usability of the annotations we apply labels in C to locate the BBs. Labels do not change functionality or timing but are maintained in the LLVM data structure. The inner condition of the *Max* example with labels can be expressed as:

```
if (a<b) {
  label1:   a = b;
} else {
  label2: ;
}
```

Using the labels, we can address the BBs in the `config.tcl` file directly, by stating

`set_parameter Cycles_Lab_label1 2`, and
`set_parameter Cycles_Lab_label2 2`,

which constrains the latency of the BBs indicated by `label1` and `label2` to 2 each.

15.4.3 Automatic Path Balancing

So far, we expected the designer to know the exact number of cycles of the BBs. In this subsection we discuss approaches that do not need that detailed low-level knowledge. Instead, we balance execution paths that follow a branch automatically. Therefore, first, we consider paths consisting a single BB only. In the second part of this section we describe an algorithm to identify and annotate paths with nested sub-branches.

15.4.3.1 Balancing Single BB-Paths

Execution time variability is generally caused by conditional branches such as

```
if cond then BBx else BBy,
```

so that either BB_x or BB_y are executed, depending on condition `cond`. Considering that binary structure, we can constrain the latency of the two BBs with the scheduling constraint:

$$latency(BB_x) - latency(BB_y) = 0.$$

This constraint enforces the faster of the two BBs to extend its latency and add idle cycles, because the BB with the higher latency cannot reduce its latency without violating other scheduling constraints. In case of the example in Fig. 15.2, we would enforce that the latency of the two BBs following the conditional statement in BB1 are equivalent, that is, $latency(BB_2) - latency(BB_3) = 0$. Since BB3 requires 3 cycles and BB2 needs 2 cycles, this constraint would force BB2 to add one idle state.

Practically, the two BBs of the two paths do not need to be addressed directly. Instead, we can label the conditional statement, since we know that the following two BBs have to be balanced. The resulting C-code for the *Max* example would be as:

```
label1:
    if (a<b) {
      a = b;
    } else { }
```

The annotation in the `config.tcl` in this case is

```
set_parameter Balance_label1,
```

stating that the branches following `label1` need to be balanced. The result in this case is that both branches now require one cycle.

15.4.3.2 Balancing Paths with Sub-branches

So far we considered only one BB in each path following a branch. However, systems can be more complex and might contain nested branches in the paths that need to be secured. The example shown in Fig. 15.5 illustrates a system in which BB1 contains a conditional statement that branches based on a secret information. The program further branches in BB2 and reconverges all secured paths in BB7.

A manual annotation in this case is not trivial, since we need to ensure that, first, the latencies of BB3 and BB4 are equal, and second, the latency of BB6 matches the latency of the path BB2-BB3-BB5. Clearly such manual annotation is error-prone and does not scale for larger systems.

Instead it is the goal to fix the timing for all paths between BB1 and BB7, while timing of BB0, BB8, BB9 is irrelevant for the security of the system. That means that all BBs between BB1 and BB7 need to be fixed so that there is no slack between adjacent BBs that would allow a BB at run-time to be executed earlier or later. Therefore, we need to

1. identify the earliest point of reconvergence (BB7),
2. identify all adjacent BBs between the secure branch (BB1) and the point of reconvergence (BB7), and
3. tie the schedule for all pairs of adjacent BBs: i.e., $sv_{beg}(BB_{drain}) - sv_{end}(BB_{source}) = 1$.

We realize the three steps in Algorithm 15.2. The input of the algorithm is a directed graph $G(B, E)$ with nodes B and edges E. B is the set of basic blocks of the system, while E is the set of direct control paths between BBs. We obtain this graph structure from the LLVM intermediate representation. In the algorithm we exploit that the identifier of BBs in the LLVM data structure is ordered and for every edge $E(BB_x, BB_y)$: from BB_x to BB_y, we can assume $id(BB_x) < id(BB_y)$. The input of the algorithm is the BB that contains the security-related branch. From that node we maintain a list of reachable nodes (R), while in every iteration we replace (line 7) the node that has the smallest id in R (b_i, line 3) with the set of directly reachable nodes (B_n) from b_i. If R contains only one node (line 9), we know that this node is the point of reconvergence, since all paths converged into that one node. Since the algorithm passes all nodes between the start and the point of reconvergence, in line 6 we can add the required scheduling constraint that ties adjacent BBs.

Considered that all BBs in the example of Fig. 15.5 require 2 cycles, but BB3 requires 3 cycles, then a possible result of the balanced schedule is shown in Fig. 15.5. Idle states are inserted in BB4 and BB6 to balance the paths between BB1 and BB7.

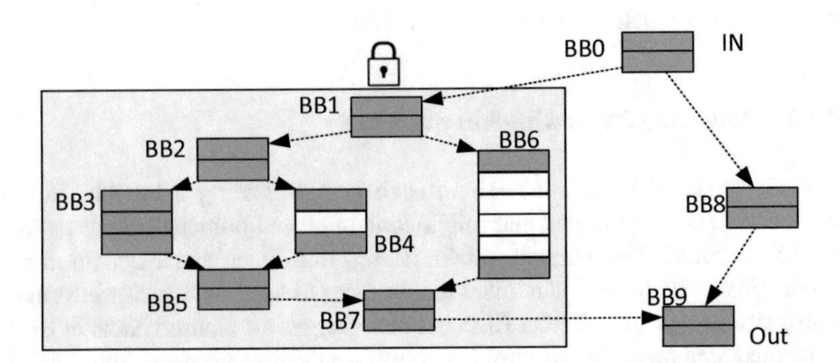

Fig. 15.5 Illustration of a CDFG with branches in the secure region. The branch in BB1 is based on sensitive information, so that the timing for all paths between BB1 and BB7 need to be fixed. The figure also illustrates one identified secured schedule. Each shaded cell in a BB is a state with scheduled operations, white cells are states with no scheduled operations. All paths between BB1 and BB7 have an invariant number of states

Algorithm 15.2 Path reconvergence detection and schedule rule insertion

Input: directed Graph $G(B, E)$, start node $b_S \in B$
Output: returns Node of minimum reconvergence, and added scheduling constraints
Procedure:

1: set $R = \{b_S\}$ *//ordered set of reachable nodes*
2: **repeat**
3: $b_i = min(R)$ *//node with minimal id in R*
4: B_n =set of reachable nodes from b_n
5: **for** $b_r \in B_n$ **do**
6: add constraint $sv_{beg}(b_r) - sv_{end}(b_i) = 1$
7: **end for**
8: $R = R + B_n - \{b_i\}$
9: **if** $B_n = \varnothing || min(R) \leq b_n$ **then**
10: **return** $\varnothing$
11: **end if**
12: **until** $|R| \leq 1$
return R

After all, each BB is executed independently, i.e., there is no global schedule. We achieve the timing invariance by extending the schedules of each separate BB, so that all feasible execution paths result in the same number states.

15.5 Verification

In this section we present a bounded model checking (BMC) [21] approach to verify the resilience or vulnerability of a synthesized design against cycle-based timing attacks. We apply the BMC on the generated scheduling model (FSMD), and aim to prove that changing sensitive data does not change the execution time for any arbitrary input signal—or to find a counterexample to show the vulnerability. BMC techniques have been successfully applied to prove that a certain condition is possible or impossible [21], and to determine the true WCET [22]. We express and solve the BMC using the satisfiability modulo theory (SMT) solver Z3 [23]. SMT solvers are capable tools to solve advanced satisfiability problems, and have proven valuable to identify preferable configurations in the domain of embedded systems [22, 24].

The basic idea of our verification approach is to encode two copies of the system FSMD to an SMT, and to query the SMT solver to find a pair of variables for which the execution time varies. If such a pair cannot be found, the system is not vulnerable against cycle-based timing attacks. First, we describe the encoding of the FSMD. In the second part of the section we describe the verification.

15.5.1 Encoding the FSMD

The basic encoding idea is similar to the encoding shown in [22]. In this approach, the FSMD with its control paths, data dependencies, and operations are mapped to a set of assertions that can be processed by SMT solvers. Registers and variables are mapped to SMT variables, ensuring static single assignment. That means that each design variable is assigned to a new instance in each state. For example, the operation $x = x + 1$ in an assumed BB2 would be encoded to the set of assertions:

```
BB2.x_1=BB2.x_in
BB2.x_2=BB2.x_1+1
BB2.x_out=BB2.x_2
```

The set of assertions is only constrained if the BB is actually active, i.e., the BB is on the execution path. Hence, we can express the assertions in the SMT as

BB2 → BB2.x_1=BB2.x_in∧BB2.x_2=BB2.x_1+1∧BB2.x_out =BB2.x2,

which means that the operation $x = x + 1$ is only asserted if BB2 is enabled. Since a BB is enabled if and only if an active control path from one of the preceding BBs exists, the BB variable is asserted in one of the preceding BBs. Assumed BB2 is active if condition e of BB1 is true, then we can write

BB1 → (BB1.e = BB2),

which means that if BB1 is active, BB2 is active if the branch condition e is true. Thus, branches activate or deactivate basic blocks and their assertions based on variables that are assigned by the solver.

In addition to state variables and BBs we encode the execution time as counter variables along the active path, starting with $t = 0$ at the start state.

15.5.1.1 An Illustrative Example

Figure 15.6 illustrates the C-code, the control path, and the operations for the *Max* example. Depending on the assigned values, exactly one path from In to Out is valid. Assumed that each block requires one cycle to process, different paths results in different timings. For instance, if enable is 0, the program requires 1 cycle, if $a \geq b$ 2 cycles, and if $a < b$ 3 cycles.

The example is encoded to the set of assertions, shown in Fig. 15.7. In Fig. 15.7, line 1 expresses that the start block is enabled and time is 0 at start. Line 2, implies (→) that if start is enabled, so BB0 is enabled and all global variables are copied to BB0. Line 3 states that if BB0 is enabled, the branch condition e is computed, time is incremented by 1, and based on e either BB1 will be activated or the time is forwarded to Out/End. Finally, the variable `end.t` is assigned to the required time that is needed to process the assigned variables.

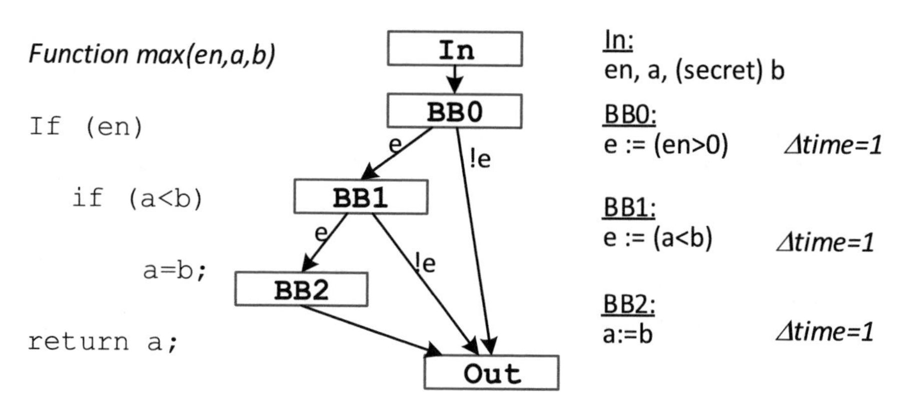

Fig. 15.6 High-level program and generated interpretation with basic blocks, operations, and timing for the *Max* example

1: Start ∧ start.t=0;
2: Start →
BB0 ∧ BB0.a=start.a ∧ BB0.b=start.b ∧ BB0.en=start.en ∧ BB0.t_0=start.t

3: BB0 → BB0.e=(BB0.en>0) ∧ BB0.t_1=BB0.t_0+1 ∧
((BB0.e ∧ BB1 ∧ BB1.a_0=bb0.a ∧ BB1.b_0=bb0.b ∧ BB1.t_0=BB0.t_1)
V (!BB0.e ∧ end.t=BB0.t_1));

4: BB1→ BB1.e=(BB1.a_0<BB1.b_0) ∧ BB1.t_1=BB1.t_0+1 ∧
((BB1.e ∧ BB2 ∧ BB2.b_0=bb1.b_0 ∧ BB2.t_0=BB1.t_1)
V(end.t=BB1.t_1));

5: BB2→
BB2.a_1=BB2.b_0 ∧ BB2.t_1=BB2.t_0+1 ∧ end.t=BB2.t_1;

Fig. 15.7 Set of SMT assertions to express the FSMD of the *Max* example

For the set of assertions, shown in Fig. 15.7, a SMT solver returns one random assignment for *en*, *a*, and *b*, as well as the time to process this set of variables in `end.t`.

15.5.2 Finding a Counterexample

To prove that the generated program is vulnerable against timing attacks, we duplicate the encoded FSMD, so that we have two models M1 and M2. Practically we duplicate all variables and states and add the prefixes M1 and M2, respectively. For the models we want to prove that for every equivalent non-critical input, changing the sensitive data does not result in different timings at the end. We do it by showing that no counterexample exists, i.e., that our request for a counterexample is not satisfiable.

For our example we can add the assertion,
$\text{M1} \bigwedge \text{M2} \bigwedge \text{M1.en} = \text{M2.en} \bigwedge \text{M1.a} = \text{M2.a} \bigwedge \text{M1.end.t} > \text{M2.end.t}.$
In this case, a counterexample can be easily found in (M1.b=1 and M2.b=0 for a=0 and en=1).

Applying our execution path balancing approach, the vulnerability is solved by adding an idle state in a new BB between BB1 and OUT. In that case, the SMT solver returns UNSAT, which means that no counterexample could be found and therefore, the design can be considered as time invariant.

15.6 Evaluation

We implemented the presented techniques, i.e., the support for the timing annotation, the balanced scheduling, and the verification for the LegUp HLS tool chain, and applied them to a range of Benchmark applications and cryptographic protocols. In the following subsection we describe the experimental setup and the result of the experiments.

15.6.1 Experimental Setup

To evaluate the effectiveness of the path balancing mechanisms, which we introduced in Sect. 15.4, we compare systems that were synthesized with and without the path balancing. For each system we conducted an empiric run-time test in addition to the timing invariance verification procedure, discussed in the previous section. We further compare synthesis time and synthesis results. The goal of the evaluation is the validation of the timing invariance of the synthesized designs, and to quantify the overheads for the applied techniques.

We tested two generic benchmark applications and two cryptographic systems:

- the *Max* example (see Sect. 15.4.2),
- the SRA algorithm [18] to compute an approximation for $\sqrt{a+b}$, while the assumed secret is b,
- the 2048-bit modular exponentiation ($x^e \ mod \ p$) of the RSA crypto algorithm, with the secret exponent e, and
- the 233-bit (kP) point multiplication of the ECC cryptographic algorithm, with the secret factor k.

For ECC and RSA we used pre-synthesized units for the algebraic operations in the finite field, which means that the complex multipliers were not synthesized in the HLS process.

Each system was synthesized with three settings:

1. No annotations (NOA), i.e., the original design,
2. Static balancing (STA), as discussed in Sect. 15.4.2, and
3. Automatic path balancing (APB), discussed in Sect. 15.4.3.

For each test case, the run-time was measured empirically by applying various random test input, including data inputs with all bits set to 0 and all bits set to 1. The verification was performed based on the generated schedule which is stored in a LegUp report file. We provide the data for the technology-dependent properties as percentage in relation to the original design. The systems were synthesized for an ALTERA Cyclone V FPGA, simulated with ModelSim 15. The area is based on the reported FPGA utilization, the power consumption is estimated using the Quartus PowerPlay tool. The synthesis time includes the required time for the HLS but does not include time for hardware synthesis and mapping. The experiments were conducted on a i7 PC with 16GB RAM.

15.6.2 Results and Discussion

The measurements and evaluation results for the tested system are summarized in Table 15.1. In this section we discuss the performance, overhead, and time invariance.

Table 15.1 Synthesis, execution time, and verification results

Design	Case	Synth time [s]	Execution time [cycles]	Verification	Area	Longest path	Power
					[%] to original design		
Max	NOA	4	10–11	Failed	–	–	–
	STA	4	11	Passed	=0.0	=0.0	−0.5
	APB	4	11	Passed	=0.0	=0.0	−0.5
SRA	NOA	4	23–26	Failed	–	–	–
	STA	4	26	Passed	=0.0	=0.0	−0.2
	APB	4	26	Passed	=0.0	=0.0	−0.2
ECC	NOA	6	12,952–13,416	Failed	–	–	–
	STA	6	13,414	Passed	+0.0	=0.0	−0.0
	APB	7	13,414	Passed	+0.0	=0.0	−0.0
RSA	NOA	8	0.1–1.6mio	Failed	–	–	–
	STA	9	1,598,356	Passed	+0.2	=0.0	−5.2
	APB	9	1,594,260	Passed	+0.2	=0.0	−5.2

15.6.2.1 Performance and Overhead

In general, the measured overheads of synthesis time, area, longest path, and power consumption are negligible. The synthesis time increases by up to 10%. The increase is expected and is caused by processing time in the added *Balancer* module, and for solving the additional constraints in the ILP solver. We expect that at tighter integration of the *Balancer* and the SDC scheduler could reduce the overhead further.

The longest combinatorial path, which also determines the maximum clock frequency of an IC, is not affected by the path balancing for any of the tested designs. This result was expected because the longest combinatorial path typically is part of the data path which is not be affected by the added complexity of the control path.

The results show a small increase of area for the synthesized design, which is caused by the added states to the control path. Overall the area is only affected marginally, which also means that, at least for our test cases, the idle-cycle-imposed reduction of resource utilization did not lead to any reduction of functional units or registers.

The power consumption is affected marginally, showing a small reduction of average power consumption, most notable for the RSA design. This behavior is expected since idle states naturally reduce the power consumption of a system.

15.6.2.2 Time Invariance

Regarding the time invariance, which is main concern of this work, we see that the execution time varies for all of the original designs. That means that all evaluated original designs show exploitable timing side channels. This vulnerability was also discovered by the automatic verification step, which identified execution path variances for all tested original designs.

The execution time for the annotated designs is invariant for all tested designs, which is validated in the automatic verification step. For the RSA case the execution time for the static annotation is slightly larger than the execution time for the automatic balancing. This is caused by the fact, that lacking complete knowledge of the basic blocks and their operations, a manual annotation leads to wrong, unsatisfiable, or non-optimal schedules. That is a known limitation of manual annotation approaches and motivated our work towards the automatic balancing. It should be noted that with manual fine tuning the static approach can result in the same performance numbers as the automated annotations.

Overall, while the worst-case execution times are similar to the original designs, the average execution time of the balanced designs is higher than the original designs. That observation is not surprising since the fundamental idea of our work is the addition of idle states to balance execution paths. Therefore, the additional runtime is a trade-off for security, that is already know for low-level countermeasures such as [8].

15.7 Conclusions

In the design of integrated circuits (IC), the use of High-Level Synthesis (HLS) plays a critical role in improving the quality of the designed circuit (in terms of performance) while substantially increasing design productivity. In this chapter we have shown that HLS can have an additional benefit of facilitating the design of ICs that have built-in guards against timing attacks. The added chip-level security strength is achieved via design automation, i.e., without the need for designers to address security vulnerabilities directly at a low-level of abstraction. We have shown how a system specification in C-code can be translated to an HDL description that provides time invariance on all security-related paths of the design.

Our key contributions include the time invariance generation of new scheduling constraints, which reflect annotations along the timing of security-related branches and execution paths. Our approach is integrated within in an HLS tool chain. Specifically, the scheduling constraints are considered in the schedule generation of an HLS flow and result in time-balanced control paths. The strength of the presented approach is the compatibility with established HLS optimizations, namely, the adding of scheduling constraints without interfering with other synthesis steps.

A verification step based on bounded model checking and implemented in satisfiability modulo theory (SMT) could prove the time invariance of the generated systems. The verification step utilizes intermediate model information from the HLS flow and bounded model checking to validate the intended timing properties for the synthesized designs.

This chapter outlined a practical implementation of the framework in the open source LegUp HLS tool chain. The practical evaluation for a range of benchmark applications as well as an Elliptic Curve Cryptography (ECC) and RSA implementation could demonstrate the practicability as well as a negligible overhead for synthesis time, as well as area overhead, longest path, and power consumption of the synthesized designs. Our evaluation also showed the importance of a careful design of the system algorithm, since unbalanced applications (like RSA) are penalized either with a significantly higher run-time. Carefully designed algorithms like the ECC implementation show an execution-time overhead of less than 5%.

While our work cannot solve all challenges regarding timing attackers for ICs, we consider our results a successful first step towards the support of side-channel resistance in HLS. As future work we want to extend the approach to reflect power attacks in addition to timing attacks. Therefore, the added idle states might be replaced with operations from parallel execution paths. Another possible future work is the integration of low-level side-channel-attack-resistant design flows such as [15] to further improve the side-channel resistance of the synthesized design.

References

1. Kocher, P.C.: Timing attacks on implementations of diffie-hellman, RSA, DSS, and other systems. In: Koblitz, N. (ed.) Advances in Cryptology—CRYPTO '96, pp. 104–113. Springer, Berlin (1996)
2. Mao, B., Hu, W., Althoff, A., Matai, J., Oberg, J., Mu, D., Sherwood, T., Kastner, R.: Quantifying timing-based information flow in cryptographic hardware. In: 2015 IEEE/ACM International Conference on Computer-Aided Design (ICCAD), pp. 552–559 (2015). https://doi.org/10.1109/ICCAD.2015.7372618
3. Fan, X., Peter, S., Krstic, M.: GALS design of ECC against side-channel attacks—A comparative study. In: 2014 24th International Workshop on Power and Timing Modeling, Optimization and Simulation (PATMOS), pp. 1–6 (2014). https://doi.org/10.1109/PATMOS.2014.6951905
4. Chakraborty, R.S., Bhunia, S.: HARPOON: An obfuscation-based SoC design methodology for hardware protection. Trans. Comp.-Aided Des. Integ. Cir. Sys. **28**(10), 1493–1502 (2009). https://doi.org/10.1109/TCAD.2009.2028166
5. Todman, T., Stilkerich, S., Luk, W.: In-circuit temporal monitors for runtime verification of reconfigurable designs. In: 2015 52nd ACM/EDAC/IEEE Design Automation Conference (DAC), pp. 1–6 (2015). https://doi.org/10.1145/2744769.2744856
6. Tiri, K., Verbauwhede, I.: A digital design flow for secure integrated circuits. IEEE Trans. Comput.-Aided Design Integr. Circ. Syst. **25**(7), 1197–1208 (2006). https://doi.org/10.1109/TCAD.2005.855939
7. López, J., Dahab, R.: High-speed software multiplication in F2m. In: Roy, B., Okamoto, E. (eds.) Progress in Cryptology —INDOCRYPT 2000, pp. 203–212. Springer, Berlin (2000)
8. Rostami, M., Koushanfar, F., Karri, R.: A primer on hardware security: models, methods, and metrics. Proc. IEEE **102**(8), 1283–1295 (2014). https://doi.org/10.1109/JPROC.2014.2335155
9. Gajski, D.D., Abdi, S., Gerstlauer, A., Schirner, G.: Embedded System Design: Modeling, Synthesis, Verification. Springer Science & Business Media, Berlin (2009)
10. Canis, A., Choi, J., Aldham, M., Zhang, V., Kammoona, A., Czajkowski, T., Brown, S.D., Anderson, J.H.: LegUp: an open-source high-level synthesis tool for FPGA-based processor/accelerator systems. ACM Trans. Embed. Comput. Syst. **13**(2) (2013). https://doi.org/10.1145/2514740
11. Meeus, W., Beeck, K.V., Goedemé, T., Meel, J., Stroobandt, D.: An overview of today's high-level synthesis tools. Des. Autom. Embedd. Syst. **16**(3), 31–51 (2012). https://doi.org/10.1007/s10617-012-9096-8
12. Peter, S., Givargis, T.: Towards a timing attack aware high-level synthesis of integrated circuits . In: 2016 IEEE 34th International Conference on Computer Design (ICCD), pp. 452–455 (2016). https://doi.org/10.1109/ICCD.2016.7753326
13. Cong, J., Zhang, Z.: An efficient and versatile scheduling algorithm based on SDC formulation. In: 2006 43rd ACM/IEEE Design Automation Conference, pp. 433–438 (2006). https://doi.org/10.1145/1146909.1147025
14. Standaert, F.X.: Introduction to Side-Channel Attacks Secure Integrated Circuits and Systems. In: Secure Integrated Circuits and Systems, pp. 27–42. Springer, Boston (2010)
15. Kulikowski, K., Smirnov, A., Taubin, A.: Automated design of cryptographic devices resistant to multiple side-channel attacks. In: Goubin, L., Matsui, M. (eds.) Cryptographic Hardware and Embedded Systems—CHES 2006, pp. 399–413. Springer, Berlin (2006)
16. Huang, Q., Lian, R., Canis, A., Choi, J., Xi, R., Brown, S., Anderson, J.: The effect of compiler optimizations on high-level synthesis for FPGAs. In: 2013 IEEE 21st Annual International Symposium on Field-Programmable Custom Computing Machines, pp. 89–96 (2013). https://doi.org/10.1109/FCCM.2013.50
17. Chen, L., Ebrahimi, M., Tahoori, M.B.: Reliability-aware operation chaining in high level synthesis. In: 2015 20th IEEE European Test Symposium (ETS), pp. 1–6 (2015). https://doi.org/10.1109/ETS.2015.7138739

18. Gajski, D.D., Zhu, J., Dömer, R., Gerstlauer, A., Zhao, S.: SpecC Specification Language and Methodology. Springer Science & Business Media, Berlin (2012)
19. Harrath, N., Monsuez, B.: SystemC waiting state automata. Int. J. Crit. Comput.-Based Syst. **3**(1/2), 60–95 (2012). https://doi.org/10.1504/IJCCBS.2012.045077
20. Genkin, D., Shamir, A., Tromer, E.: RSA key extraction via low-bandwidth acoustic cryptanalysis. In: Annual Cryptology Conference, pp. 444–461. Springer, Berlin (2014)
21. Cordeiro, L., Fischer, B., Marques-Silva, J.: SMT-based bounded model checking for embedded ANSI-C software. In: 2009 IEEE/ACM International Conference on Automated Software Engineering, pp. 137–148 (2009). https://doi.org/10.1109/ASE.2009.63
22. Henry, J., Asavoae, M., Monniaux, D., Maïza, C.: How to compute worst-case execution time by optimization modulo theory and a clever encoding of program semantics. SIGPLAN Not. **49**(5), 43–52 (2014). https://doi.org/10.1145/2666357.2597817
23. Bjørner, N., Phan, A.D., Fleckenstein, L.: νZ - an optimizing SMT solver. In: Baier, C., Tinelli, C. (eds.) Tools and Algorithms for the Construction and Analysis of Systems, pp. 194–199. Springer, Berlin (2015)
24. Peter, S., Givargis, T.: Component-based synthesis of embedded systems using satisfiability modulo theories. ACM Trans. Des. Autom. Electron. Syst. **20**(4) (2015). https://doi.org/10.1145/2746235

Chapter 16
Integrating Information Flow Tracking into High-Level Synthesis Design Flow

Wei Hu, Armaiti Ardeshiricham, Lingjuan Wu, and Ryan Kastner

16.1 Introduction

High-level synthesis (HLS) enables hardware designers to write an untimed circuit description allowing them to focus on architectural design optimizations like pipelining, task level parallelism, and array partitioning [1]. By removing the burden of describing cycle accurate behaviors, HLS designers can perform a more comprehensive design space exploration to better find an architecture that meets the desired power, performance, and area (PPA) constraints.

While HLS tools are effective at exploring tradeoffs and optimizations related to PPA, security has largely been an after-thought. The emergence of hardware security flaws and threats [2–5] has brought a demand for hardware security verification tools. Due to the high cost (or even technical impossibility) to patch hardware security vulnerabilities after chip fabrication, identifying and eliminating security flaws in the early design phase is crucial.

Information flow tracking (IFT) is a fundamental technique for hardware security verification [6–8]. IFT allows the designer to verify security properties related to confidentiality, integrity, and availability. An important first step is to create security enhanced circuit models for accurate description of security-related design behaviors and formal verification of security properties [9–12]. Some recent

W. Hu
Northwestern Polytechnical University, Xi'an, China
e-mail: weihu@nwpu.edu.cn

A. Ardeshiricham · R. Kastner (✉)
UC San Diego, La Jolla, CA, USA
e-mail: aardeshi@ucsd.edu; kastner@ucsd.edu

L. Wu
Huazhong Agricultural University, Wuhan, China
e-mail: wulj@mail.hzau.edu.cn

S. Katkoori, S. A. Islam (eds.), *Behavioral Synthesis for Hardware Security*,
https://doi.org/10.1007/978-3-030-78841-4_16

works [13–15] incorporate security models in HLS to allow automated synthesis of secure hardware accelerators. Ideally, these security models can be seamlessly integrated into the standard EDA flow in order to allow security to be verified alongside traditional design constraints during design space exploration [16, 17] without incurring additional design burdens (e.g., learning new design languages and tools) on hardware designers. However, securing hardware accelerators is still a significant challenge for HLS [18].

We aim to answer the question of "how do we best integrate security into HLS hardware design flow?" To better understand this question, we describe a security aware HLS flow that integrates into a property driven hardware security verification flow [6]. We enhance the HLS design flow by integrating information flow tracking to allow the verification of security properties. We discuss the value of performing information flow security verification at the register transfer level (RTL); we develop precise IFT methods; we present fine-granularity information flow model formalizations, and we illustrate how hardware security properties related to confidentiality, integrity, isolation, constant time, and malicious design modification could be formally verified using standard EDA verification tools. Specifically, we make the following contributions:

1. Proposing a method to enhance the backend of HLS by employing information flow security verification using standard EDA tools;
2. Developing precise hardware IFT methods at the register transfer level and deriving security enhanced circuit model formalizations in a standard HDL;
3. Presenting experimental results to demonstrate the effectiveness of our security verification techniques in proving hardware security properties and identifying security vulnerabilities.

The remainder of this chapter is organized as follows. Section 16.2 provides a background discussion on how to integrate security into the HLS design flow. We lay out a basic methodology for security enhanced HLS design flow. Section 16.3 illustrates how hardware security properties can be modeled and verified from the perspective of information flow—a frequently used technique for enforcing security in hardware designs. In Sect. 16.4, we elaborate on our efforts in developing standardized hardware security models for security verification at the RTL. Section 16.5 presents experimental results to demonstrate the effectiveness of our security verification solution in identifying security flaws. We conclude the book chapter in Sect. 16.6.

16.2 Background

HLS allows designers to specify circuits in a more abstract manner. The key feature is that HLS specifications are *algorithmic* or *untimed*—designers do not need to describe circuit behaviors on a cycle-by-cycle basis. This allows them to more efficiently build and verify their hardware designs.

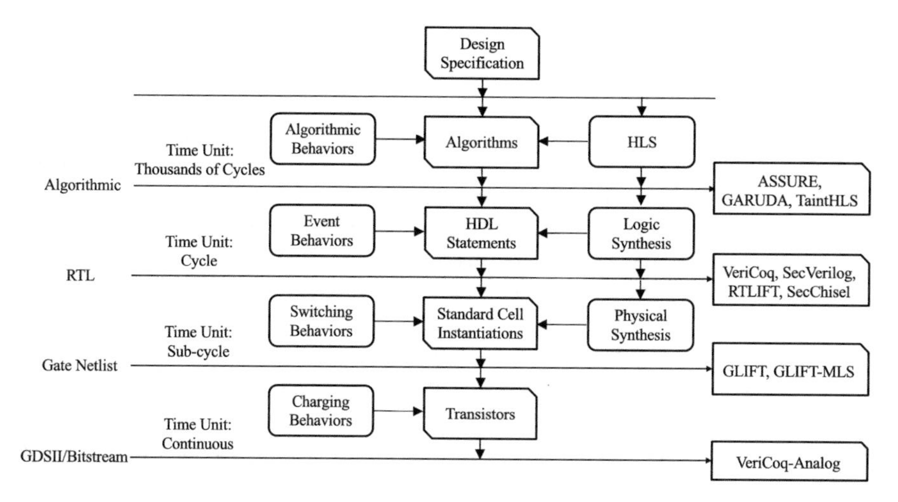

Fig. 16.1 Hardware IFT at different levels of abstraction

An HLS designer provides a specification in a high-level language (e.g., C, C++, SystemC). The designer then goes through an optimization process to create different architectures, and hopefully find one that suits their needs with respect to the traditional circuit metrics of performance, power, and area. This process involves adding HLS optimizations involving pipelining, data partitioning, and data representation to find a final design that best fits their design goals. HLS tools translate the high-level algorithmic specification into a *register transfer level (RTL)*. RTL provides more details about the functionality of the circuit. In particular, it has cycle accurate behaviors. This RTL is then gradually translated down to lower abstraction levels and eventually a physical layout (GDSII).

Figure 16.1 shows the hardware design flow using HLS as a design entry point. At higher of abstraction, we use models of computation that describe hardware design behaviors, e.g., algorithms and functions. While these high-level entities are more concise, they abstract away a huge amount of information that is needed for the final hardware implementation. As the synthesis process proceeds to RTL and gate level, the hardware design is represented using more concrete circuit models such as function units and standard macro cells. The timing behavior of hardware designs also becomes more accurate as the design description is refined; RTL provides cycle level accuracy while gate level and lower abstractions yield sub-cycle timing information. Additionally, area models are better understood as more details about gate sizes, their locations, and wire lengths become available.

In a similar sense, hardware security verification can be performed at different abstraction levels. There are projects that perform verification on algorithmic specifications [13–15]. Other hardware security techniques work at RTL [11, 12, 19–21]. Some techniques perform verification at the gate level [9, 22] and even some take into account analog characteristics [23].

It is important to select the right abstraction when employing secure information flow analysis for hardware design [24]. *The key question that we aim to answer is: "What is the best level of abstraction for information flow analysis for an HLS design flow?"* To try to better understand these tradeoffs, we perform a comparison of hardware IFT techniques. We focus on the HLS, RTL, and gate levels which are commonly considered in the hardware security verification flow.

At the algorithmic level, the hardware design is described using highly abstractive design models. As a result, we need to make very conservative assumptions about design behaviors and employ conservative rules for security label propagation. In addition, it is difficult to model timing flows (see Sect. 16.3) due to the lack of accurate timing information. Additionally, distinguishing implicit flows from explicit ones is a challenging task at this level. The effect of conditional operations can spread across a wide range of operations, which are hard to track. The benefit of modeling IFT at HLS is that the verification is typically faster than that at lower levels due to the simplified circuit models and utilization of conservative label propagation rules.

Performing information flow analysis at the RTL allows analysis that requires timing accurate behaviors. This includes timing side channels, which would not be exposed when considering IFT using an algorithmic (HLS) abstraction. Additionally, interactions related to sharing resources are also apparent at the RTL but are not visible at higher abstraction levels. This includes shared registers, shared functional units, memories, and interfaces. Understanding the implications of resource sharing is a particularly challenging aspect of hardware security and performing IFT at the algorithmic level would abstract away some important aspects related to this.

Moving to the gate level for IFT analysis provides some additional benefits since more details of the circuit are available. Sub-cycle timing and switching behaviors are better understood which can provide better analysis of security concerns related to timing or switching. Unfortunately, as the hardware design refines, it is translated into a significantly larger number of gates. This poses big challenges in scalability of IFT techniques and security verification performance. It is generally known to the EDA community that RTL verification is much easier and faster than gate-level verification methods [25]. That is an important reason why recent hardware IFT techniques have gravitated towards the RTL for better performance and scalability.

Clearly there are tradeoffs for performing analysis at these different abstraction levels. The lack of details about cycle level timing and resource sharing severely limit the type of information flow analysis that can be done using an algorithmic description. This requires the IFT logic to be very conservative which forces the designer to be overly conservative in their security decisions. This points to the need to perform the analysis at a lower level of abstraction.

Gate-level analysis has sub-cycle timing information and a clear notion of resource sharing. Additionally, the information flow analysis is simplified in some regards as the basic units are logical operations; analyzing flows of information is a lot easier when computation is broken done at this level of granularity [22]. Yet, some higher-level information about control is lost in the translation to gates. For example, it is hard to differentiate between control and data flow since everything

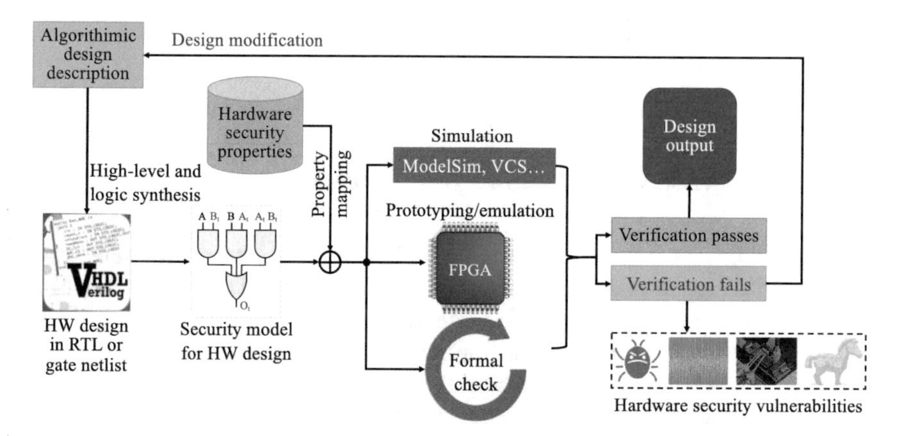

Fig. 16.2 Secure hardware design flow that employs information flow security verification for secure vulnerability detection

is just Boolean gates. This means that timing flows cannot be separated from functional flows [20, 26].

We argue that IFT analysis performed at the RTL provides the best tradeoff between scalability and ability to differentiate different flows of information including timing side channels, vulnerabilities related to resource sharing, and those related to control and data path interactions. We make a case for this in subsequent sections. But before we do that, we need to describe the hardware security verification process and some background on information flow analysis.

Figure 16.2 describes a general framework for integrating information flow analysis into an HLS security verification flow, which follows a property based approach to hardware security [6]. The framework takes an algorithmic hardware description and a set of security properties. The description is synthesized in a typical fashion using HLS, logic, and physical synthesis. The key question that we consider is when to generate the security models and perform the security analysis. Our experiments and discussions attempt to understand value and tradeoffs of performing this analysis at different levels.

We use an IFT analysis method that generates a security model for analysis. The security model is derived from the original circuit; it is fully synthesizable, but separate circuit, that can be analyzed using existing EDA verification tools. The security model is used to verify information flow security properties specified using standard property specification languages such as SystemVerilog Assertion (SVA). Since both the formal circuit model and security properties are described with standard HDLs, the verification process can be performed through simulation, FPGA prototyping or emulation as well as formal proof. If the hardware design adheres to all desired security properties, it is ready for design output. Otherwise, the security verification fails, indicating potential existence of unintentional design flaws or intended malicious design modifications. In such a case, the design process should iterate until the design passes security verification. The security model is not required to be added to the final circuit though it can be if one desires run-time security violation checking.

16.3 Hardware Information Flow Tracking

Information can flow through hardware designs in a variety of different ways. This includes *logical flows* and *physical flows*. This book chapter primarily aims to understand the flow of information in hardware designs during the early design phase. Thus, we only account for the logical information flows.

Logical information flows can be further classified into *explicit* and *implicit*. Explicit information flows happen when data is explicitly assigned to a destination operand while implicit flows usually occur when some operands are conditionally updated. The following code snippet illustrates the difference between these two types of information flows.

```
1:  key_hash := hash(key)
2:  if(key_hash == CORRECT_HASH_RECORD)
3:      unlock := 1
4:  else
5:      unlock := 0
6:  end if
```

In this example, there is an explicit flow from *key* (or more accurately) *hash*(*key*) to *key_hash* resulting from the first explicit assign statement (*line* 1). There is also a piece of implicit information flow from *key_hash* to *unlock* even if *key_hash* is not directly assigned to *unlock* (*lines* 2–5). This is due to the fact that by observing the status of *unlock*, we can learn if the *key_hash* matches the record.

From the example, explicit flows are easy to capture while implicit flows are more difficult to determine. An even more subtle case of implicit flow is *timing flow*, which is caused by conditional updates of stateful elements [20]. The following code snippet shows a case of timing flow caused by fast paths (*lines* 1–2 and *lines* 3–4) in the exponentiation unit. An unauthorized process may be able to infer the exponent by observing the amount of time take to calculate the exponentiation.

```
1:  if(exponent == 0)
2:      power := 1
3:  else if(exponent == 1)
4:      power := base
5:  else
6:      power := exp(base, exponent)
7:  end if
```

From the above examples, information flows in hardware design can lead to leakage of protected information and cause security violation. In order to understand

and further prevent such leakage, we need to specify and enforce information flow security properties, which will be discussed in the following subsection.

16.3.1 Information Flow Security Properties

Undesired flows of information could violate different security policies such as *confidentiality*, *integrity*, and *isolation*. For instance, in the *key verifier* example from the previous subsection, flow of secret information from the *key_hash* variable to the *unlock* signal violates the confidentiality property if the adversary has access to the value of *unlock* signal. To detect such violations, information flow security properties are added to the design in forms of logical assertions. An assertion in the form of $assert(key \not\rightarrow unlock)$ detects the confidentiality breach in the *key verifier* example. Here, the "$\not\rightarrow$" operator indicates the absence of information flows from the left-hand side variable to the right-hand side variable.

Information flow properties capture security relevant design behavior which cannot be expressed by existing property specification languages for functional verification. In the following we review the major security policies which are specified using the model of information flow.

- **Confidentiality:** The confidentiality property analyzes the relation between secret data and publicly observable ports. To preserve confidentiality, we need to constrain the flow of information from data objects which contain secret information. For instance, in a cryptographic core, confidentiality is stated as $assert(key \not\rightarrow pub)$, where *pub* represents public ports that are not encrypted.
- **Integrity:** The Integrity property is the dual of confidentiality and refers to the information flow from untrusted data to critical components in the design. For instance, to preserve integrity of the program counter (PC) register in a processor, there should be no flow of information from the public inputs such as the Ethernet port to the PC. This property can be specified as $Ethernet\_port \not\rightarrow PC$.
- **Isolation:** The isolation property denotes eliminating information flow between two entities. As an example, consider a SoC where the AES core and the IIR filter should be isolated. This expectation is modeled by $assert(AES\_out \not\rightarrow IIR\_in \;\&\&\; IIR\_out \not\rightarrow AES\_in)$. Note that isolation is a two-way policy and is enforced on both cores.
- **Timing Side Channels:** Information flow leakage through timing side channel can be used to break confidentiality even in cases where the secret data is not directly readable. Hence, to avoid timing side channel attacks, we need to eliminate timing leakage. For instance, to detect timing leakage in a cryptographic core, we need to specify that the secret key does not flow to the cipher via timing channels. This is stated as $assert(key \not\rightarrow_{time} cipher)$. Here, "$\not\rightarrow_{time}$" operator represents absence of timing flows.
- **Hardware Trojans:** Information flow properties can detect certain class of hardware Trojans where a malicious circuitry is inserted to generate unauthorized

flow of information [27]. For example, in the Trust-Hub benchmarks [28], hardware Trojans are added to cryptographic cores to transfer the secret key to the output "*Antena*". The information flow property to detect these Trojans can be formulated as $assert(key \not\rightarrow Antena)$.

The information flow security policies can be translated to a set of SystemVerilog assertions written over the instrumented design. More specifically, an IFT policy modeled as $assert(A \not\rightarrow B)$ is translated to the following properties, where A_t and B_t are the security labels of A and B, respectively.

```
assume (A_t == 1);
assume (B_t == 0);
```

In this example we assume both A and B have single bit labels. A timing side channel policy modeled as $assert(A \not\rightarrow_{time} B)$ is verified using the following properties, where B_time is the timing security label of B.

```
assume (A_t == 1);
assume (B_time == 0);
```

16.3.2 Fundamentals of Hardware IFT

Hardware IFT is a commonly used technique for measuring the flow of information in circuit designs. The core idea behind hardware IFT is to associate data objects in the hardware design with a label for encoding security attribute, e.g., sensitive information can be labeled as `secret` while information from an open computing environment should be marked as `untrusted`. These meta data will be processed along with the data objects to determine the security attribute of the outputs. The output label will be updated according to the flow of information. Specifically, an input flows to the output *if and only if* the input has an influence on the value of the output. In this case, the security label of the input will be involved in determining the security label of the output. Thus, information flows can be measured by observing the relation between the input and output security labels. Figure 16.3 uses streaming cipher as an example to illustrate how hardware IFT is performed.

Apart from the *original logic* that XORs the plaintext (i.e., m_i) and key (i.e., k_i) streams shown in Fig. 16.3a, we need to associate the inputs with security labels, i.e., m_t_i and k_t_i, respectively. Additional *IFT logic* is then instantiated to process these metadata and calculate the output security label as shown in Fig. 16.3b. The way in which the IFT logic is implemented depends on the label propagation policy employed. This example simply takes the logical OR (i.e., the upper bound) of input security labels as the security label for the output. There are more complex but also more precise label propagation policies as we will illustrate in Sect. 16.4. Before that, we briefly cover the fundamental aspects of hardware IFT.

It is possible to succinctly describe the flow relationships on the Boolean operations. IFT logic formalizations and generation algorithms lay the theoretic

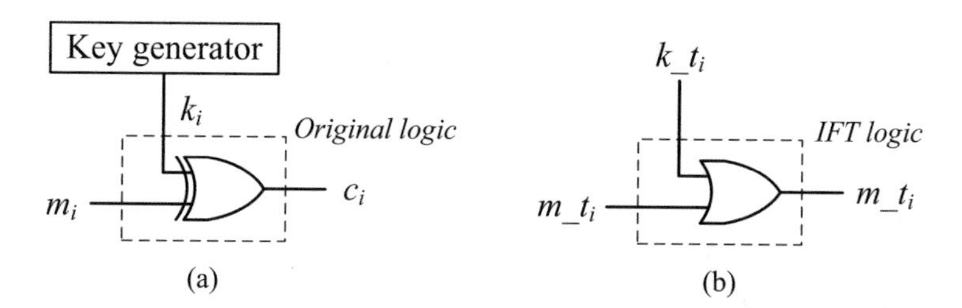

Fig. 16.3 Hardware information flow tracking uses IFT logic for label propagation. (**a**) The data encryption logic for stream ciphers. (**b**) The IFT logic for the encryption operator

Table 16.1 IFT logic for Boolean operations

Gate	Boolean function	IFT logic
AND-2	$f = g \cdot h$	$\mathcal{L}(f) = g \cdot \mathcal{L}(h) + h \cdot \mathcal{L}(g) + \mathcal{L}(g) \cdot \mathcal{L}(h)$
OR-2	$f = g + h$	$\mathcal{L}(f) = \overline{g} \cdot \mathcal{L}(h) + \overline{h} \cdot \mathcal{L}(g) + \mathcal{L}(g) \cdot \mathcal{L}(h)$
XOR-2	$f = g \oplus h$	$\mathcal{L}(f) = \mathcal{L}(g) + \mathcal{L}(h)$
INV	$f = \overline{g}$	$\mathcal{L}(f) = \mathcal{L}(g)$
AND-N	$f = f_1 \cdot f_2 \cdots f_n$	$\mathcal{L}(f) = \prod_{i=1}^{n}(f_i + \mathcal{L}(f_i)) - f$
OR-N	$f = f_1 + f_2 + \cdots + f_n$	$\mathcal{L}(f) = \prod_{i=1}^{n}(\overline{f_i} + \mathcal{L}(f_i)) - \overline{f}$
XOR-N	$f = f_1 \oplus f_2 \oplus \cdots \oplus f_n$	$\mathcal{L}(f) = \sum_{i=1}^{n} \mathcal{L}(f_i)$

groundwork for hardware IFT [29]. Table 16.1 shows the flow tracking logic for Boolean operations where $\mathcal{L}(\cdot)$ denotes the function for calculating security label, sum represents logical OR while product means logical AND. The minus operator means excluding that term from the equation.

The IFT logic shown in Table 16.1 can be extended to Boolean operations of variable widths, e.g., using the generate feature of HDL. In this way, we construct an IFT library for deriving IFT methods as we will introduce in the following section. The role of the IFT library in creating IFT logic for large circuits is similar to technology library in technology mapping.

16.4 Register Transfer Level Information Flow Tracking

RTLIFT software accepts a hardware design implemented in the Verilog language and generates functionally equivalent Verilog code which is instrumented with information flow tracking logic. The outputted code is described at the same abstraction level as the input design. This is achieved by defining label propagation rules for RTL language constructs eliminating the need to synthesize the input design to a netlist. The code generated by RTLIFT can be analyzed by standard EDA verification tools and allows leveraging decades of research on functional testing to assess security properties of hardware designs. If the instrumented design passes

```
input [7:0] a,b,c;
output [8:0] o;
assign o = a + b&c;
```

(a)

```
module and_IFT
(Z, Z_t, X, X_t, Y, Y_t);
parameter w1,w2,w3;
output [w1-1:0] Z, Z_t;
input [w2-1] X, X_t;
input [w3-1] Y, Y_t;
assign out = X & Y;
assign out_t = X_t | Y_t;
endmodule
```

(b)

```
input [7:0] a,b,c;
input [7:0] a_taint,b_taint,c_taint;
output [8:0] o;
output [8:0] o_taint;
wire [7:0] temp, temp_taint;
and_IFT #(8,8,8) and1
(temp, temp_taint, b, b_taint, c, c_taint );
add_IFT #(9,8,8) add1
(o, o_taint, temp, temp_taint, a, a_taint);
```

(c)

Fig. 16.4 Tracking explicit flows via RTLIFT. (**a**) Input Verilog code. (**b**) RTLIFT library for *and* operation. (**c**) Instrumented Verilog code

the security verification, the original code can be used for fabrication. Otherwise, the original code should be modified, instrumented, and verified again.

RTLIFT enables tracking both explicit and implicit flows. Flow tracking starts by extending each design component (e.g., wires and registers in a given Verilog code) with a label that carries out information regarding the security properties of the data. After extending the variables with security labels, every HDL operation is replaced with an IFT-enhanced operation. An IFT-enhanced operation is functionally equivalent to the original operation, but it also includes the logic for tracking explicit flows through that operation. The IFT-enhanced operations are defined in RTLIFT library for all valid Verilog operations. Figure 16.4 shows an example of tracking explicit flows at RTL.

Tracking only explicit flows might inaccurately report the absence of information flow by ignoring existence of implicit flows through conditional statements. To capture these flows, we need to obtain a list of variables which affect the execution of each statement. The logic for tracking the implicit flows can be generated using this list with different levels of precision. To implement a conservative IFT approach, any usage of tainted conditions should yield a tainted output. To have more precise flow tracking, we need to figure out if different outcomes are possible for the right-hand side of an assignment, assuming the conditions were flipped. Using this approach, we can track the implicit flow through each conditional statement by modeling it as a multiplexer.

To illustrate the idea, we show implicit flow tracking for a simple example shown in Fig. 16.5. Here, `e1_t` and `e2_t` represent the explicit flows from expressions `e1` and `e2`. The imprecise approach, as shown in Fig. 16.5b, marks the output of a conditional statement as tainted whenever the condition is tainted. The precise IFT logic specifies that information flows from the condition signal to the output only in cases where the condition is tainted while both inputs are tainted or they have different Boolean values.

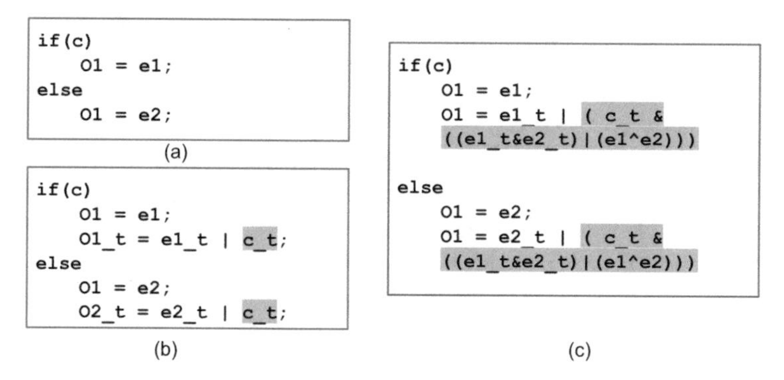

Fig. 16.5 Tracking implicit flows via RTLIFT. (**a**) Input Verilog code. (**b**) Verilog code instrumented with imprecise logic. (**c**) Verilog code instrumented with precise logic. Tracking logic for implicit flows are highlighted

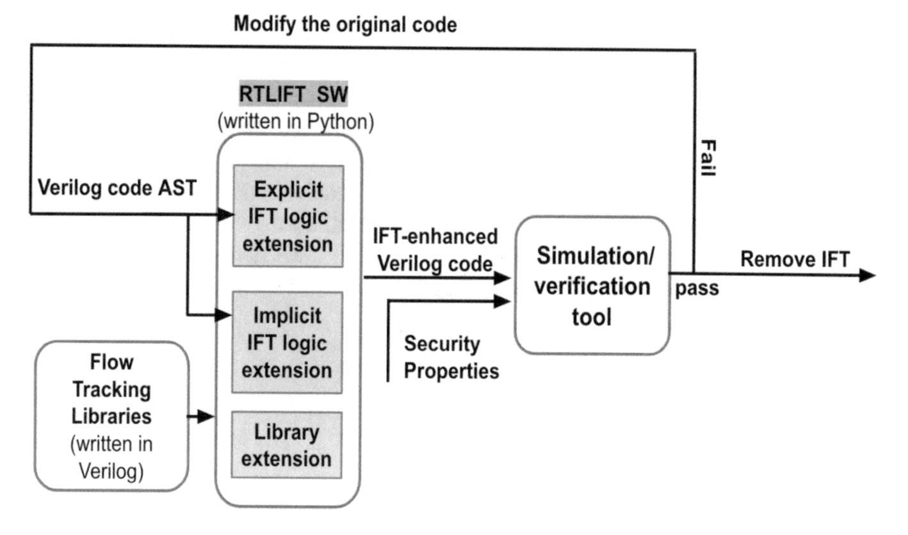

Fig. 16.6 AST based RTLIFT overview

Figure 16.6 gives an overview of an Abstract Syntax Tree (AST) Based RTLIFT tool. The IFT instrumentation is done by analyzing the data flow graph of the input design. The data flow graph is collected by using *Yosys frontend compiler* [30] to transform the code to its AST representation. RTLIFT analyzes the node of each assignment statement via in-order traversal. And for each operation, it adds a module from the IFT library. RTLIFT considers a single bit label for each variable bit. Here, a high value indicates either sensitive or untrusted value, depending on the property to be verified.

The AST generated by high-level HDL frontends are further converted into the RTLIL main internal data format for further design optimization in Yosys. Afterwards, the RTLIL representation can be converted into various formats including

Verilog and ILANG. We provide another RTL hardware IFT method that targets RTLIL to leverage the synthesis optimizations in Yosys.

The RTLIL defines design objects such as *module*, *cell*, *wire*, *process*, and *memory*. For the *module* and *wire* objects, we only need to extend the original design variable list with security labels. We also assign one-bit security label for each data bit. Similarly, the IFT logic for *cell* design objects can also be created by mapping these cells into a standard IFT logic library. The *process* design objects represent the *if–then–else* and *case* statements. These are the most challenging steps in RTL tracking logic generation. We take an approach similar to AST based RTLIFT to create IFT logic for processes.

16.5 Experimental Results

This section presents experimental results to better understand the tradeoffs of when to perform IFT analysis. We start with some concrete design examples to demonstrate the effectiveness of our information flow security verification method in detecting design flaw, timing channel and malicious design modification in Sect. 16.5.1. Section 16.5.2 performs an information flow security verification performance analysis. Section 16.5.3 performs an analysis of hardware IFT logic generated from different RTL intermediate representations and gate-level netlist in terms of complexity and precision. In Sect. 16.5.4, we demonstrate our design methodology that integrates IFT into HLS flow for hardware security verification using a RSA example.

16.5.1 Security Verification Results

16.5.1.1 Design Flaw

We use a Present encryption core from Opencores [31] to demonstrate how our security verification method can be used to detect design flaws that can cause security issues. Figure 16.7 shows the architecture of the core.

We create RTL IFT logic for the core using the IFT logic generation method introduced in Sect. 16.4 and label the lowest plaintext bit (i.e., *message*[0]) as `secret`. We then formally verify the diffusion security property stating that *each plaintext bit should affect (or flow to) multiple bits in the ciphertext*. The property can be formalized as follows:

```
assume message_t = 64'h01
assume key_t = 80'h00
assert cipher_t[0] == cipher_t[1]
```

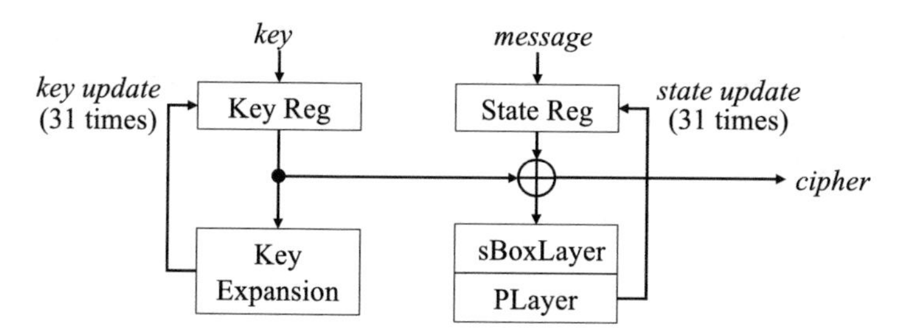

Fig. 16.7 A Present cipher core from Opencores [31]

The security property requires that when the lowest plaintext bit flows to the lowest ciphertext bit (i.e., *cipher*[0]), it should also flow to the second bit of the ciphertext (i.e., *cipher*[1]). Or the lowest plaintext bit should have not yet flowed to either of these two ciphertext bits, e.g., after core reset.

We translate the security property into security constraints and formally verify if the Present core adheres to such a property using *Mentor Graphics Questa Formal*. Formal proof result indicates that the security property can be violated under certain conditions. The counter example returned by the formal verification tool shows that the security property fails to hold right after the *message* is loaded into the state register. This is because the state and key registers are shared and updated during different rounds of encryption. The $key \oplus state$ is assigned directly to *cipher* at all times. Loading the state register will allow *message* to flow to *cipher*, rendering the security labels of *cipher*[0] and *cipher*[1] logical `1` and `0`, respectively.

In this example, our security verification method has successfully identified the design flaw that feeds intermediate encryption results to the observable ciphertext port.

16.5.1.2 Timing Channel

Timing variations in hardware designs have been repeatedly exploited by attackers to break software implementations of ciphers such as RSA and AES. Many of these timing side channel attacks target timing variations through caches and cipher implementations that use pre-computed values that are indexed based on the value of the secret key [32, 33]. In such scenarios, the attacker can collect information regarding the secret key by extracting the cache access pattern of the process running the encryption. To mitigate such attacks, several cache architectures have been developed to eliminate the correlation between the index value of sensitive cache accesses and the time that it takes for the cache to retrieve data in later cycles.

We use hardware IFT to show existence of timing flows in a traditional cache implementation (i.e., with no mitigation technique in place) and the random permutation cache (RPcahce) introduced in [34]. To write the security properties, we consider two processes with isolated address spaces that share the cache. We mark indexes of accesses made by one process (with pid i) as tainted and check if the data read by the other process (with pid j) contains timing variation. The IFT properties for detecting timing flows in an instrumented cache are written as follows.

```
if (pid == i)
  assume(index_t == 16'hFFFF);
if (pid == j)
  assert(data_rd_proc_time == 32'b0);
```

The IFT verification fails for the conventional cache as expected. This is due to the fact that the address which is used to access the cache (i.e., *index_t*) influences the data which is being evicted from the cache. Once process j accesses the evicted data, this access takes longer and is distinguishable through repeated measurements.

We next test the RPcache that eliminates any relation between the cache collisions by randomly permuting the mapping of memory to cache addresses, and randomly choosing a cache line for eviction. This randomization disables the attacker from observing the victim's cache patterns. The IFT verification passes for the RPcache assuming that the random number generator is untainted.

16.5.1.3 Stealthy Hardware Trojan

We use a satisfiability Trojan example proposed by Hu et al. [35] to demonstrate how our security verification method can be used to detect malicious design modifications. The Trojan uses a signal pair that cannot be logical `1` at the same time from AES S-Box as Trojan trigger and adds two multiplexers to multiplex the AES key to the ciphertext output port, as shown in Fig. 16.8. This Trojan design will be activated when both signals are logical `1` and thus the Trojan will never be activated under normal operation. As a consequence, the Trojan cannot be detected using functional testing or even formal equivalence checking.

We also use the method introduced in Sect. 16.4 to create RTL IFT logic for the AES design and label the *key* as `secret`. We declassify at the last add round key operation (i.e., *cipher*) and manually set *cipher* to `unclassified`. This declassification operation is generally regarded as safe and thus allowed. We then

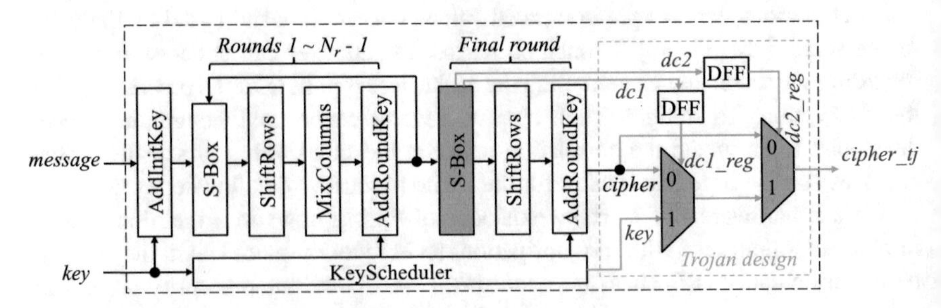

Fig. 16.8 A satisfiability hardware Trojan that leaks the AES key [35]

formally prove that the public output port of the AES core should not take a `secret` label after the declassification operation. We specify the following security property for this proof:

```
assume key_t = {128{1'b1}}
assume cipher_t = 128'h00
assert cipher_tj == 128'h00
```

We then formally prove the above security property under the open source Yosys proof tool. Proof result indicates that the security property does not hold under certain conditions. The counter example returned by Yosys shows that the $cipher\_tj$ can be non-zero when $dc1$ and $dc2$ are both logical `1`. Thus, the formal security verification has precisely captured the trigger condition of the Trojan.

16.5.2 Verification Performance Analysis

We use several design examples and benchmarks for verification performance analysis. We use the IFT logic generated by our RTLIFT tool as the security verification model and verify information flow security properties on these security models. In our test, we use the *SAT* solver in *Yosys* to prove security properties. Table 16.2 shows the security properties proved and verification performance results.

As an example, it takes 384.55 s to run formal verification and detect the Trojan for the example discussed in Sect. 16.5.1.3. From Table 16.2, RTLIFT provides an approach for constructing security models that allow security properties to be verified within acceptable amount of time.

16.5.3 Complexity and Precision Analysis

We use several IWLS benchmarks [37] for IFT complexity and precision analysis. We use the number of cells in synthesized IFT circuits as a measure for complexity while the number of simulated information flows as a measure of precision. We use combinational benchmarks in complexity and precision analysis to more accurately measure input-output flow relations, eliminating the complex flow relations over multiple clock cycles. Figure 16.9 shows our test flow as well as tools used to create different IFT logic circuits.

We use five different test flows for IFT logic generation. The *ABC-resyn2* flow first uses the *resyn2* script in *ABC* [38] to synthesize the benchmarks to gate-level netlists and then uses our GLIFT script to create GLIFT logic. The *ABC-dc* test flow uses the *resyn2rs*, *compress2rs*, *dc2*, *dch* and *mfs3* scripts to synthesize the design, which yields higher optimization effort and also enables don't care based optimization. The *AST2-conservative* and *AST2-precise* test flows first dump

Table 16.2 Proof time (*sec*) for verifying security properties on several design examples and benchmarks

Designs	Security properties	Proof time
AES-DPA [36]	assume key_t = {127{1'b1}, 1'b0}; assert	13.01
	cipher_t[0] == cipher_t[1]	
PresentEncryptor [31]	assume key_t = {79{1'b1}, 1'b0}; assert cipher_t[0] == cipher_t[1]	1.31
AES-T400 [28]	assume key_t = {128{1'b1}}; assert Antena_t == 1'b0	474.26
AES-T1700 [28]	assume key_t = {128{1'b1}}; assert Antena_t == 1'b0	703.76
RSA-T400 [28]	assume indata_t = 32'h0; assume indata_t = {32{1'b1}}; assert count_t[0] == 1'b0	4.59
AES-DC-TJ [35]	assume key_t = {128{1'b1}}; assume cipher_t = 128'h0; assert cipher_tj == 128'h0	384.55

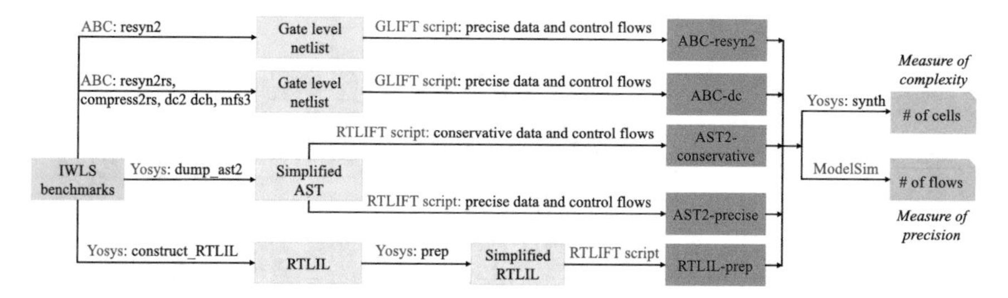

Fig. 16.9 Different test flows for IFT logic generation

simplified AST (i.e., -dump_ast2) using the Verilog frontend in Yosys and then create IFT logic from the AST extracted. The difference lies in that the *AST2-conservative* test flow employs a conservative policy to measure data and control flows while *AST2-precise* uses a precise one. The *RTLIL-prep* flow first constructs the RTLIL intermediate presentation for the benchmark and then invokes the *prep* synthesis script in Yosys to optimize the RTLIL representation. The IFT logic is generated from the optimized RTLIL.

After generating different versions of IFT circuits at the gate level and RTL for each benchmark, we use the *synth* script in *Yosys* [30] to synthesize these IFT logic circuits and report the number of cells in the synthesized IFT circuits. We also use *ModelSim* to test the IFT logic circuits under 2^{20} random test vectors and observe the total number of information flows. Table 16.3 shows the test results.

From Table 16.3, the *AST2-precise* and *RTLIL-prep* test flows perform identical optimizations and produce identical IFT logic. This is formally verified by equivalence checking of IFT logic circuits created by these two test flows. The *AST2-conservative* flow employs conservative policy to track data and control flows. It significantly reduces the complexity of IFT logic circuits at the side effect of a larger number of false positives in information flow measurement. These additional information will not actually happen. The *ABC-dc* test flow that performs don't care based logic optimization on the original design leads to smaller number of cells and information flows. By comparison of the *ABC-resyn* and *AST2-precise/RTLIL-prep* test flows, AST and RTLIL based optimizations tend to lead to larger number of cells while close number of information flows. The test results reveal the tradeoff between complexity and precision of hardware IFT.

For a better understanding, we visualize the number of cells and flows normalized to those for the *GLIFT-resy2* test flow. The normalized results are shown in Figs. 16.10 and 16.11, respectively.

Table 16.3 Precision and complexity of hardware IFT logic generated using different methods

Benchmarks	GLIFT-resyn2		GLIFT-dc		AST2-conservative		AST2-precise		RTLIL-prep	
	Cells	Flows	Cells	Flows	Cells	Flows	Cells	Flows	Cells	Flows
C1355	656	33546328	1398	33545816	334	33554368	3099	33546328	3101	33546328
C1908	1879	20954234	2028	21074948	425	26214350	2437	21074890	2434	21074890
C2670	3406	83584759	2891	83528247	921	97009773	3836	84786443	3820	83747198
C3540	6152	20982566	6324	21099557	1675	22864600	6205	20913238	6510	20913238
C432	1285	6552124	1219	6307165	353	7340006	1019	6553500	1008	6553500
C499	1625	33546253	2644	33545521	334	33554368	739	33546253	743	33546253
C5315	9224	92555711	9175	91789720	2498	112689371	9482	90829295	9522	90829295
C7552	7758	78802237	7426	78931128	2689	87655782	12661	79899472	12661	79899472
C880	2140	16049183	1909	16021046	523	25711253	1961	16011824	1972	16011824
frg2	5604	71020251	5312	70543284	3537	134929650	11081	70389291	11078	70389291
i10	12057	124753233	11910	128373586	3455	221522371	13438	121632233	13379	121632233
i5	1864	42869700	1566	42869700	712	68502275	2211	42869700	2221	42869700
i6	2731	54712363	2717	54712363	577	67152895	2379	54719267	2404	54719267
i7	3659	58060003	3297	58634475	922	68584377	3205	58134101	3189	58134101
i8	5474	69087423	5944	71595869	3257	84915387	10094	71866588	10077	71866588
i9	3460	60539033	3633	60326644	1060	66048068	3925	60573112	3925	60573112
k2	7845	23911899	6858	23630132	3637	45044927	13532	22074582	13296	22074582
too_large	26864	3127215	20924	3021917	23762	3145722	72424	1494259	72430	1494259
vda	3693	28572052	4024	26112629	1272	40758299	4999	27998546	4917	27998546
x1	4028	19340837	2285	18783256	3635	32155816	11026	17633240	11015	17633240
x4	4028	41280920	2102	41788569	1511	68221032	4614	42186099	4608	42186099

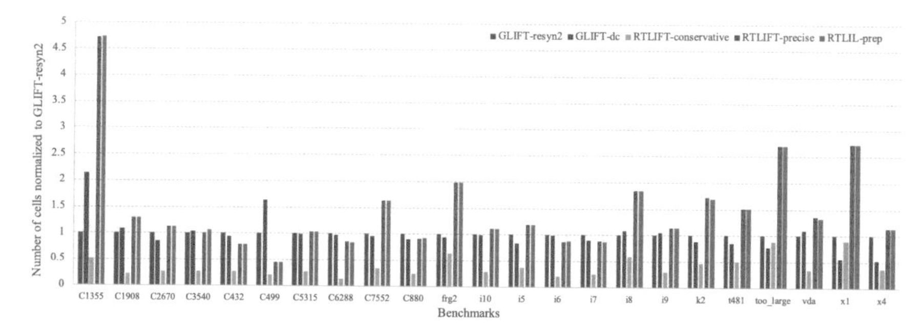

Fig. 16.10 The number of cells normalized to GLIFT-resyn2

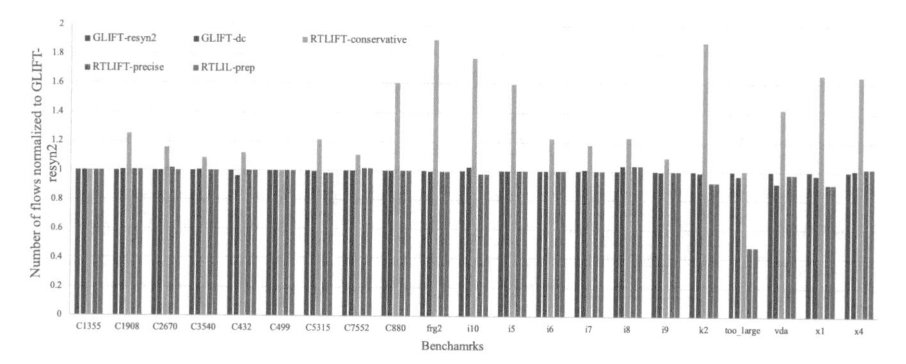

Fig. 16.11 The number of flows normalized to GLIFT-resyn2

16.5.4 An HLS Design and Verification Example

We use a RSA example to demonstrate our design methodology that integrates IFT into HLS flow for hardware security verification. Figure 16.12 shows our test flow. We implement a 32-bit instance of the right-to-left repeated squaring algorithm for calculating modular exponentiation (i.e., the basic operation of RSA) in HLS C. We use *Xilinx Vivado* to synthesize the C code into Verilog design. The resulting RSA design is then converted into IFT model using our RTLIFT tool.

The IFT model is combined with the following security property, which labels the secret exponent d as `high` while all other inputs as `low` and asserts that the encryption result ready output *ap_ready* should be `low`, to perform formal verification of the security property under *Mentor Graphics Questa Formal*. In our proof, we constrain the secret exponent d and modulus n to allow constant values so that the prover only needs to search on the message c, which minimizes the search state space and in turn accelerates the proof process. Such constraint should be applied since the RSA has its key generation rules and only allowed values can be used as legal RSA key pairs.

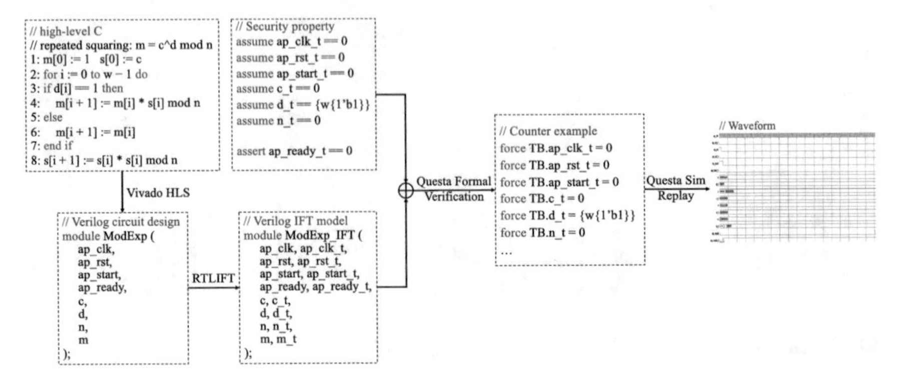

Fig. 16.12 An HLS design and verification example

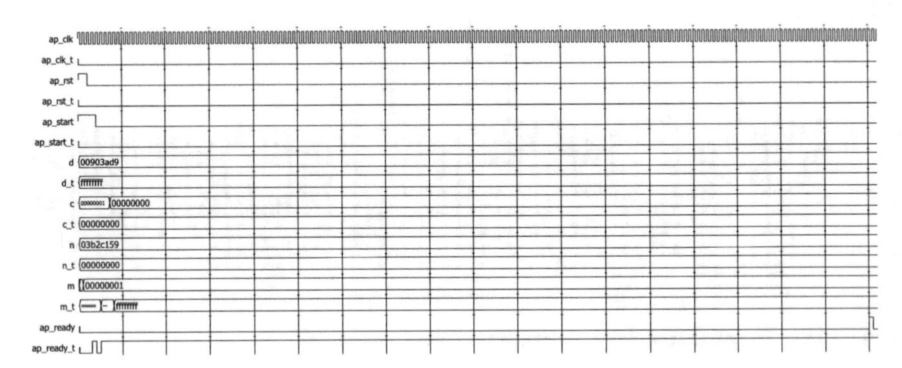

Fig. 16.13 Replay waveform of the counter example from RSA security verification

```
assume ap_clk_t == 0
assume ap_rst_t == 0
assume ap_start_t == 0
assume c_t == 0
assume d_t = {w{1'b1}}
assume n_t = 0

assert ap_ready_t == 0
```

Formal verification result shows that the specified security property can be violated and *Questa Formal* provides a counter example to show when such violation can happen. We replay the counter example under *Mentor Graphics Questa Sim* and the replay waveform is shown in Fig. 16.13.

From Fig. 16.13, the *ap_ready* output can be `high` (i.e., $ap\_ready\_t = 1$), indicating that it can contain information about the secret exponent d since we only labeled d as `high`. This is because the right-to-left repeated squaring RSA implementation contains a timing channel that leaks the secret exponent to *ap_ready* in that the exponent d is used to control a timing-unbalanced conditional branch, i.e., the *if-else* branch statement shown in the high-level C code in Fig. 16.12. This creates timing information flows from d to ap_ready since the secret exponent dominates the encryption time. Our design methodology has detected the timing channel in this RSA implementation using standard verification tools.

16.6 Conclusion

We introduce a methodology to integrate information flow analysis into the HLS design flow. We argue that the RTL provides an optimal place to perform information flow analysis. We describe RTLIFT—a precise information flow tracking method at the RTL for secure hardware design. We provide IFT logic formalization, information flow security property specification, and verification methodology. We demonstrate how our security verification method can be employed to enhance the EDA flow and identify hardware security vulnerabilities in the early design phase.

Acknowledgments This work was supported in part by the Natural Science Foundation of Shaanxi Province under Grant 2019JM-244, NSF award 1718586, and the Semiconductor Research Corporation Task 2770.001.

References

1. Kastner, R., Matai, J., Neuendorffer, S.: Parallel programming for FPGAs (2018). Preprint, arXiv:1805.03648
2. Bulck, J.V., Minkin, M., Weisse, O., Genkin, D., Kasikci, B., Piessens, F., Silberstein, M., Wenisch, T.F., Yarom, Y., Strackx, R.: Foreshadow: extracting the keys to the Intel SGX kingdom with transient out-of-order execution. In: 2018 27th USENIX Security Symposium (USENIX Security 18), pp. 991–1008. USENIX Association, Baltimore, MD (2018). https://www.usenix.org/conference/usenixsecurity18/presentation/bulck
3. Weisse, O., Bulck, J.V., Minkin, M., Genkin, D., Kasikci, B., Piessens, F., Silberstein, M., Strackx, R., Wenisch, T.F., Yarom, Y.: Foreshadow-NG: breaking the virtual memory abstraction with transient out-of-order execution (2018). https://foreshadowattack.eu/foreshadow-NG.pdf
4. Skorobogatov, S., Woods, C.: Breakthrough Silicon Scanning Discovers Backdoor in Military Chip, pp. 23–40. Springer, Heidelberg (2012)
5. Andreou, A., Bogdanov, A., Tischhauser, E.: Cache timing attacks on recent microarchitectures. In: 2017 IEEE International Symposium on Hardware Oriented Security and Trust (HOST), pp. 155–155 (2017)
6. Hu, W., Althoff, A., Ardeshiricham, A., Kastner, R.: Towards property driven hardware security. In: 2016 17th International Workshop on Microprocessor and SOC Test and Verification (MTV), pp. 51–56. IEEE, Piscataway (2016)
7. Hu, W., Ardeshiricham, A., Gobulukoglu, M.S., Wang, X., Kastner, R.: Property specific information flow analysis for hardware security verification. In: 2018 IEEE/ACM International Conference on Computer-Aided Design (ICCAD), pp. 1–8 (2018)
8. Ma, H., He, J., Liu, Y., Zhao, Y., Jin, Y.: CAD4EM-P: security-driven placement tools for electromagnetic side channel protection. In: 2019 Asian Hardware Oriented Security and Trust Symposium (AsianHOST), pp. 1–6 (2019)

9. Tiwari, M., Wassel, H.M., Mazloom, B., Mysore, S., Chong, F.T., Sherwood, T.: Complete information flow tracking from the gates up. In: the 14th International Conference on Architectural Support for Programming Languages and Operating Systems (ASPLOS), pp. 109–120 (2009)
10. Bidmeshki, M., Makris, Y.: Toward automatic proof generation for information flow policies in third-party hardware IP. In: 2015 IEEE International Symposium on Hardware Oriented Security and Trust (HOST), pp. 163–168 (2015)
11. Zhang, D., Wang, Y., Suh, G.E., Myers, A.C.: A hardware design language for timing-sensitive information-flow security. In: Proceedings of the Twentieth International Conference on Architectural Support for Programming Languages and Operating Systems (ASPLOS), pp. 503–516. ACM, New York, NY (2015)
12. Ardeshiricham, A., Hu, W., Marxen, J., Kastner, R.: Register transfer level information flow tracking for provably secure hardware design. In: Design, Automation & Test in Europe Conference & Exhibition (DATE), pp. 1691–1696 (2017)
13. Sefton, S., Siddiqui, T., Amour, N.S., Stewart, G., Kodi, A.K.: GARUDA: designing energy-efficient hardware monitors from high-level policies for secure information flow. IEEE Trans. Comput. Aided Des. Integr. Circuits Syst. **37**(11), 2509–2518 (2018)
14. Jiang, Z., Dai, S., Suh, G.E., Zhang, Z.: High-level synthesis with timing-sensitive information flow enforcement. In: Proceedings of the International Conference on Computer-Aided Design (ICCAD), pp. 88:1–88:8. ACM, New York, NY (2018)
15. Pilato, C., Wu, K., Garg, S., Karri, R., Regazzoni, F.: TaintHLS: high-level synthesis for dynamic information flow tracking. IEEE Trans. Comput. Aided Des. Integr. Circuits Syst. **38**(5), 798–808 (2019)
16. Ravi, P., Najm, Z., Bhasin, S., Khairallah, M., Gupta, S.S., Chattopadhyay, A.: Security is an architectural design constraint. Microprocess. Microsyst. **68**, 17–27 (2019)
17. Knechtel, J., Kavun, E.B., Regazzoni, F., Heuser, A., Chattopadhyay, A., Mukhopadhyay, D., Dey, S., Fei, Y., Belenky, Y., Levi, I., Güneysu, T., Schaumont, P., Polian, I.: Towards Secure Composition of Integrated Circuits and Electronic Systems: On the Role of EDA. In: Design, Automation & Test in Europe Conference & Exhibition (DATE), pp. 508–513 (2020).
18. Pilato, C., Garg, S., Wu, K., Karri, R., Regazzoni, F.: Securing hardware accelerators: a new challenge for high-level synthesis. IEEE Embed. Syst. Lett. **10**(3), 77–80 (2018)
19. Deng, S., Gümüşoğlu, D., Xiong, W., Sari, S., Gener, Y.S., Lu, C., Demir, O., Szefer, J.: SecChisel framework for security verification of secure processor architectures. In: Proceedings of the 8th International Workshop on Hardware and Architectural Support for Security and Privacy (HASP), pp. 7:1–7:8. ACM, New York, NY (2019)
20. Ardeshiricham, A., Hu, W., Kastner, R.: Clepsydra: modeling timing flows in hardware designs. In: 2017 IEEE/ACM International Conference on Computer-Aided Design (ICCAD), pp. 147–154 (2017)
21. Jin, Y., Guo, X., Dutta, R.G., Bidmeshki, M., Makris, Y.: Data secrecy protection through information flow tracking in proof-carrying hardware IP–Part I: framework fundamentals. IEEE Trans. Inf. Forensics Secur. **12**(10), 2416–2429 (2017)
22. Hu, W., Oberg, J., Irturk, A., Tiwari, M., Sherwood, T., Mu, D., Kastner, R.: Theoretical fundamentals of gate level information flow tracking. IEEE Trans. Comput. Aided Des. Integr. Circuits Syst. **30**(8), 1128–1140 (2011)
23. Bidmeshki, M., Antonopoulos, A., Makris, Y.: Information flow tracking in analog/mixed-signal designs through proof-carrying hardware IP. In: Design, Automation & Test in Europe Conference & Exhibition (DATE), pp. 1703–1708 (2017)
24. Li, X., Tiwari, M., Hardekopf, B., Sherwood, T., Chong, F.T.: Secure information flow analysis for hardware design: using the right abstraction for the job. In: Proceedings of the 5th ACM SIGPLAN Workshop on Programming Languages and Analysis for Security (PLAS), pp. 8:1–8:7. ACM, New York, NY (2010)
25. Stroud, C.E., Wang, L.T., Chang, Y.W.: Introduction. In: Wang, L.T., Chang, Y.W., Cheng, K.T.T. (eds.) Electronic Design Automation, Chap. 1, pp. 1–38. Morgan Kaufmann, Boston (2009)

26. Oberg, J., Meiklejohn, S., Sherwood, T., Kastner, R.: Leveraging gate-level properties to identify hardware timing channels. IEEE Trans. Comput. Aided Des. Integr. Circuits Syst. **33**(9), 1288–1301 (2014)
27. Hu, W., Mao, B., Oberg, J., Kastner, R.: Detecting hardware trojans with gate-level information-flow tracking. Computer **49**(8), 44–52 (2016)
28. Shakya, B., He, T., Salmani, H., Forte, D., Bhunia, S., Tehranipoor, M.: Benchmarking of hardware trojans and maliciously affected circuits. J. Hardware Syst. Secur. **1**, 85–102 (2017)
29. Hu, W., Oberg, J., Irturk, A., Tiwari, M., Sherwood, T., Mu, D., Kastner, R.: On the complexity of generating gate level information flow tracking logic. IEEE Trans. Inf. Forensics Secur. **7**(3), 1067–1080 (2012)
30. Wolf, C., Glaser, J.: Yosys - a free Verilog synthesis suite (2013). http://www.clifford.at/yosys/
31. Ameli, R.: Present Cipher Encryption IP Core (2011). https://opencores.org/ocsvn/present_encryptor/present_encryptor/trunk
32. Percival, C.: Cache missing for fun and profit. In: Proc. of BSDCan 2005 (2005)
33. Bernstein, D.J.: Cache-timing attacks on AES. VLSI Des. IEEE Comput. Soc. **51**(2), 218–221 (2005)
34. Wang, Z., Lee, R.B.: New cache designs for thwarting software cache-based side channel attacks. SIGARCH Comput. Archit. News **35**(2), 494–505 (2007)
35. Hu, W., Zhang, L., Ardeshiricham, A., Blackstone, J., Hou, B., Tai, Y., Kastner, R.: Why you should care about don't cares: exploiting internal don't care conditions for hardware trojans. In: IEEE/ACM International Conference on Computer-Aided Design (ICCAD), pp. 707–713 (2017)
36. Satoh, A.: AES Encryption/Decryption Macro (2007). http://www.aoki.ecei.tohoku.ac.jp/crypto/
37. IWLS: IWLS Benchmarks Ver. 3.0 (2005). http://iwls.org/iwls2005/benchmarks.html
38. Berkeley Logic Synthesis and Verification Group: ABC: A System for Sequential Synthesis and Verification (2020). http://www.eecs.berkeley.edu/~alanmi/abc

Index

S. Katkoori, S. A. Islam (eds.), *Behavioral Synthesis for Hardware Security*,
https://doi.org/10.1007/978-3-030-78841-4

Printed in the United States
by Baker & Taylor Publisher Services